# Exploring IoT Sensors and Their Applications: Advancements, Challenges, and Opportunities in Smart Environments

# Exploring IoT Sensors and Their Applications: Advancements, Challenges, and Opportunities in Smart Environments

Editors

Lei Jing
Yoshinori Matsumoto
Zhan Zhang

Basel • Beijing • Wuhan • Barcelona • Belgrade • Novi Sad • Cluj • Manchester

*Editors*

Lei Jing
Graduate School of Computer
Science and Engineering
The University of Aizu
Aizuwakamatsu
Japan

Yoshinori Matsumoto
Department of Applied
Physics and Physico-Informatics
Keio University
Yokohama
Japan

Zhan Zhang
School of Computer Science
and Engineering
Harbin Institute of Technology
Harbin
China

*Editorial Office*
MDPI AG
Grosspeteranlage 5
4052 Basel, Switzerland

This is a reprint of articles from the Special Issue published online in the open access journal *Micromachines* (ISSN 2072-666X) (available at: www.mdpi.com/journal/micromachines/special_issues/IYE40N4A39).

For citation purposes, cite each article independently as indicated on the article page online and as indicated below:

Lastname, A.A.; Lastname, B.B. Article Title. *Journal Name* **Year**, *Volume Number*, Page Range.

ISBN 978-3-7258-2024-5 (Hbk)
ISBN 978-3-7258-2023-8 (PDF)
doi.org/10.3390/books978-3-7258-2023-8

© 2024 by the authors. Articles in this book are Open Access and distributed under the Creative Commons Attribution (CC BY) license. The book as a whole is distributed by MDPI under the terms and conditions of the Creative Commons Attribution-NonCommercial-NoDerivs (CC BY-NC-ND) license.

# Contents

**About the Editors** . . . . . . . . . . . . . . . . . . . . . . . . . . . . . . . . . . . . . . . . . . . . . . . . . vii

**Preface** . . . . . . . . . . . . . . . . . . . . . . . . . . . . . . . . . . . . . . . . . . . . . . . . . . . . . . . . . ix

**Lei Jing, Yoshinori Matsumoto and Zhan Zhang**
Editorial for the Special Issue on Exploring IoT Sensors and Their Applications: Advancements, Challenges, and Opportunities in Smart Environments
Reprinted from: *Micromachines* **2024**, *15*, 1048, doi:10.3390/mi15081048 . . . . . . . . . . . . . . . 1

**Zeping Yu, Chenghong Lu, Yunhao Zhang and Lei Jing**
Gesture-Controlled Robotic Arm for Agricultural HarvestingUsing a Data Glove with Bending Sensor and OptiTrack Systems
Reprinted from: *Micromachines* **2024**, *15*, 918, doi:10.3390/mi15070918 . . . . . . . . . . . . . . . . 4

**Pontakorn Sonchan, Neeranut Ratchatanantakit, Nonnarit O-Larnnithipong, Malek Adjouadi and Armando Barreto**
Robust Orientation Estimation from MEMS Magnetic, Angular Rate, and Gravity (MARG) Modules for Human–Computer Interaction
Reprinted from: *Micromachines* **2024**, *15*, 553, doi:10.3390/mi15040553 . . . . . . . . . . . . . . . . 22

**Heyu Yin, Sylmarie Dávila-Montero and Andrew J. Mason**
Geometry Scaling for Externally Balanced Cascade Deterministic Lateral Displacement Microfluidic Separation of Multi-Size Particles [†]
Reprinted from: *Micromachines* **2024**, *15*, 405, doi:10.3390/mi15030405 . . . . . . . . . . . . . . . . 57

**Yoshiro Tajitsu, Saki Shimda, Takuto Nonomura, Hiroki Yanagimoto, Shun Nakamura, Ryoma Ueshima, et al.**
Application of Braided Piezoelectric Poly-l-Lactic Acid Cord Sensor to Sleep Bruxism Detection System with Less Physical or Mental Stress
Reprinted from: *Micromachines* **2024**, *15*, 86, doi:10.3390/mi15010086 . . . . . . . . . . . . . . . . 75

**Myoungsu Chae, Doowon Lee and Hee-Dong Kim**
Low-Power Consumption IGZO Memristor-Based Gas Sensor Embedded in an Internet of Things Monitoring System for Isopropanol Alcohol Gas
Reprinted from: *Micromachines* **2024**, *15*, 77, doi:10.3390/mi15010077 . . . . . . . . . . . . . . . . 93

**Madiha Javeed, Maha Abdelhaq, Asaad Algarni and Ahmad Jalal**
Biosensor-Based Multimodal Deep Human Locomotion Decoding via Internet of Healthcare Things
Reprinted from: *Micromachines* **2023**, *14*, 2204, doi:10.3390/mi14122204 . . . . . . . . . . . . . . . 105

**Tianlin Li, Guanglu Sun, Linsen Yu and Kai Zhou**
HRBUST-LLPED: A Benchmark Dataset for Wearable Low-Light Pedestrian Detection
Reprinted from: *Micromachines* **2023**, *14*, 2164, doi:10.3390/mi14122164 . . . . . . . . . . . . . . . 125

**Liufeng Fan, Zhan Zhang, Biao Zhu, Decheng Zuo, Xintong Yu and Yiwei Wang**
Smart-Data-Glove-Based Gesture Recognition for Amphibious Communication
Reprinted from: *Micromachines* **2023**, *14*, 2050, doi:10.3390/mi14112050 . . . . . . . . . . . . . . . 145

**Chi Cuong Vu, Jooyong Kim and Thanh-Hai Nguyen**
Health Monitoring System from Pyralux Copper-Clad Laminate Film and Random Forest Algorithm
Reprinted from: *Micromachines* **2023**, *14*, 1726, doi:10.3390/mi14091726 . . . . . . . . . . . . . . . 168

**Thanh-Hai Nguyen, Ba-Viet Ngo, Thanh-Nghia Nguyen and Chi Cuong Vu**
Flexible Pressure Sensors and Machine Learning Algorithms for Human Walking Phase Monitoring
Reprinted from: *Micromachines* **2023**, *14*, 1411, doi:10.3390/mi14071411 . . . . . . . . . . . . . . . . **179**

# About the Editors

**Lei Jing**

Lei Jing (Member, IEEE) received his Ph.D. degree in computer science and engineering from The University of Aizu in Japan in 2008. He is currently a senior associate professor at UoA, where he leads the Human Intelligent Sensing Group and the team dedicated to human motion tracking. His particular interest lies in hand motion detection, employing multimodal sensor fusion methods. The applications of his work include activity abnormality detection, sign language recognition, and skill heritage. He has published over 100 papers and holds six patents in related areas. Lei Jing has been a member of IEEE and ACM since 2010 and a member of IPSJ since 2012.

**Yoshinori Matsumoto**

Yoshinori Matsumoto received a B.S. degree and Ph.D. degree in Electronic Engineering from Tohoku University, Sendai. He has been a research associate at the department of Electric and Electronic Engineering of Toyohashi University of Technology since 1993. He worked on SOI MEMS sensors and integrated circuits. Since 2010, he has been a professor at the Department of Applied Physics and Physico-Informatics, Faculty of Science and Technology, Keio University. His research interests are smart sensors and IoT environmental measurement systems.

**Zhan Zhang**

Zhang Zhan received a Ph.D. degree in computer science and technology from the Harbin Institute of Technology (HIT), China. He is now a professor with the Research Center of Fault-Tolerance and Mobile Computing, School of Computer Science and Technology, HIT. His research interests include wearable computers, fault-tolerant computer, and system evaluation theory and technology.

# Preface

It is with great pleasure that we introduce this volume, which delves into the ever-evolving world of IoT sensors and their transformative applications in smart environments. As the integration of IoT sensors becomes increasingly central to innovations in healthcare, industry, and everyday life, this reprint serves as a comprehensive guide to understanding both the advancements and the ongoing challenges within this dynamic field.

The rapid progress in IoT technologies has been nothing short of remarkable, yet the field continues to face significant challenges. These include the need for sensors that are not only more accurate and reliable but also energy-efficient and capable of being integrated into systems that can handle vast amounts of data with precision. This volume addresses these challenges by presenting cutting-edge research and developments that push the boundaries of what is possible with IoT sensors.

Inside, you will find a collection of carefully selected contributions that showcase the latest innovations in sensor technology, from flexible pressure sensors used for human motion detection to advanced systems for health monitoring and environmental sensing. These contributions not only highlight related technical advancements but also explore their practical applications, offering insights into how these technologies can be utilized in real-world scenarios.

Moreover, this reprint emphasizes the importance of interdisciplinary collaboration in driving forward the next generation of IoT solutions. By integrating AI, data science, and sensor technology, we can create smarter, more autonomous systems that can revolutionize industries and improve the quality of life.

As you explore the pages of this reprint, I encourage you to consider the broader implications of these technologies. The future of IoT lies not just in the innovations themselves but also in how we apply them to solve the complex challenges of our time.

I hope this volume inspires new ideas, fosters collaboration, and contributes to the continued growth and success of the IoT field.

**Lei Jing, Yoshinori Matsumoto, and Zhan Zhang**
*Editors*

 micromachines

*Editorial*

# Editorial for the Special Issue on Exploring IoT Sensors and Their Applications: Advancements, Challenges, and Opportunities in Smart Environments

Lei Jing [1,*], Yoshinori Matsumoto [2,*] and Zhan Zhang [3,*]

1. Graduate School of Computer Science and Engineering, The University of Aizu, Tsuruga, Ikki-machi, Aizuwakamatsu 965-8580, Japan
2. Department of Applied Physics and Physico-Informatics, Keio University, Yokohama 223-8522, Japan
3. School of Computer Science and Technology, Harbin Institute of Technology, Harbin 150001, China
* Correspondence: leijing@u-aizu.ac.jp (L.J.); matsumoto@appi.keio.ac.jp (Y.M.); zhangzhan@hit.edu.cn (Z.Z.)

As the editor of the Special Issue on "Exploring IoT Sensors and Their Applications: Advancements, Challenges, and Opportunities in Smart Environments", I am delighted to present this collection of groundbreaking research that addresses the emerging needs and challenges in the field of IoT sensors and smart environments. This Special Issue brings together ten contributions that showcase the significant advancements and innovative applications of IoT sensors, providing valuable insights into the current state of the art and highlighting future research directions.

The field of IoT sensors has rapidly evolved, driven by the increasing demand for smart solutions in healthcare, industry, and everyday life. Recent studies, such as Deng et al. [1], have highlighted the growing importance of edge computing in reducing latency and improving the efficiency of IoT systems. Similarly, Li et al. [2] have explored the integration of wireless sensor networks (WSNs) with IoT, emphasizing the need for low-power, high-reliability communication protocols.

However, several challenges remain, including the need for more accurate, reliable, and energy-efficient sensors, as well as the integration of these sensors into comprehensive systems that can effectively process and analyze the vast amounts of data they generate. For instance, Deng et al. [3] underscore the ongoing efforts to develop power management strategies that extend the operational life of IoT devices, which is critical for applications in remote and resource-constrained environments.

This Special Issue addresses these gaps by presenting research that not only introduces new sensor technologies but also explores novel algorithms and systems that enhance the performance and utility of these sensors in various applications.

The following contributions in this Special Issue cover a broad range of applications and technologies, each addressing specific gaps in current knowledge:

Nguyen et al. [4] present the development of flexible pressure sensors designed to monitor human walking phases. The sensors are integrated with machine learning algorithms, specifically tailored to accurately capture and interpret the complex biomechanical processes involved in human gait. This work addresses the challenge of creating wearable devices that can provide real-time, reliable data for applications in healthcare and rehabilitation. By focusing on the integration of flexible materials and advanced algorithms, the authors have laid the groundwork for future innovations in wearable health monitoring systems.

Vu et al. [5] propose a health monitoring system that leverages Pyralux copper-clad laminate film as the sensor material, combined with a random forest algorithm for data analysis. This system is designed to enhance the accuracy and reliability of health monitoring, particularly in environments where traditional sensors may be less effective. The paper highlights the potential of combining flexible sensor materials with robust machine learning techniques to create more effective health monitoring systems, paving the way for broader applications in remote health care and wearable devices.

Fan et al. [6] introduce a smart-data glove designed for gesture recognition in amphibious communication scenarios. The glove integrates multiple sensors to capture hand movements and gestures, which are then processed to enable effective communication in challenging environments, such as underwater. The authors address the gap in reliable communication methods for environments where traditional devices fail, demonstrating how wearable technology can be adapted for specialized use cases.

Li et al. [7] present HRBUST-LLPED, a benchmark dataset specifically designed for the development and testing of wearable low-light pedestrian detection systems. The dataset addresses the critical need for reliable pedestrian detection in low-visibility conditions, such as at night or in poorly lit environments. The authors provide a comprehensive analysis of the dataset and demonstrate its potential for improving pedestrian safety through enhanced detection algorithms.

In this paper, Javeed et al. [8], the authors explore the use of biosensors in multimodal deep human locomotion decoding, facilitated by the Internet of Healthcare Things (IoHT). This research focuses on integrating multiple biosensor data streams to create a comprehensive understanding of human locomotion patterns, which is crucial for applications in health monitoring, rehabilitation, and sports science. The study highlights the potential of IoHT to enhance the accuracy and utility of wearable health monitoring systems by leveraging deep learning techniques.

In research by Chae et al. [9], the authors introduce a low-power IGZO memristor-based gas sensor designed for detecting isopropanol alcohol gas, embedded within an IoT monitoring system. The authors address the challenge of energy consumption in sensor networks by developing a gas sensor that is both highly sensitive and energy-efficient. This work is particularly relevant for applications in industrial safety and environmental monitoring, where continuous, reliable detection of volatile compounds is essential.

In a paper by Tajitsu et al. [10], the authors present an innovative application of braided piezoelectric poly-l-lactic acid (PLLA) cord sensors in a sleep bruxism detection system. The authors focus on developing a sensor that minimizes physical and mental stress, which is crucial for ensuring user compliance and comfort during long-term health monitoring. The sensor's application in detecting bruxism—a condition that can lead to severe dental issues if untreated—demonstrates the potential of wearable IoT sensors in providing non-invasive, stress-free health monitoring solutions.

Yin et al. [11] investigate the geometry scaling of microfluidic devices used for deterministic lateral displacement (DLD) to separate multi-size particles. The authors introduce an externally balanced cascade design that enhances the separation efficiency of microfluidic systems. This work addresses the challenge of precisely controlling particle separation in lab-on-a-chip devices, which has significant implications for biomedical research and diagnostics.

Sonchan et al. [12] investigate robust orientation estimation using MEMS MARG modules, which integrate magnetic, angular rate, and gravity sensors. The authors develop algorithms that improve the accuracy and reliability of orientation estimation, which is critical for human-computer interaction systems, including virtual reality (VR) and augmented reality (AR) applications. By enhancing the performance of MARG modules, this research contributes to more immersive and responsive user experiences in HCI systems.

Finally, Yu et al. [13] demonstrate the application of gesture-controlled robotic arms in agricultural harvesting, utilizing a data glove with bending sensors and OptiTrack systems, which underscores the growing importance of IoT sensors in precision agriculture. This innovative application of IoT sensors in agriculture addresses the challenge of improving precision and efficiency in harvesting operations. The integration of gesture recognition technology with robotic systems represents a significant step forward in the development of smart agricultural solutions.

While the advancements presented in this Special Issue represent significant progress, there remain several avenues for future research. Interdisciplinary collaboration will be crucial in further integrating IoT sensors with AI and data science to create smarter, more autonomous systems. Additionally, the scalability and miniaturization of sensors, coupled with advancements in battery life and energy harvesting, will be vital for the continued expansion of IoT applications. Moreover, ensuring the security and privacy of the vast amounts of data generated by IoT systems will require ongoing attention, particularly as these technologies become more embedded in our daily lives.

In conclusion, this Special Issue has provided a comprehensive overview of the current advancements in IoT sensors and their applications in smart environments. The contributions have addressed critical gaps in the field and set the stage for future innovations. I look forward to witnessing the continued evolution of this dynamic field, driven by the collaborative efforts of researchers and practitioners worldwide.

**Author Contributions:** Conceptualization, L.J., Y.M. and Z.Z.; methodology, L.J., Y.M. and Z.Z.; software, L.J., Y.M. and Z.Z.; validation, L.J., Y.M. and Z.Z.; formal analysis, L.J., Y.M. and Z.Z.; investigation, L.J., Y.M. and Z.Z.; resources, L.J., Y.M. and Z.Z.; data curation, L.J., Y.M. and Z.Z.; writing—original draft preparation, L.J.; writing—review and editing, L.J., Y.M. and Z.Z.; visualization, L.J., Y.M. and Z.Z.; supervision, L.J., Y.M. and Z.Z.; project administration, L.J., Y.M. and Z.Z.; funding acquisition, L.J., Y.M. and Z.Z. All authors have read and agreed to the published version of the manuscript.

**Conflicts of Interest:** The authors declare no conflict of interest.

# References

1. Deng, S.; Zhao, H.; Fang, W.; Yin, J.; Dustdar, S.; Zomaya, A.Y. Edge Intelligence: The Confluence of Edge Computing and Artificial Intelligence. *IEEE Internet Things J.* **2020**, *7*, 7457–7469. [CrossRef]
2. Li, J.; Mahmoodi, T. The Entanglement of Communication and Computing in Enabling Edge Intelligence. *IEEE Internet Things J.* **2024**, *11*, 19278–19302. [CrossRef]
3. Teng, L.; Wang, H.; Liu, Y.; Fu, M.; Liang, J. A Three-Transistor Energy Management Circuit for Energy-Harvesting-Powered IoT Devices. *IEEE Internet Things J.* **2024**, *11*, 1301–1310. [CrossRef]
4. Nguyen, T.-H.; Ngo, B.-V.; Nguyen, T.-N.; Vu, C.C. Flexible Pressure Sensors and Machine Learning Algorithms for Human Walking Phase Monitoring. *Micromachines* **2023**, *14*, 1411. [CrossRef] [PubMed]
5. Vu, C.C.; Kim, J.; Nguyen, T.-H. Health Monitoring System from Pyralux Copper-Clad Laminate Film and Random Forest Algorithm. *Micromachines* **2023**, *14*, 1726. [CrossRef] [PubMed]
6. Fan, L.; Zhang, Z.; Zhu, B.; Zuo, D.; Yu, X.; Wang, Y. Smart-Data-Glove-Based Gesture Recognition for Amphibious Communication. *Micromachines* **2023**, *14*, 2050. [CrossRef] [PubMed]
7. Li, T.; Sun, G.; Yu, L.; Zhou, K. HRBUST-LLPED: A Benchmark Dataset for Wearable Low-Light Pedestrian Detection. *Micromachines* **2023**, *14*, 2164. [CrossRef] [PubMed]
8. Javeed, M.; Abdelhaq, M.; Algarni, A.; Jalal, A. Biosensor-Based Multimodal Deep Human Locomotion Decoding via Internet of Healthcare Things. *Micromachines* **2023**, *14*, 2204. [CrossRef] [PubMed]
9. Chae, M.; Lee, D.; Kim, H.-D. Low-Power Consumption IGZO Memristor-Based Gas Sensor Embedded in an Internet of Things Monitoring System for Isopropanol Alcohol Gas. *Micromachines* **2024**, *15*, 77. [CrossRef] [PubMed]
10. Tajitsu, Y.; Shimda, S.; Nonomura, T.; Yanagimoto, H.; Nakamura, S.; Ueshima, R.; Kawanobe, M.; Nakiri, T.; Takarada, J.; Takeuchi, O.; et al. Application of Braided Piezoelectric Poly-l-Lactic Acid Cord Sensor to Sleep Bruxism Detection System with Less Physical or Mental Stress. *Micromachines* **2024**, *15*, 86. [CrossRef] [PubMed]
11. Yin, H.; Dávila-Montero, S.; Mason, A.J. Geometry Scaling for Externally Balanced Cascade Deterministic Lateral Displacement Microfluidic Separation of Multi-Size Particles. *Micromachines* **2024**, *15*, 405. [CrossRef] [PubMed]
12. Sonchan, P.; Ratchatanantakit, N.; O-Larnnithipong, N.; Adjouadi, M.; Barreto, A. Robust Orientation Estimation from MEMS Magnetic, Angular Rate, and Gravity (MARG) Modules for Human–Computer Interaction. *Micromachines* **2024**, *15*, 553. [CrossRef] [PubMed]
13. Yu, Z.; Lu, C.; Zhang, Y.; Jing, L. Gesture-Controlled Robotic Arm for Agricultural Harvesting Using a Data Glove with Bending Sensor and OptiTrack Systems. *Micromachines* **2024**, *15*, 918. [CrossRef] [PubMed]

**Disclaimer/Publisher's Note:** The statements, opinions and data contained in all publications are solely those of the individual author(s) and contributor(s) and not of MDPI and/or the editor(s). MDPI and/or the editor(s) disclaim responsibility for any injury to people or property resulting from any ideas, methods, instructions or products referred to in the content.

Article

# Gesture-Controlled Robotic Arm for Agricultural Harvesting Using a Data Glove with Bending Sensor and OptiTrack Systems

Zeping Yu [1], Chenghong Lu [1], Yunhao Zhang [1] and Lei Jing [2,*]

[1] Graduate School of Computer Science and Engineering, University of Aizu, Aizuwakamatsu 965-8580, Japan; d8251103@u-aizu.ac.jp (Z.Y.); d8222104@u-aizu.ac.jp (C.L.); m5272029@u-aizu.ac.jp (Y.Z.)
[2] Department of Computer Science and Engineering, University of Aizu, Aizuwakamatsu 965-8580, Japan
* Correspondence: leijing@u-aizu.ac.jp

**Abstract:** This paper presents a gesture-controlled robotic arm system designed for agricultural harvesting, utilizing a data glove equipped with bending sensors and OptiTrack systems. The system aims to address the challenges of labor-intensive fruit harvesting by providing a user-friendly and efficient solution. The data glove captures hand gestures and movements using bending sensors and reflective markers, while the OptiTrack system ensures high-precision spatial tracking. Machine learning algorithms, specifically a CNN+BiLSTM model, are employed to accurately recognize hand gestures and control the robotic arm. Experimental results demonstrate the system's high precision in replicating hand movements, with a Euclidean Distance of 0.0131 m and a Root Mean Square Error (RMSE) of 0.0095 m, in addition to robust gesture recognition accuracy, with an overall accuracy of 96.43%. This hybrid approach combines the adaptability and speed of semi-automated systems with the precision and usability of fully automated systems, offering a promising solution for sustainable and labor-efficient agricultural practices.

**Keywords:** gesture control; robotic arm; agricultural harvesting; data glove; bending sensors; OptiTrack; machine learning; CNN+BiLSTM; spatial tracking; ergonomic design

**Citation:** Yu, Z.; Lu, C.; Zhang, Y.; Jing, L. Gesture-Controlled Robotic Arm for Agricultural Harvesting Using a Data Glove with Bending Sensor and OptiTrack Systems. *Micromachines* **2024**, *15*, 918. https://doi.org/10.3390/mi15070918

Academic Editor: Muhammad Ali Butt

Received: 2 June 2024
Revised: 6 July 2024
Accepted: 11 July 2024
Published: 16 July 2024

**Copyright:** © 2024 by the authors. Licensee MDPI, Basel, Switzerland. This article is an open access article distributed under the terms and conditions of the Creative Commons Attribution (CC BY) license (https:// creativecommons.org/licenses/by/ 4.0/).

## 1. Introduction

With the continuous growth of the global population, food issues around the world are becoming increasingly severe. The United Nations Department of Economic and Social Affairs reported that the world population reached 7.942 billion in 2022, with projections indicating increases to 8.512 billion by 2030 and 9.687 billion by 2050 [1]. Concurrently, data from the International Labor Organization reveal a downward trend in the employment in agriculture, forestry, and fishing, decreasing from approximately 968.475 million in 2010 to about 855.386 million by 2020 [2]. This decline is further compounded by demographic shifts in developed nations, particularly the aging population issue, which is steadily eroding the agricultural workforce. A case in point is Japan, where, based on Agricultural Census data, the number of individuals engaged in agriculture dropped from 1.757 million in 2015 to 1.363 million in 2020, and the percentage of those aged 60 and above rose from 78.7% to 79.9% [3]. The confluence of a shrinking and aging agricultural workforce, against the backdrop of a burgeoning global population, portends a looming food crisis.

Mechanization and automation have emerged as pivotal solutions to the crisis faced in agriculture. Statistical analysis reveals a significant reduction in labor requirements for China's three principal cereal crops, decreasing from 13.80 labor days per mu (a unit of area measurement commonly used in China equivalent to approximately 666.67 m$^2$) in 1998 to 4.81 labor days per mµ in 2018, marking a substantial decline of 65.14%. In contrast, advancements in the mechanization and automation of the fruit sector lag behind those of the grain industry. Labor inputs per mu for apple production saw a decrease from 48.70 labor days in 1998 to 33.85 labor days in 2018 (a reduction of 14.85 labor days), translating to a modest decline of 30.49%. The Ministry of Agriculture and Rural Affairs of China estimates that the comprehensive mechanization rate for major cereal crops surpassed 80% in 2018 [4].

At this moment, fruit production is still significantly dependent on human labor, despite the potential for mechanization in specific segments of the process. The stages that demand the highest amount of labor include pruning, pollination, bagging, and harvesting, with the latter being the most labor-intensive. The necessity for such extensive manual labor stems, mainly due to the irregular spatial growth patterns of many fruits, hinders the widespread adoption and effectiveness of mechanization and automation technologies. Although numerous studies have been aimed at developing automated systems for fruit production, the operational speed of these systems significantly lags behind human performance, and they often come with a steep learning curve [5]. Therefore, this research focuses on the development of a user-friendly robotic arm controller that enables efficient fruit harvesting, among other agricultural tasks, through gesture recognition for remote control, effectively mimicking human actions.

To achieve the aforementioned objectives, efforts must be concentrated on three fronts. First, we must define and recognize easily executable human hand gestures to control certain behaviors of the robotic arm, such as power on/off or opening/closing of the end effector. Second, it is imperative to accurately capture the movement distance of the human hand and replicate its coordinates for the robotic arm. Third, a system must be established that processes the aforementioned data and sends control commands to the robotic arm.

Initially, to identify human hand gestures, we fabricated a data glove equipped with bending sensors at the joints of the hand. This glove is capable of measuring the voltage changes in the bending sensor detection circuit and transmitting the data to a server. Subsequently, by placing three reflective markers on the glove, we employed an OptiTrack system to acquire motion data. The associated Motive (Body 3.0.1 Final) software computes the coordinate data and dispatches them to the server. Finally, by executing custom-developed data transmission and processing software on the server, machine learning and deep learning algorithms are utilized to recognize the bending sensor data transmitted by the data glove. This recognition process identifies hand gestures and sends the corresponding control commands to the robotic arm. Concurrently, the coordinate data received from the OptiTrack system are converted and relayed as coordinate control commands to the robotic arm.

Existing solutions and their limitations can be summarized as follows:

- Fully automated systems: These systems promise reduced human labor but suffer from slow operational speeds, inefficiencies in adapting to various crop types and environmental conditions, and high initial and maintenance costs.
- IMU-based data gloves: While enhancing interactivity between human operators and robotic systems, these gloves are prone to drift errors and require complex calibration, affecting precision crucial for tasks like fruit harvesting.
- OptiTrack systems: Despite offering superior accuracy and lower latency, they face challenges in environments with potential obstructions to the line of sight, leading to inaccuracies in gesture recognition.

The contributions in this paper can be summarized as follows:

- An integrated system combining bending sensors and an OptiTrack system was developed for precise gesture recognition and spatial tracking.
- The convenience and accuracy of robotic control were enhanced through advanced hand gesture recognition.
- By leveraging the complementary strengths of bending sensors and the OptiTrack system, issues associated with IMU-based data gloves, such as spatial coordinate drift, were mitigated, and the keypoint loss problem in OptiTrack's hand movement tracking was addressed.

The remainder of this paper is structured as follows. This work is compared with other data gloves and robotic arm control systems, discussing the systematic design choices made in both hardware and software aspects. Moreover, the paper evaluates the gesture accuracy of several research subjects and the response speed of the robotic arm to movement and gesture recognition, followed by a discussion of the results. The final chapter summarizes the main findings in this work and provides an outlook on future research.

## 2. Related Works

The automation of fruit harvesting has seen various technological interventions, primarily categorized into fully automated systems and semi-automated systems that incorporate human operators. The development of fully automated fruit harvesting robots, such as those discussed by Yoshida et al. (2022) [5] and Majeed et al. (2022) [6], has primarily focused on providing a complete mechanization solution that promises to reduce human labor. However, these systems are often hampered by slow operational speeds and inefficiencies, particularly when adapting to diverse types of crops or varying environmental conditions. The high initial setup and maintenance costs, coupled with their limited flexibility, render these systems less feasible for widespread adoption.

In contrast, data gloves equipped with Inertial Measurement Units (IMUs) represent a significant advancement in enhancing the interactivity between human operators and robotic systems. Lu et al. (2023) [7] provided detailed hand-tracking capabilities such devices, which offer real-time spatial tracking and have been utilized in various applications, ranging from virtual reality to interactive robotics. Despite their versatility, IMUs are prone to drift errors and often require complex calibration procedures to maintain accuracy, as highlighted by Lin et al. (2018) [8] and Rodić et al. (2023) [9]. These limitations significantly impact the precision required for tasks like fruit harvesting, where delicate handling and exact positioning are crucial.

To address the spatial tracking issues inherent in IMU systems, OptiTrack systems have been employed due to their use of high-precision cameras and reflective markers for motion tracking. This technology offers superior accuracy and lower latency compared to IMU-based systems, making it suitable for applications requiring high precision. However, as noted by the Comparative Analysis of OptiTrack Motion Capture Systems (2018) [10], these systems can face challenges in environments where the line of sight can be obstructed, leading to inaccuracies in gesture recognition. Such occlusions, common in outdoor agricultural settings, can significantly reduce the efficacy of tracking systems.

Table 1 below provides a comparison of various data gloves based on their performance metrics for hand gesture recognition. It includes details about the type of sensors used for detecting hand gestures and positioning, the cost of the gloves, and the corresponding references. While commercial systems such as HaptX, Manus Meta, and TactGlove DK2 may appear more robust, they are not necessarily smaller or cheaper. Our glove costs approximately USD 100 and could be even more affordable in mass production. Additionally, it is important to note that TactGlove DK2, while being the least expensive option, does not inherently possess any gesture recognition or positioning capabilities. It relies entirely on external VR cameras for these functions, which contributes to its lower price.

Table 1. Comparison of data gloves for hand gesture recognition.

| Data Glove | Hand Gesture Sensor | Hand Position Sensor | Glove Costs | Reference |
|---|---|---|---|---|
| MIMU data glove | IMU sensor | IMU sensor | USD 200 | [7] |
| Inertial sensor-based data glove | IMU sensor | IMU sensor | USD 500 | [8] |
| HaptX Gloves G1 (HaptX Inc., Seattle, WA, USA) | IMU sensor | Magnetic capture system | USD 5495 | [11] |

Table 1. *Cont.*

| Data Glove | Hand Gesture Sensor | Hand Position Sensor | Glove Costs | Reference |
|---|---|---|---|---|
| XSENS PRIME 3 (MANUS Meta., Eindhoven, The Netherlands) | IMU and bending sensor | VR device camera | USD 4000 | [12] |
| TactGlove DK2 (bHaptics, Daejeon, Republic of Korea) | VR device camera | VR device camera | USD 250 | [13] |
| Bending sensor and OptiTrack-based data glove | Bending sensor | OptiTrack system | USD 100 | This work |

In recent years, deep learning algorithms have made significant breakthroughs in sensor-based gesture recognition. For instance, Guan Yuan et al. [14] proposed a hand gesture recognition system using a deep feature fusion network based on wearable sensors. This glove includes two armbands and an integrated three-dimensional bending sensor capable of capturing fine-grained movements of the entire arm and all finger joints, with an LSTM model using fused feature vectors as input, yielding excellent results. Yongfeng et al. [15] proposed a dynamic gesture recognition algorithm (DGDL-GR), which achieved promising results by capturing finger movements and bending data. Jiawei Wu et al. [15] further advanced this field by introducing a gesture recognition method that combines Convolutional Neural Networks (CNNs) and Bidirectional Long Short-Term Memory networks (BiLSTMs), incorporating an attention mechanism to enhance recognition accuracy and robustness. Additionally, Yang Song et al. [16] utilized a wearable wrist sensor made from flexible pressure sensors integrating CNN and BiLSTM models for gesture recognition, demonstrating the potential of this approach. These methods not only capture complex finger movements and bending degrees but also achieve precise gesture recognition through deep learning algorithms. The combination of the powerful feature extraction capabilities of CNNs and the sensitivity of BiLSTMs to temporal information significantly enhances the performance of gesture recognition systems based on bending sensors. These advancements highlight the crucial role of deep learning algorithms in improving the accuracy and robustness of gesture recognition systems.

Furthermore, recent advancements in machine learning and sensor technology have opened new avenues for enhancing these hybrid systems. As demonstrated by Ran Bi (2023) [17], integrating machine learning algorithms with sensor data can significantly improve the adaptability and efficiency of gesture recognition systems, paving the way for more responsive and intuitive control mechanisms in agricultural robotics. This hybrid approach addresses the critical shortcomings of fully automated systems, such as adaptability and speed, while overcoming the spatial accuracy and occlusion issues prevalent in traditional data glove systems. For example, combining vision and bending sensor data can enhance recognition performance. The multi-modal fusion gesture recognition system proposed by Lu et al. (2021) [18] successfully integrates camera data and data glove data, improving the recognition rate of gestures under occlusion. However, the introduction of video data results in higher computational costs. The integration of these technologies presents a promising avenue for developing more efficient and flexible robotic solutions for fruit harvesting, potentially transforming agricultural practices to be more sustainable and less labor-intensive.

Moreover, the human–robot interaction (HRI) aspect of semi-automated systems, which is crucial for tasks requiring high levels of precision and adaptability, has not been fully explored. Current systems often do not account for the ergonomic and cognitive loads placed on human operators, which can affect the overall efficiency and adoption of these technologies. As highlighted by Rodić et al. (2023) [9], enhancing the intuitive aspects of human–machine interfaces and reducing the cognitive burden through better design and integration of feedback mechanisms are critical areas needing attention.

In light of these challenges, this study proposes a novel hybrid approach that combines the OptiTrack system with bending sensors integrated into a data glove. This method not only leverages the high spatial accuracy of OptiTrack but also incorporates the flexibility and resilience of bending sensors to provide robust gesture recognition. even in complex and dynamic agricultural environments. This approach effectively bridges the gap between the adaptability and speed of fully automated systems and the accuracy and usability of semi-automated systems.

## 3. System Architecture and Design

### 3.1. Overview

Figure 1 shows the system architecture, which consists of the following three main components: the robotic arm controller, the server, and the robotic arm. The robotic arm controller consists of the data glove, six OptiTrack cameras, and a video monitor. The data glove has ten bending sensors for gesture recognition and three reflective markers for captured hand coordinates. The server side consists of three different programs; Motive is responsible for processing the video data from the OptiTrack camera and converting it into coordinate data, the gesture recognition program is responsible for decoding the received data and sending the recognized results to the data processing program, the data processing program sends the control commands to the robotic arm based on the recognized results, and the data between the programs are exchanged through sockets. The arm is controlled by a built-in Raspberry Pi, which operates six joint motors and end effectors after receiving the control commands. The robotic arm is also equipped with a Wi-Fi camera, which is used to remotely transmit the real-time image to the video monitor of the controller.

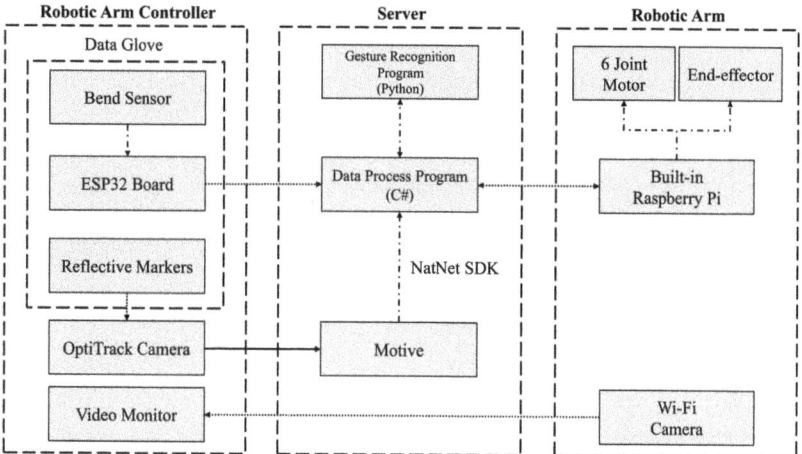

**Figure 1.** System block diagram. Solid arrows means that the data are transmitted through a wire, dashed arrows mean that the data are transmitted wirelessly, and dotted arrows mean that the data are transmitted within the system.

### 3.2. Hardware Design

#### 3.2.1. Data Glove

The data glove contains ten BS-65 bending sensors from Sensia Technology, three reflective markers provided by the OptiTrack system, and a custom-built board with resistance detection and charging/discharging circuits powered by an ESP32-S3 microcontroller (Espressif Systems, Shanghai, China) .

The response curve of the BS-65 bending sensor is shown in Figure 2. Its resistance is 20 k$\Omega$ ± 10% when unbent and changes from about +250% to −40% when bent. We followed the method described by A. K. Bose et al. [19] to test the impact of stress on this bending sensor. The calculated results are presented as follows: at 5 N, $\Delta R/R = 0.585$; at 10 N, $\Delta R/R = 1.269$; at 15 N, $\Delta R/R = 2.253$; at 20 N, $\Delta R/R = 3.194$. To avoid the influence of stress on the bending sensor's readings, we fixed only one end of the sensors to the glove during its fabrication, allowing the sensor to move freely within a certain range.

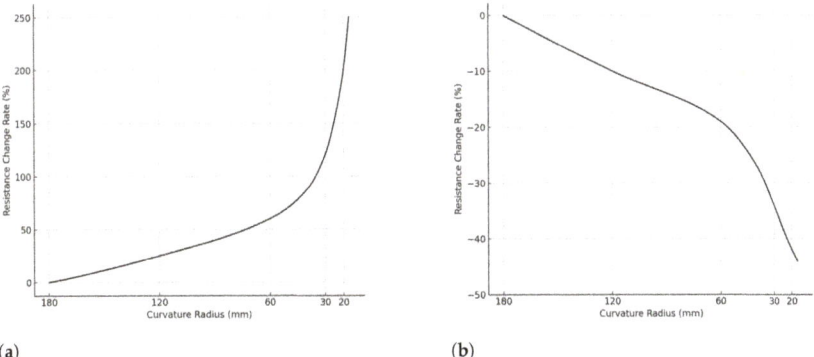

(a) (b)

**Figure 2.** BS-65 bending sensor response curve. When bent in the direction that stretches the sensor surface, the resistance increases. When bent in the direction that compresses the sensor surface, the resistance decreases. (**a**) Tensile strength of the sensor surface. (**b**) Compressive strength of the sensor surface.

Figure 3 shows the custom-designed ESP32 board, which includes battery charging/discharging management, Wi-Fi data transmission, Real-Time Clock (RTC), and Analog-to-Digital Conversion (ADC) functions. The board's through holes are specially designed. It has larger, oval-shaped solder pads and holes to facilitate easy attachment to the data glove through sewing.

**Figure 3.** The custom-designed ESP32 board. The board features an ESP32S3-WROOM1 module (left) for Wi-Fi and ADC functionalities, operational amplifiers (U7, U8, and U9), a real-time clock (U5), and a battery management system (U3).

Figure 4 shows the data glove equipped with bending sensors and OptiTrack reflective markers. The reflective markers, small spheres made of reflective material, efficiently reflect infrared light emitted by the OptiTrack system's cameras, allowing for precise position tracking. These markers are attached to the data glove using Velcro straps. The glove itself is integrated with bending sensors sewn into the fabric to accurately measure hand movements.

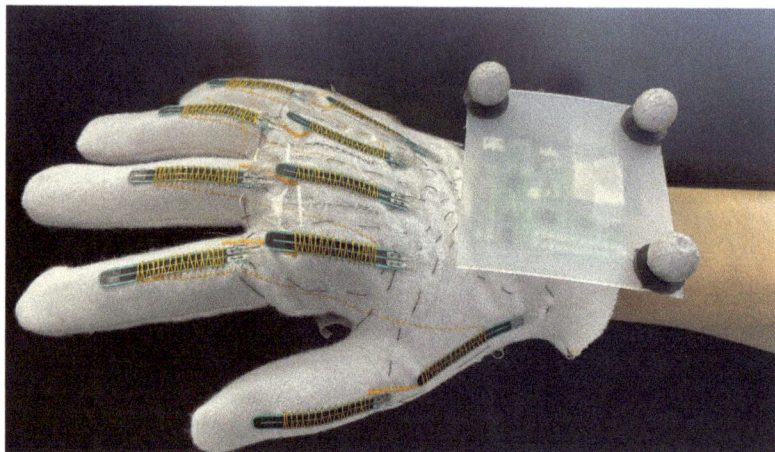

**Figure 4.** Bending sensor data glove, consisting of ten bend sensors fixed on cloth.

Figure 5 shows the resistance measurement circuit on the ESP32 board; Figure 5a is an inverted signal amplifier circuit used to convert the change in measured resistance to a change in voltage for easy measurement with the following formula:

$$V_{out} = -\frac{R_{fb}}{R_{bs}} \times V_{ref} \quad (1)$$

where $V_{out}$ represents the output voltage, which is directly connected to the AD conversion pin of the ESP32; $R_{bs}$ is the resistance of the bending sensor; $R_{fb}$ is the feedback resistor of the inverting signal amplification circuit; and $V_{ref}$ is the reference voltage at the output of the circuit shown in Figure 5b.

The reference voltage circuit consists of a resistor divider circuit and a voltage follower, and the value of $V_{ref}$ is calculated by the following formula:

$$V_{ref} = \frac{R_1}{R_1 + R_2} \times V_{dd} \quad (2)$$

where $V_{dd}$ is 3.3 V, serving as the system's supply voltage. The system is powered by a battery and includes an LDO chip (Texas Instruments TLV75733). $R_1$ and $R_2$ are the resistors in the voltage divider circuit, with resistance values of 1.8 kΩ and 200 Ω, respectively.

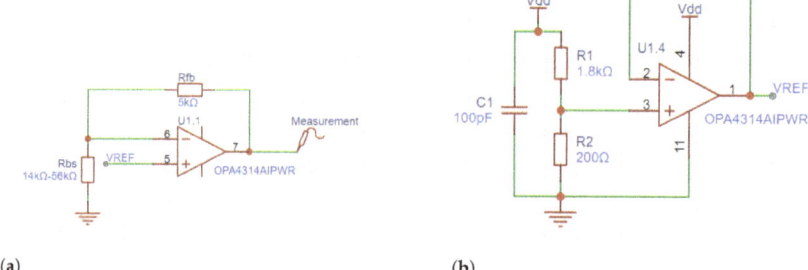

(a) (b)

**Figure 5.** The resistance measurement circuit; these circuits are used to measure and convert resistance changes in the bending sensor to voltage changes for easy measurement. (**a**) The inverting signal amplifier circuit. (**b**) The reference voltage circuit.

The circuit depicted in Figure 5 was simulated using PSpice for TI 17.4-2023, with the results presented in Figure 6. The simulations varied the feedback resistor ($R_{fb}$) at values of 90 kΩ, 110 kΩ, and 130 kΩ, while the bending sensor resistance ($R_{bs}$) ranged from 10 kΩ to 60 kΩ. The output voltage ($V_{out}$) exhibited significant differences under these conditions. Given that the minimum resistance of $R_{bs}$ is approximately 14 kΩ, utilizing a 130 kΩ $R_{fb}$ might position the operating point within a cutoff region, while a 90 kΩ $R_{fb}$ could result in insufficient sensitivity when $R_{fb}$ exceeds 50 kΩ. After a comprehensive evaluation, an $R_{fb}$ value of 110 kΩ was determined to optimize both the measurement range and sensitivity, ensuring a balanced operational profile.

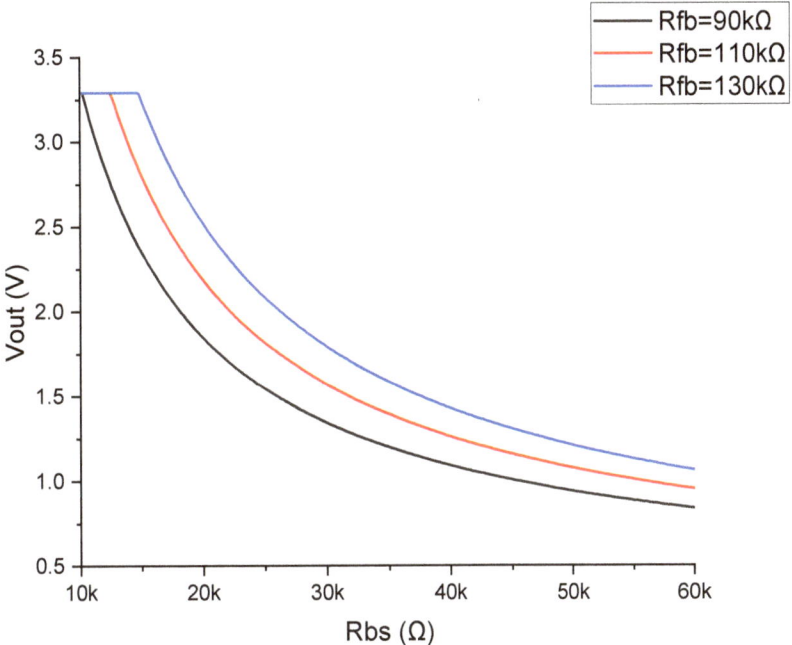

**Figure 6.** The output voltage ($V_{out}$) versus the resistance of the bending sensor ($R_{bs}$) for different feedback resistor values ($R_{fb}$) in the inverting signal amplifier circuit. The graph shows the response curves for $R_{fb}$ values of 90 kΩ (black), 110 kΩ (red), and 130 kΩ (blue).

### 3.2.2. Robotic Arm

The utilized robotic arm is a commercially available model from Elephant Robotics. As illustrated in Figure 7, myCobot 320 Pi adopts a Raspberry Pi microprocessor. Its body weights 3 kg with a load of 1 kg and a working radius of 320 mm. A total of six degrees of freedom can be achieved using this robotic arm. The end effector shown in Figure 10 would be considered an additional degree of freedom, as it can close its claws there by cutting in the desired way.

### 3.3. Software Design

As illustrated in Figure 8, due to variations in sampling rates across different devices, linear processing methods can lead to significant latency. To reduce the overall system delay, the software design is segmented into four specialized programs.

The ESP32 board program (C#, Visual Studio 2022 IDE) leverages the ADC functionality of the ESP32 and the circuit detailed in Figure 5 to read voltage values from bending sensors and transmit them to the server. The deep learning program(Python 3.12.4) is tasked with importing real-time data into a pre-trained model and exporting the outcomes. The data process program(C, Arduino IDE 2.3.2) oversees data processing and exchange within the entire system, transmitting sensor data from the data glove to the deep learning program and converting the results from the deep learning program into control commands for the robotic arm program. Additionally, it translates coordinate data from OptiTrack into further control commands. The robotic arm program forwards these commands to the motor and end effector, calculating delays via a connected RTC. All four components are designed to operate simultaneously upon system startup.

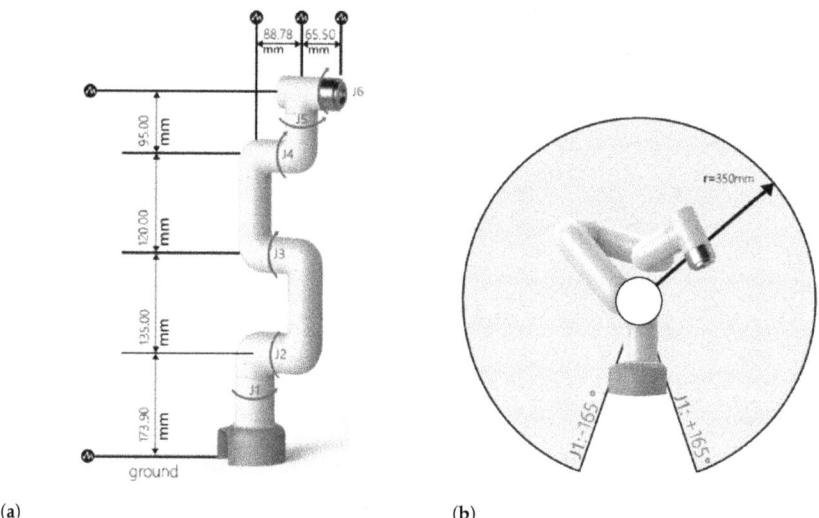

(a) (b)

**Figure 7.** Schematic diagram of the robotic arm showing its dimensions and range of motion. (a) The figure illustrates the lengths of each segment of the robotic arm and their respective joints (J1 to J6), with measurements provided in millimeters. (b) The figure illustrates the range of motion.

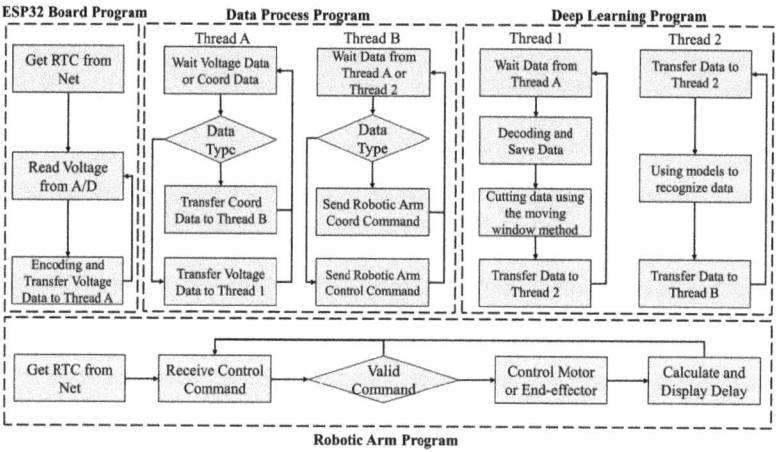

**Figure 8.** System architecture and data flow for the gesture-controlled robotic arm. The architecture comprises the following four main programs: the ESP32 board program, data processing program, deep learning program, and robotic arm program.

### 3.4. Proportion Calculation for Elimination of Systematic Errors

To address and eliminate systematic errors in the robotic arm, it is essential to account for any potential delay between the coordinates of the human hand and the robotic arm. This delay can be calculated based on the sampling frequency and time differences observed in the data.

First, the data are aligned by shifting the robotic arm's coordinate data forward by the calculated number of frames. This temporal alignment ensures that the data from the robotic arm correspond accurately to the data from the human hand.

To further eliminate systematic errors, the proportion factors between the coordinates of the robotic arm and the human hand are computed using linear regression. This method involves finding the best-fitting line that minimizes the sum of squared differences between the observed values and the values predicted by the model. The linear regression equations are

$$x_m = k_x \cdot x_h + b_x$$
$$y_m = k_y \cdot y_h + b_y \qquad (3)$$
$$z_m = k_z \cdot z_h + b_z,$$

where $x_m$, $y_m$, and $z_m$ are the coordinates of the robotic arm; $x_h$, $y_h$, and $z_h$ are the coordinates of the human hand; $k_x$, $k_y$, and $k_z$ are the proportion factors; and $b_x$, $b_y$, and $b_z$ are the biases for the X, Y, and Z coordinates, respectively.

The linear regression model is fitted to determine the proportion factors and biases. This fitting process helps to precisely model and correct the relationship between the coordinates of the robotic arm and the human hand, thereby eliminating systematic errors in the robotic arm's movements.

*3.5. Deep Learning Models for Gesture Recognition*

In the development of the hand gesture recognition system for robotic control, experiments were conducted with three deep learning structures, namely LSTM, BiLSTM, and a combination of CNN with BiLSTM.

LSTMs are ideal for this application because they handle sequences, such as the readings from the glove's bend sensors, by retaining information for extended durations. This capability is crucial for predicting sequences of hand positions.

BiLSTMs build on LSTMs by analyzing sequences both from the beginning to the end and vice-versa. This two-way analysis is better for recognizing complex gestures, as it considers what comes before and after in a sequence.

This combined model, CNN+BiLSTM, starts with CNN layers that pull out important spatial features from sensor data, then passes these on to BiLSTM layers. This mix is particularly effective at capturing patterns related to both space and time, which helps make recognition more accurate and reliable.

As shown in Figure 9, the sensor data first pass through a 1D CNN layer, which detects important spatial patterns, specifically the relationship between the curved sensors in a frame. Batch normalization ensures that the model learns efficiently and consistently.

Next comes a Leaky ReLU activation function, which introduces non-linearity. This means the network can learn more complex patterns, which is essential for differentiating subtle hand movements.

Two BiLSTM layers form the core of the setup. They are adept at understanding long-term patterns in the sensor data, considering what comes both before and after in a sequence. This provides a strong foundation for accurately classifying gestures.

Another batch normalization and Leaky ReLU set up the final BiLSTM layer. This is crucial when recognizing complex gestures that change significantly over time. This hybrid approach effectively captures both spatial and temporal features, significantly improving the recognition accuracy and robustness against noisy data. However, due to the increase in model expressiveness, the risk of overfitting is exacerbated.

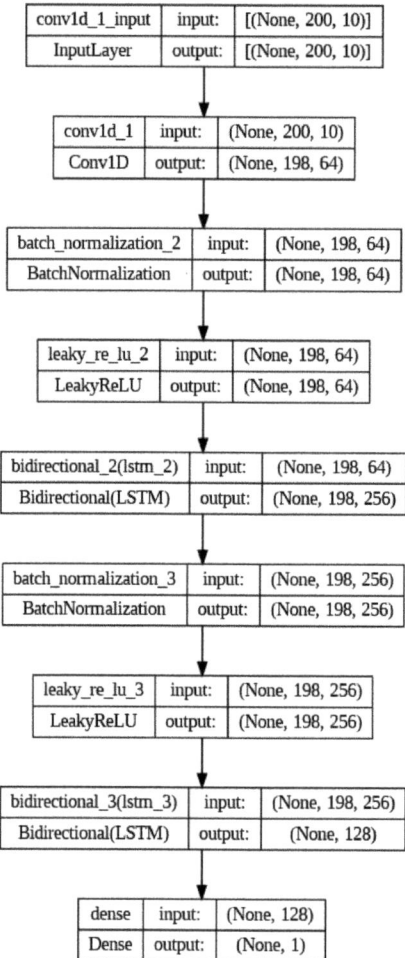

**Figure 9.** CNN+BiLSTM network architecture.

## 4. Experiments and Results

The system primarily achieves the following two functions: following the trajectory of hand movements with the end effector of the robotic arm and recognizing hand gestures to execute corresponding functions. Therefore, the validation of the two experiments focuses on the spatial trajectory error between the hand and the robotic arm and the accuracy of hand gesture recognition.

*4.1. Experiment 1: Spatial Trajectory Error*

In this experiment, data were collected for one minute of back-and-forth movement along each motion axis, as shown in Figure 10, totaling 3 min and including 1800 frames of data. For the spatial trajectory, the displacement ratio between the hand and the robotic arm was adjusted to be one to one. A marker was also placed on the end-effector of the robotic arm to track its trajectory. The errors between the trajectories of the hand and the robotic arm were then compared.

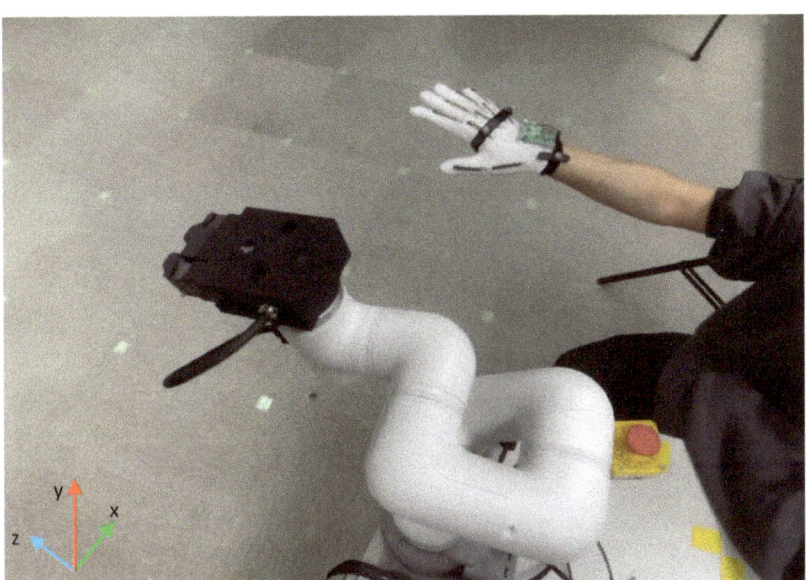

**Figure 10.** Experimental setup showing a hand wearing the data glove and the robotic arm.

To address and eliminate systematic errors in the robotic arm, as shown in Figure 11, it is essential to evaluate the stability of the delay between the coordinates of the human hand and the robotic arm. Based on the collected delay data, several statistical measures were computed to assess this stability. The mean delay was found to be 0.325 s, representing the average time difference between the movements of the human hand and the corresponding response of the robotic arm. The standard deviation of the delays was 0.0625 s, indicating the amount of variation or dispersion from the average delay. A lower standard deviation suggests that the delays are more consistent. Additionally, the interquartile range (IQR) of the delays was calculated to be 0.09 s, reflecting the range within which the middle 50% of the delay values fall. The IQR is a robust measure against outliers, providing a clear picture of the variability in the delays. The first quartile (Q1) was 0.275 s, and the third quartile (Q3) was 0.365 s. These statistical measures collectively demonstrate the effectiveness of the delay correction process in achieving stable and predictable synchronization between the human hand and the robotic arm.

To evaluate the spatial trajectory error between the hand and the mechanical hand, the following three metrics were calculated: mean Euclidean distance, the standard deviation of the Euclidean distance, and root mean square error (RMSE). These metrics provide a quantitative assessment of the accuracy and precision of the mechanical hand's movements in replicating the intended hand gestures.

The Euclidean distance measures the straight-line distance between corresponding points on trajectories of the hand and the mechanical hand. It is a direct measure of the deviation at each time point. The Euclidean distance is calculated using the following formula:

$$E = \sqrt{(x_1 - x_2)^2 + (y_1 - y_2)^2 + (z_1 - z_2)^2} \tag{4}$$

RMSE provides an aggregated measure of the overall error by considering the squared differences between the trajectories of the hand and mechanical hand, thereby giving more weight to larger errors. RMSE is calculated using the following formula:

$$\text{RMSE} = \sqrt{\frac{1}{n} \sum_{i=1}^{n} [(x_{1_i} - x_{2_i})^2 + (y_{1_i} - y_{2_i})^2 + (z_{1_i} - z_{2_i})^2]} \tag{5}$$

where $(x_1, y_1, z_1)$ are the coordinates of the hand and $(x_2, y_2, z_2)$ are the coordinates of the end effector.

The results of this experiment demonstrate that the mean Euclidean distance between the hand and the mechanical hand is 0.0131 m, while the root mean square error (RMSE) is 0.0095 m. Additionally, the standard deviation of the Euclidean distance is 0.0100 m.

These results indicate a high level of precision in the mechanical hand's ability to replicate the intended hand movements, with minimal deviation and error.

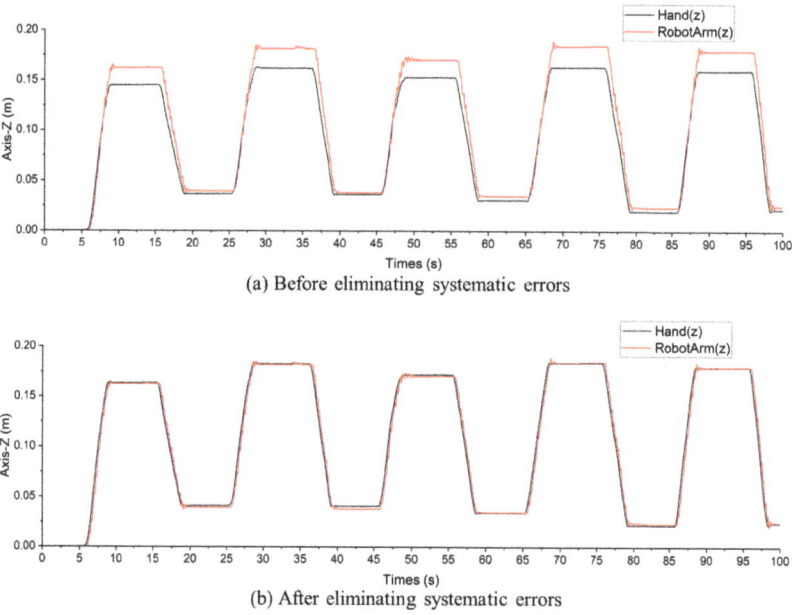

**Figure 11.** Data comparison before and after eliminating systematic errors. (**a**) Discrepancies between the robotic arm and human hand coordinates. (**b**) Corrected data demonstrating improved alignment after addressing systematic errors.

### 4.2. Experiment 2: Hand Gesture Recognition Accuracy

In the second experiment, the accuracy of hand gesture recognition was evaluated. A dataset was amassed comprising seven types of hand gestures with a temporal dimension, as shown in Figure 12. These gestures include rest, show 1 (index finger up), show 2 (index and middle fingers up), claw, fist, pinch with index finger and thumb, and all-finger pinch. Each gesture was recorded at a sampling rate of 100 Hz for a 2 s duration, ensuring a variety of temporal states and initiating positions were captured. Each gesture was performed ten times under ten distinct conditions, resulting in 100 data samples per gesture and a comprehensive total of 1400 s of data.

The collected data were meticulously divided into training, validation, and testing sets with a ratio of 6:2:2. This split was designed to provide a robust training framework while retaining sufficient data for effective model validation and testing.

In terms of model training, a low learning rate of 0.00001 was set to fine tune the network's adjustments during learning. 'Sparse categorical cross entropy' was utilized as the loss function due to its suitability for multi-class classification tasks. The Adam optimizer facilitated the learning process over 300 epochs, with dropout implemented to combat overfitting. The model learned to discern the subtleties between different hand gestures, which is critical for accurate classification.

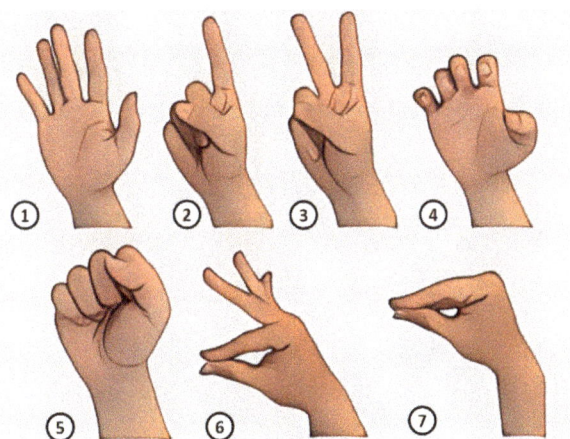

**Figure 12.** Gestures: ① rest, ② show 1, ③ show 2, ④ claw, ⑤ fist, ⑥ pinch with index finger and thumb, and ⑦ all-finger pinch.

This meticulous approach to data collection and the deliberate choice of model parameters were fundamental in developing a hand gesture recognition system capable of interpreting nuanced human gestures for the control of robotic hands.

In this experimental setup, each model was trained and validated on a dataset comprising seven distinct hand gestures. The CNN+BiLSTM model outperformed the standalone LSTM and BiLSTM models in terms of accuracy and processing speed. This superior performance can be attributed to its ability to leverage both spatial and temporal dynamics, which is critical for dynamic and accurate gesture recognition in real-time applications.

The performance of three different models for hand gesture recognition was compared, namely CNN+BiLSTM, BiLSTM, and LSTM. The CNN+BiLSTM model exhibited the highest performance, achieving an accuracy of 0.9500, a precision of 0.9531, a recall of 0.9500, and an F1 score of 0.9503. As shown in Table 2, this model, with 365,511 total parameters, effectively recognized and classified the various hand gestures. The combination of convolutional neural networks (CNNs) for feature extraction and bidirectional long short-term memory (BiLSTM) networks for temporal dependencies allowed the model to capture the intricate details of hand gestures, thus demonstrating its superior capability in this task.

**Table 2.** Performance metrics of different models for hand gesture recognition.

|  | Accuracy | Precision | Recall | F1-Score | Total Parameters |
| --- | --- | --- | --- | --- | --- |
| CNN+BiLSTM | 0.9643 | 0.9669 | 0.9643 | 0.9642 | 365,511 |
| BiLSTM | 0.8857 | 0.8908 | 0.8857 | 0.8821 | 80,071 |
| LSTM | 0.6929 | 0.7229 | 0.6929 | 0.6878 | 19,655 |

The BiLSTM model was less complex than the CNN+BiLSTM model yet still maintained high classification performance. The reduction in model complexity did not significantly compromise its accuracy, making it a viable option for applications requiring a balance between performance and computational efficiency.

In contrast, the LSTM model, having only 19,655 parameters, demonstrated the limitations of a simpler architecture in accurately recognizing hand gestures. While it required less computational power, its significantly lower performance highlights the necessity of more sophisticated models like CNN+BiLSTM or BiLSTM for tasks demanding high precision and reliability in gesture recognition.

The confusion matrix shows that the CNN+BiLSTM model achieves high accuracy across most hand gesture categories, with gestures 0, 4, 5, and 6 classified correctly 100% of the time. Minor misclassifications occur in gestures 1, 2, and 3, indicating similarities that make them harder to distinguish. Overall, the model demonstrates robust performance, with a majority of gestures being accurately classified.

As shown in Figures 13 and 14, the confusion matrix shows that the CNN+BiLSTM model achieves high accuracy across most hand gesture categories, with gestures 0, 4, 5, and 6 classified correctly 100% of the time. Minor misclassifications occur for gestures 1, 2, and 3, indicating similarities that make them harder to distinguish. Overall, the model demonstrates robust performance, with a majority of gestures being accurately classified.

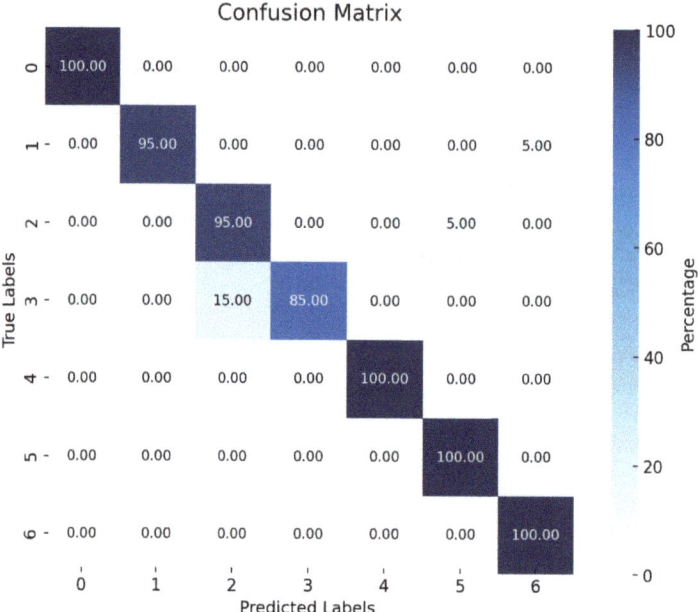

**Figure 13.** CNN+BiLSTM confusion matrix. (The darker the blue, the higher the recognition rate percentage).

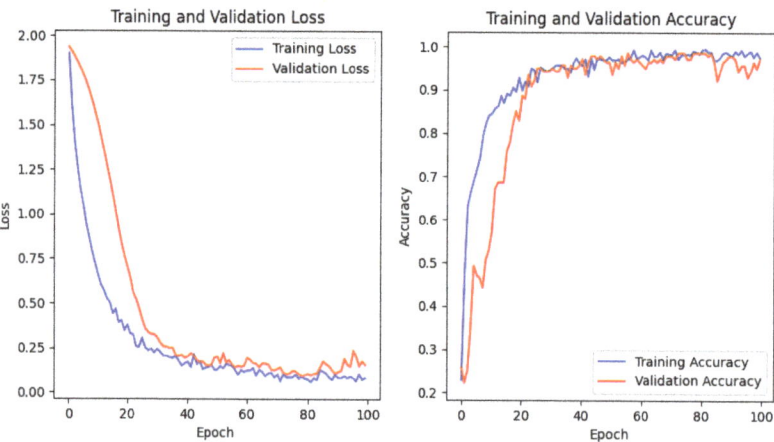

**Figure 14.** CNN+BiLSTM loss and accuracy.

The training and validation loss and accuracy graphs indicate effective learning with minimal overfitting. Both training and validation loss decrease smoothly, while the accuracy curves rise and stabilize close to each other, reflecting strong generalization to unseen data. These results highlight the model's capability for accurate and reliable hand gesture recognition, making it suitable for real-world applications.

## 5. Discussion and Conclusions

The results of this study indicate that combining bending sensors and OptiTrack systems in a data glove to control a robotic arm in agricultural harvesting offers significant advantages over traditional methods. This hybrid approach addresses the critical shortcomings of fully automated systems, such as adaptability and speed, while overcoming the spatial accuracy and occlusion issues prevalent in traditional data glove systems. The integration of these technologies presents a promising avenue for developing more efficient and flexible robotic solutions for fruit harvesting, potentially transforming agricultural practices to be more sustainable and less labor-intensive.

Experiment 1 demonstrated that the spatial trajectory error between the hand and the robotic arm was minimal. By evaluating the Euclidean distance and root mean square error (RMSE), it was found that the robotic arm was able to closely follow the hand movements with a mean Euclidean distance of 0.0131 m, a standard deviation of the Euclidean distance of 0.0100 m, and an RMSE of 0.0095 m. This high level of precision indicates that this system can accurately replicate intended hand movements, which is crucial for delicate tasks like fruit harvesting.

The CNN+BiLSTM model outperformed both the standalone LSTM and BiLSTM models, achieving the highest accuracy, precision, recall, and F1-score in Experiment 2. This superior performance can be attributed to the model's ability to leverage both spatial and temporal dynamics, which is crucial for dynamic and accurate gesture recognition in real-time applications. Confusion matrix analysis revealed high accuracy across most hand gesture categories, with minor misclassifications occurring in gestures that are inherently similar, indicating the robustness of the model.

One of the key insights from this study is the necessity to balance model complexity and computational efficiency. While the CNN+BiLSTM model demonstrated the highest performance, the BiLSTM model, despite being less complex, maintained high classification performance, making it a viable option for applications where computational resources are limited. In contrast, the LSTM model, with its simpler architecture, was less effective in accurately recognizing hand gestures, highlighting the need for more sophisticated models in tasks demanding high precision and reliability.

The ergonomic aspect of the data glove was also a critical consideration. By integrating OptiTrack systems with bending sensors in a user-friendly design, the goal was to minimize the cognitive load on operators while maximizing the system's adaptability and efficiency. This integration not only improved gesture recognition accuracy in complex environments but also enhanced the ergonomic experience, making it more practical for everyday use by agricultural workers.

In conclusion, this research demonstrates that a gesture-controlled robotic arm using a data glove with bending sensors and OptiTrack systems is a feasible and effective solution for agricultural harvesting. The hybrid approach leverages the strengths of both technologies, offering high precision, adaptability, and user-friendly operation. The CNN+BiLSTM model proved to be the most effective in gesture recognition, underscoring the importance of combining spatial and temporal analysis for accurate and reliable performance.

Future research should focus on further refining the ergonomic design of the data glove to reduce operator fatigue and enhance usability. Additionally, exploring other machine learning models and integrating advanced sensor technologies could further improve the system's accuracy and efficiency. Future work should aim to integrate OptiTrack with IMUs, enabling the system to switch to IMU-based operation when visual localization is lost, ensuring continuous functionality. In addition to this integration, the system might face challenges such as occlusion and varying light conditions, affecting the reflective markers within the indoor environment. For instance, equipment or other objects might block the line of sight to the markers, or shadows cast by moving equipment could impact marker visibility. To overcome these challenges, employing multi-sensor fusion technology, such as by combining data from IMUs and cameras, could enhance the system's robustness. Furthermore, integrating temperature and humidity sensors to provide real-time calibration of the MEMS bending sensors can help eliminate the influence of environmental factors on sensor accuracy, ensuring consistent performance. Finally, field trials in diverse agricultural environments are essential to validate the system's performance under real-world conditions and identify any areas for improvement.

By addressing these aspects, more sophisticated, reliable, and user-friendly robotic solutions can be developed that will significantly contribute to the automation of agricultural tasks, thereby addressing labor shortages and increasing the efficiency and sustainability of food production.

**Author Contributions:** Conceptualization, Z.Y. and L.J.; methodology, Z.Y.; software , Z.Y., C.L., and Y.Z.; validation, Z.Y. and C.L.; data curation, C.L.; writing—original draft preparation, Z.Y. and C.L.; writing—review and editing, L.J.; visualization, Z.Y.; supervision, L.J.; project administration, L.J.; funding acquisition, L.J. All authors have read and agreed to the published version of the manuscript.

**Funding:** This work was supported by JSPS KAKENHI Grant Number 22K12114, the JKA Foundation, and NEDO Intensive Support for Young Promising Researchers Grant Number JPNP20004.

**Data Availability Statement:** The data presented in this study are available from the corresponding author upon request.

**Conflicts of Interest:** The authors declare no conflicts of interest.

## References

1. United Nations, Department of Economic and Social Affairs, Population Division. World Population Prospects 2022: Summary of Results. 2022. Available online: https://www.un.org/development/desa/pd/sites/www.un.org.development.desa.pd/files/wpp2022_summary_of_results.pdf (accessed on 13 February 2024).
2. International Labour Organization. ILO Modelled Estimates Database. 2022. Available online: https://ilostat.ilo.org/data/ (accessed on 14 February 2024).
3. Ministry of Agriculture Forestry and Fisheries. Census of Agriculture and Forestry in Japan Census Results Report. 2020. Available online: https://www.e-stat.go.jp/en/stat-search/files?stat_infid=000032172307 (accessed on 14 February 2024).
4. Zhang, X.; Du, L.; Zhu, Q. The Changes in the Patterns of Labor Use, Supervision over Hired Labor, and Agricultural Transformation: A Comparison between Grain and Fruit Production. *Rural China* **2022**, *19*, 159–180. [CrossRef]
5. Yoshida, T.; Onishi, Y.; Kawahara, T.; Fukao, T. Automated harvesting by a dual-arm fruit harvesting robot. *Robomech J.* **2022**, *9*, 19. [CrossRef]
6. Majeed, Y.; Zhang, Y. In-Depth Evaluation of Automated Fruit Harvesting in Unstructured Environments. *Machines* **2022**, *12*, 151. [CrossRef]
7. Lu, C.; Dai, Z.; Jing, L. Measurement of Hand Joint Angle Using Inertial-Based Motion Capture System. *IEEE Trans. Instrum. Meas.* **2023**, *72*, 9503211. [CrossRef]
8. Lin, J.; Lee, J. Design of an Inertial-Sensor-Based Data Glove for Hand Function Evaluation. *Sensors* **2018**, *18*, 1545. [CrossRef]
9. Rodic, A.; Stancic, A. Analysis of Sensor Data and Machine Learning Models for Gesture Recognition. In Proceedings of the 2023 8th International Conference on Smart and Sustainable Technologies (SpliTech), Split/Bol, Croatia, 20–23 June 2023. [CrossRef]
10. Furtado, J.S.; Liu, H.H.T.; Lai, G.; Lacheray, H.; Desouza-Coelho, J. Comparative Analysis of OptiTrack Motion Capture Systems. In Proceedings of the Symposium on Mechatronics, Robotics, and Control (SMRC'18)—Canadian Society for Mechanical Engineering International Congress 2018, Toronto, ON, Canada, 27–30 May 2018.
11. HaptX Inc. HaptX Gloves G1. 2024. Available online: https://haptx.com/gloves-g1/ (accessed on 30 June 2024).
12. MANUS Meta. XSENS PRIME 3. 2023. Available online: https://www.manus-meta.com/products/prime-3-haptic-xr (accessed on 30 June 2024).
13. bHaptics Inc. TactGlove DK2. 2024. Available online: https://www.bhaptics.com/tactsuit/tactglove/ (accessed on 30 June 2024).
14. Yuan, G.; Liu, X.; Yan, Q.; Qiao, S.; Wang, Z.; Yuan, L. Hand Gesture Recognition Using Deep Feature Fusion Network Based on Wearable Sensors. *IEEE Sensors J.* **2021**, *21*, 539–547. [CrossRef]
15. Dong, Y.; Liu, J.; Yan, W. Dynamic Hand Gesture Recognition Based on Signals from Specialized Data Glove and Deep Learning Algorithms. *IEEE Trans. Instrum. Meas.* **2021**, *70*, 2509014. [CrossRef]

16. Song, Y.; Liu, M.; Wang, F.; Zhu, J.; Hu, A.; Sun, N. Gesture Recognition Based on a Convolutional Neural Network–Bidirectional Long Short-Term Memory Network for a Wearable Wrist Sensor with Multi-Walled Carbon Nanotube/Cotton Fabric Material. *Micromachines* **2024**, *15*, 185. [CrossRef]
17. Bi, R. Sensor-based gesture recognition with convolutional neural networks. In Proceedings of the 3rd International Conference on Signal Processing and Machine Learning, Wuhan, China, 31 March–2 April 2023; Volume 4.
18. Lu, C.; Kozakai, M.; Jing, L. Sign Language Recognition with Multimodal Sensors and Deep Learning Methods. *Electronics* **2023**, *12*, 4827. [CrossRef]
19. Bose, A.K.; Zhang, X.; Maddipatla, D.; Masihi, S.; Panahi, M.; Narakathu, B.B.; Bazuin, B.J.; Williams, J.D.; Mitchell, M.F.; Atashbar, M.Z. Screen-Printed Strain Gauge for Micro-Strain Detection Applications. *IEEE Sens. J.* **2020**, *20*, 12652–12660. [CrossRef]

**Disclaimer/Publisher's Note:** The statements, opinions and data contained in all publications are solely those of the individual author(s) and contributor(s) and not of MDPI and/or the editor(s). MDPI and/or the editor(s) disclaim responsibility for any injury to people or property resulting from any ideas, methods, instructions or products referred to in the content.

*Article*

# Robust Orientation Estimation from MEMS Magnetic, Angular Rate, and Gravity (MARG) Modules for Human–Computer Interaction

Pontakorn Sonchan, Neeranut Ratchatanantakit, Nonnarit O-Larnnithipong, Malek Adjouadi and Armando Barreto *

Electrical and Computer Engineering Department, Florida International University, Miami, FL 33174, USA
* Correspondence: barretoa@fiu.edu

**Abstract:** While the availability of low-cost micro electro-mechanical systems (MEMS) accelerometers, gyroscopes, and magnetometers initially seemed to promise the possibility of using them to easily track the position and orientation of virtually any object that they could be attached to, this promise has not yet been fulfilled. Navigation-grade accelerometers and gyroscopes have long been the basis for tracking ships and aircraft, but the signals from low-cost MEMS accelerometers and gyroscopes are still orders of magnitude poorer in quality (e.g., bias stability). Therefore, the applications of MEMS inertial measurement units (IMUs), containing tri-axial accelerometers and gyroscopes, are currently not as extensive as they were expected to be. Even the addition of MEMS tri-axial magnetometers, to conform magnetic, angular rate, and gravity (MARG) sensor modules, has not fully overcome the challenges involved in using these modules for long-term orientation estimation, which would be of great benefit for the tracking of human–computer hand-held controllers or tracking of Internet-Of-Things (IoT) devices. Here, we present an algorithm, GMVDμK (or simply GMVDK), that aims at taking full advantage of all the signals available from a MARG module to robustly estimate its orientation, while preventing damaging overcorrections, within the context of a human–computer interaction application. Through experimental comparison, we show that GMVDK is more robust to magnetic disturbances than three other MARG orientation estimation algorithms in representative trials.

**Keywords:** MEMS MARG orientation; orientation for human–computer interaction; magnetic distortion; GMVDμK algorithm; GMVDK algorithm

**Citation:** Sonchan, P.; Ratchatanantakit, N.; O-Larnnithipong, N.; Adjouadi, M.; Barreto, A. Robust Orientation Estimation from MEMS Magnetic, Angular Rate, and Gravity (MARG) Modules for Human–Computer Interaction. *Micromachines* **2024**, *15*, 553. https://doi.org/10.3390/mi15040553

Academic Editors: Lei Jing, Zhan Zhang and Yoshinori Matsumoto

Received: 28 February 2024
Revised: 12 April 2024
Accepted: 15 April 2024
Published: 21 April 2024

**Copyright:** © 2024 by the authors. Licensee MDPI, Basel, Switzerland. This article is an open access article distributed under the terms and conditions of the Creative Commons Attribution (CC BY) license (https://creativecommons.org/licenses/by/4.0/).

## 1. Introduction

Lee et al. [1] propose that "The first micromachined accelerometer appeared in 1979 at Stanford University", referring to the device published by Roylance and Angell [2]. However, Lee et al. also point-out that "it took over 15 years before such devices became accepted mainstream products for large volume applications. In the late 1980s, surface micromachining emerged as a perceived low-cost alternative for accelerometers, aimed primarily at automotive applications" [1]. It would not be until 2009 that the motion sensing capabilities of micro electro-mechanical systems (MEMS) would be supplemented with commercial three-axis gyroscope chips [3]. Soon, the combination of a tri-axial accelerometer and a tri-axial gyroscope would be made available into packages designated as "MEMS inertial measurement units" (MEMS IMUs), which prompted the expectation that, perhaps, it would be possible to use them to track the position and orientation of objects in ways similar to the methods used to track ships and aircraft from signals generated by full-size, navigation-grade accelerometers and gyroscopes usually present in those vehicles. Those large vehicles are commonly tracked using the principles of "strapdown inertial navigation" [4–6].

Unfortunately, the past two decades have shown that the traditional signal processing approaches that are successful in tracking vehicles from navigation-grade accelerometers and gyroscopes could not merely be re-adopted when the signals are being produced by MEMS IMUs, as detailed in the next section.

## 1.1. Need for a Different Approach

The mere re-adoption of algorithms used to track large-scale vehicles is not feasible because the performance indices of low-cost MEMS accelerometers and gyroscopes are significantly below those of their full-size, navigation-, or tactical-grade counterparts. This is, essentially, the conclusion reached by Foxlin [7] after studying the motion tracking requirements for virtual environments. Foxlin clearly identifies one of the major shortcomings of MEMS IMUs: "The problem with tracking orientation using only gyros is drift", where drift is the gradually increasing error in the orientation estimate. Further, Foxlin reveals the gyroscope bias (non-zero gyroscope output when no rotation is taking place) and the gyroscope bias instability (shift in the amount of bias as time progresses) as important causes for the orientation drift error, pointing out that "Bias stability is often the critical parameter for orientation drift performance". Table 1 shows the values (extracted from Figure 8.4 in [7]) of Gyro bias stability and Gyro bias initial uncertainty that Foxlin presents as typical of the different categories (grades) of IMUs.

**Table 1.** Parameters of gyroscopes in different categories (grades) of IMUs *.

| Parameter | Commercial Grade | Tactical Grade | Navigation Grade | Strategic Grade |
|---|---|---|---|---|
| Gyro bias initial uncertainty | **150°/h** | 1.5°/h | 0.0015°/h | 0.0000015°/h |
| Gyro bias stability | **1500°/h/$\sqrt{h}$** | 15°/h/$\sqrt{h}$ | 0.015/h/$\sqrt{h}$ | 0.000015/h/$\sqrt{h}$ |

* Values extracted from Figure 8.4 in [7]. "Commercial Grade" is underlined and bolded because it is the type that would be used in an instrumented glove for human-computer interaction.

It is noticeable in Table 1 how the critical parameters of "commercial-grade" gyroscopes (such as miniature MEMS gyroscopes) have values that are orders of magnitude worse than those of strategic, navigation, or even tactical IMUs. This lower performance of MEMS IMUs prevents their direct use in the "strapdown" configuration previously employed to estimate both the position and the orientation of a rigid body, with respect to an external, immobile "reference" or "inertial" frame of coordinates. In the strapdown approach, the accelerometer and gyroscope of the IMU are affixed to the moving object, i.e., to its "body frame". Every time the gyroscope measurements are read, they are added to a running tally that indicates, therefore, the total rotation (orientation change) of the object from its initial orientation. That is, in this "dead reckoning" approach, the integration of gyroscope signals estimates the current orientation of the "body frame" with respect to its initial orientation in the "inertial frame". The accuracy of this first estimate of orientation is paramount because it is relied upon to project or "resolve" the readings from the accelerometer (along the directions of the "body frame") to the axes of the inertial frame where the known acceleration of gravity is subtracted from them, to leave only the so-called "linear" acceleration components (in the inertial frame). If the intended operation of the system were correct until this point, each of the linear acceleration components could then be "double-integrated" (in ways similar to the integration of gyroscope readings) to yield the displacement components along the inertial axes, i.e., the position of the moving object from its initial location, in terms of the inertial frame. The stringent demand for gyroscope accuracy in this scheme is underscored by the realization that any error in the orientation estimated from the gyroscope will lead to an inaccurate resolution of accelerations to the inertial frame, which will introduce errors in the inertial frame accelerations upon the incorrect removal of gravity. Through the necessary double integration of accelerations, the resulting position errors will grow as a quadratic function of time. Accordingly, Foxlin states that, for this approach, "An even more critical cause of error in position measurement is error in the orientation determined by the gyros" [7]. Based on simulations of strapdown configurations with accelerometers and gyroscopes of different grades (commercial, navigation, tactical, strategic), he concluded, referring to the use of MEMS IMUs, that "Therefore, human-motion tracking systems that can maintain position to a few centimeters for more than a minute without external correction are not on the horizon."

Despite improvements in MEMS accelerometers and gyroscopes made in the years after Foxlin's simulation, their successful operation in a strapdown configuration continued to be unfeasible. Woodman, in his 2007 report concluded "It is not currently possible to construct an INS which maintains sub-meter accuracy for more than 60 s using MEMS devices" [8].

The bias level and variability of commercial-grade MEMS gyroscopes are typically higher in comparison to other types of gyroscopes. For example, a white paper published in 2014 by KVH Industries, Inc. (Middletown, RI, USA) [9], a manufacturer of high-end inertial systems, characterizes the "bias offset" of MEMS gyroscopes (at 25 °C) as "$\pm 250°/h$" with "bias instability (at constant temperature)" of "less than $1°/h$, $1\sigma$", as compared to only "$\pm 2°/h$" and "less than $0.05°/h$, $1\sigma$", respectively, for fiber optic gyros (FOGs), which are more expensive, larger, tactical-grade gyroscopes.

Technological advances in recent years have achieved important reductions in the offset level and instability of some MEMS gyroscopes, although the most significant improvements are mainly found in research devices or specialized commercial devices with costs that would be prohibitive for instrumenting a computer interaction glove with 10 or more of them. For example, Wu et al. published their development of a "sub-$0.1°/h$ bias-instability MEMS gyroscope" in 2021 [10]. In their device, they pursue the strategy of "resonant constant frequency" (RCF) to ensure "that the excitation frequency and resonant frequency are equal and constant" in order to "eliminate effects of excitation-frequency instability and drift on gyroscope output performance." More specifically, this same group combined RCF with real-time automatic mode-matching, achieving a measurement of "bias instability of $0.09°/h$" in an experimental device [11]. More recently, Bu et al. [12] proposed an "online compensation method for ZRO drift based on multiparameter fusion". Upon application of their approach, they observed that "after online compensation, the BI [bias instability] reached $0.23°/h$". This demonstrates that technological improvements can yield much smaller offset levels and instability, but sophisticated approaches are mainly commercially available in high-cost MEMS gyros, such as the Tronics/TDK (Crolles, France) GYPRO3300 one-axis (yaw) MEMS Gyroscope, which retails for more than USD 600 per unit. According to its datasheet [13], this device has a bias instability specified as typical = $0.8°/h$, maximum = $3°/h$. On the other end of the cost spectrum for commercial MEMS chips that integrate three-axial gyroscope, tri-axial accelerometers (and possibly tri-axial magnetometers), such as the Bosch (Gerlingen, Germany) BMI088 MEMS 6 DOF IMU (retailing for less than USD 10 per unit), continue to have large values bias instability (e.g., the "zero-rate offset" is specified as "$\pm 1°/s$" in the 2024 datasheet for the Bosch BMI088 MEMS IMU [14]). Our focus throughout this article is on MEMS MARG modules in the cost range that would still make it affordable to instrument an interaction glove with 10 or more of the devices. In particular, we experimented with the 3-Space™ Micro USB MARG module from Yost Labs (Portsmouth, OH, USA) specified with a "Gyro bias stability @ 25 °C $2.5°/h$ average for all axes". In addition to the relative affordability of these modules, we selected them for our experimentation due to their ease of connection to a computer (via a USB cable) and the availability of the C#-language-based application programming interface (API) that facilitated the software development in our experiments. Importantly, the API is directly compatible for the use of the "Nano" version of the 3-Space MARG (retailing for USD 39 per unit). The complete set of specifications offered by the manufacturer is transcribed in Appendix A.

In view of the difficulties for obtaining reliable tracking estimates from just gyroscope and accelerometer readings, some manufacturers also included a tri-axial magnetometer in packages advertised as "9-degrees-of-freedom IMUs" or magnetic and inertial measurement units (MIMUs)", or, the designation we prefer, "magnetic, angular rate, and gravity (MARG)" sensor modules. One of the earliest of these MEMS sensors modules was InvenSense's MPU-9150, initially released in 2011 [15].

This paper presents our algorithm to obtain orientation estimations for a low-cost MEMS MARG module in real time, guarding against the estimate degradation that could occur in environments with large ferromagnetic objects, for the purpose of human–computer interaction.

*1.2. Prediction–Correction of MARG Orientation*

The unavoidable presence of time varying bias in the signals from the gyroscopes contained in low-cost, miniature MEMS MARG modules, along with the single and double integrations required as part of the strapdown approach, have led to re-focusing the goal of MARG tracking to estimation of the orientation of the module alone (which, in any case, would be a prerequisite to the implementation of the strapdown approach for position tracking).

This has prompted attempts to implement classical methods for orientation estimation that combine an initial estimation generated from gyroscope data, followed by a correction phase frequently achieved with the involvement of a different sensor modality, for example, accelerometer data, and also potentially involving magnetometer readings.

One of the most popular prediction–correction algorithms used in the navigation arena is the Kalman filter [16], which is an optimal linear estimator in the sense that it minimizes the variance of the estimates obtained [17]. However, application of Kalman filtering for orientation tracking of low-cost MEMS MARG modules has resulted in limited success, as we discuss further in Section 1.6.

*1.3. Scope of Our Approach to Prediction–Correction MARG Orientation for HCI*

From the previously presented considerations, we sought to define an alternative way to make use of the three types of signals available from a low-cost MEMS MARG module in the context of a hand-tracking application for human–computer interaction. This means that we would restrict our focus to uses of the commercial-grade, low-cost MARG modules, considering them attached to hand-held human–computer interaction devices, or embedded in an interaction glove, to be worn by the computer user. It should be noted that the same kind of context surrounds the utilization of MEMS MARG modules in many Internet-of-Things (IoT) applications.

Our specific focus on human–computer interaction applications allows us to establish two important expectations for the operation of the MARG:

(a) As we expect the MARG to be affixed, in one way or another, to the hand of a human computer user, we can expect that there will be intervals in which the MARG will be static (or very close to it), occurring frequently.

(b) The overall travel of a MARG in any particular use run will be constrained to the scale of meters. This, in fact, implies that objects around the MARG (and, in particular, large ferromagnetic objects) will remain in static or slowly moving relative positions with respect to the MARG.

(Both these expectations are also plausible in many IoT potential uses of the low-cost MEMS MARG modules.)

*1.4. New View of the Information Sources Onboard a MARG Module*

For application in HCI studies, the focus, then, will be to estimate the orientation of the coordinate frame considered attached to the body of the MARG module (the "body frame") with respect to an external, immobile frame of reference, which we will call here the "inertial frame". The extent of the overall movement of a MARG attached to a segment of the human body will be small enough so that a corner of the floor in a room (or the intersection of two perpendicular streets) can be proposed as the origin of the (Cartesian) inertial coordinate frame, with the third axis passing through that corner and following the orientation of a plumb line. To achieve the estimation of the MARG orientation, our algorithm seeks to make comprehensive use of the readings of all three sensor modalities available in the MARG module, but must, simultaneously, scale down the involvement of any given sensor modality when the assumptions made for its utilization are not sufficiently met. Using information from a sensor modality in an unvetted, unbridled fashion may result in effectively adding error to the recurrent estimates of MARG orientation obtained. The assumptions made for each of the three kinds of sensors are explained next.

*1.5. Estimation of Orientation Differences*

Orientation is, in effect, a relative measurement. Much like the measurement of electrical potential (voltage), which is expressed as a difference in potential from a test point to a reference point, we seek to measure the difference of orientation between the MARG body frame and the inertial frame. Therefore, we describe the orientation of the MARG as the rotation that, if applied to the body frame, would take it from an initial state, where it was aligned with the inertial frame (no orientation difference), to its current orientation.

It should be noted that this paper assumes that both the inertial frame and the MARG body frame are Cartesian coordinate systems, adhering to the "left hand rule" (this means that if the left hand is put in a flat configuration, with the fingers initially pointing in the direction of the positive X axis, and then the pinky, ring, middle, and index finger are bent 90 degrees to match the direction of the positive Y axis, the extended thumb will point in the direction of the positive Z axis). It is also important to indicate here that we (as many contemporary researchers in the field) have chosen to express the MARG orientation estimate resulting from our algorithm as a quaternion, which is a representation of 3D rotation [18–20]. That is, we identify the estimated MARG orientation with the rotation that would turn the MARG from its initial orientation to its "current orientation", at any time. Further, we invoke the assumption (without loss of generality; if a known rotation mediates between the desired inertial reference frame and the startup orientation of the MARG, it would just be necessary to "compound" that rotation with the $q_{OUT}$ result from our algorithm after each iteration (see Section 2.1)) that the initial orientation (at "t = 0" or at "startup") of the "body frame" of the MARG coincides with the fixed orientation of the inertial frame.

1.5.1. Inferring Orientation from Gyroscope Signals

The tri-axial gyroscope in a MARG reports the angular speed at which the MARG has rotated (with respect to the X, Y, and Z axes of its body frame) in the last interval of observation, which we denote as $\Delta T$. Under the assumption that the rotations (in the three body axes) took place at an approximately constant angular speed during the interval of observation, the change of orientation completed during the interval of observation is the product of the angular speed multiplied by $\Delta T$. Tracking systems keep a running tally of the total orientation change over multiple observation intervals by accumulating the newest orientation change to the previous total that had been reached prior to the latest observation interval. Then, this tally represents the rotation that would be necessary to take the MARG from its initial orientation (coincident with the orientation of the inertial frame) to its current orientation, i.e., it estimates the orientation of the MARG.

However, we have already mentioned the unavoidable imperfections present in the operation of MEMS gyroscopes, characterized by their bias (non-zero reading obtained when no rotation is taking place), and, most critically, their bias instability, which expresses the magnitude of change in bias that can be expected in a gyroscope, as time progresses. This variability of the gyroscope bias makes it impossible to achieve a single bias compensation (when the sensor is initialized) that would continue to completely eliminate the effective bias for any significant interval of time. More specifically, Aggarwal et al. identify at least two aspects of the bias uncertainty. First, the "run-to-run" gyroscope bias is such that "every time the sensor is switched on, a slightly different bias or scale factor is observed" [21].

Even more critically, the "in-run" gyroscope bias stability represents an "error that occurs due to change in bias or scale factor during a run" [21]. Further, the behavior of the bias and its evolution has been found to be significantly affected by temperature differences. "The temperature-dependent variations can be quite pronounced in very low-cost MEMS sensors" [22].

Given that the mechanism for orientation estimation from gyroscopic measurements unavoidably requires the accumulation, i.e., the integration, of the rates of rotation read from the gyroscope at every sampling interval, any slowly varying uncompensated remnant of bias in the gyroscope will result in the growing orientation error referred to as "drift". To the best of our knowledge, there are no bias compensation algorithms, currently, which can completely eliminate the bias of low-cost MEMS gyroscopes on a permanent basis. Therefore, the orientation estimates resulting from gyroscope signal integration must be corrected periodically to prevent the drift error from growing to levels that will render the orientation estimations irrelevant.

1.5.2. Inferring Orientation Information from Accelerometer Measurements

Each axis of the tri-axial accelerometer in a MARG really measures forces acting on a known test mass. Therefore, each accelerometer channel (axis) responds to the "linear acceleration" associated with change of speed of the MARG in the corresponding direction but also responds to the component of the gravitational acceleration in that direction. While the "strapdown" inertial measurement systems focused on detecting the total accelerations, resolving them to the inertial frame and then removing the components due to gravity, most contemporary MEMS MARG orientation estimation approaches seek to detect the components of gravity (exclusively) to use it as a "vector observation" [23]. This means that the gravitational acceleration is viewed as a uniform vector field in the space where the MARG is operating. In consequence, the way in which gravitational acceleration measurements along the X, Y, and Z body axes change, from the original orientation at startup (when the body frame and the inertial frame were aligned) to the current orientation of the MARG, must be the result of the rotation that mediates between the two frames. As such, the difference in X, Y, and Z axes readings contain information about the rotation of the MARG from startup to its current orientation. Observation of just one vector field would not be enough to determine the rotation between the inertial frame and the current MARG body frame, as indicated by Shuster and Oh [24]. (Nonetheless, the information obtained from observing even a single vector field may be used to correct a preliminary orientation estimate, as we do in our approach.) However, a critical assumption to extract correcting information from the difference between accelerometer axes now and at startup is that the readings are dominated by the gravitational acceleration. That is, the assumption will only be practically verified when the MARG is static or close to it, such that the non-gravitational "linear acceleration" contributions included in the accelerometer readings are negligeable with respect to the gravitational contributions. During intervals when this assumption is not verified, applying strong orientation corrections based on accelerometer readings may introduce errors in the orientation estimates.

1.5.3. Inferring Orientation Information from Magnetometer Measurements

The inference of orientation information from magnetometer readings follows the same basic rationale as for the accelerometer readings. That is, if the geomagnetic field in the operating space of the MARG were considered uniform, orientation information can be derived from the magnetometer readings as one "vector observation". Fortunately, the magnetometer readings are not altered if the MARG is moving or static. However, in this case, the magnetic field in the area of operation of the MARG may, itself, be non-uniform in certain regions, with some areas exhibiting permanent distortions. This will happen in the neighborhood to ferromagnetic items, such as furniture parts made of steel or iron in offices, classrooms, and laboratories [25], or even structural elements involved in the construction of most contemporary buildings [26]. Due to their higher magnetic permeability, ferromagnetic objects will provide an "easier" path for lines of magnetic flow. Because of this, the magnetic field vector in the neighborhood of these objects may point in a different direction than it does in the regions not affected by neighboring ferromagnetic objects. The permanent or slow varying distortion of the magnetic field produced by these objects is particularly disrupting to the assumption of a uniform (in magnitude and direction) magnetic field (more so than other rapidly alternating variations of the field). In consequence, the inference of MARG orientation from magnetometer readings also has a precondition for it to be valid (the magnetic field at the current location of the MARG should not be distorted).

*1.6. Relevant Previous Work*

The potential to track the motion of diverse segments of the human body has been highly attractive to several research communities for a long time. The biomechanics research community might have been one of the earliest to try to take advantage of inertial sensors for their work [27,28], among others. Similarly, the human–computer interaction community saw in MEMS MARG modules the possibility to develop an unobtrusive mechanism to achieve real-time monitoring of the motions of a computer user to provide input to computer systems. This was particularly attractive at the time in which the technologies driving virtual reality were emerging.

Unfortunately, the sobering findings of Foxlin [7] and Woodman [8], which we have already mentioned, among others, made it clear that the full the potential of MEMS IMU and MARG modules for human movement tracking would not be immediately realized. In fact, there has been a continuous tug-of-war between those sensor limitations and the enthusiasm for developing promising motion tracking systems with MEMS MARG modules. As an example, there have been continued efforts to develop a system, based on MEMS MARG modules, that could track the movement of the hand in real time. In particular, the group from University of Twente has continued to develop a system for real-time ambulatory assessment of hand kinematics that involves the attachment of multiple MARG modules to each finger [29,30], which were later consolidated in their "Power Glove" [31]. That type of system continues to be pursued by multiple groups, such as the one from National Taipei University, whose glove combines MEMS MARG modules with force sensing resistors (FSRs) for the assessment of hand function [32]. Any of these applications, however, requires the availability of a reliable and robust algorithm to estimate the orientation of each individual MARG used.

Many approaches to achieve the desired robustness in the orientation estimates have been attempted. A systematic and comprehensive review of those approaches was recently carried out by Nazarahari and Rouhani [33]. In that broad review, they identify three major groups of approaches: those based on vector observations (VO); those that are fundamentally a complementary filter (CF); and those that stem from the Kalman filter, as one of the most popular state estimators for the study of dynamic systems. Nazarahari and Rouhani also performed a critical comparison of 36 of the algorithms mentioned in their survey [34], which they organized into seven major groups: the linear (LCF) and nonlinear (NLCF) complementary filters, and five varieties of Kalman filters: linear Kalman filter (LKF); complementary Kalman filter (CKF); extended Kalman filter (EKF); square-root cubature Kalman filter (SRCKF); and square-root unscented Kalman filter (SRUKF).

Indeed, several variants of the Kalman filter have been used for orientation estimation on the basis of inertial measurements since the 1960s [35,36], but they have been less successful when applied to sensors (e.g., MEMS MARG modules) that have lower levels of performance than the aeronautical and space inertial sensors.

In our view, one aspect that may contribute significantly to the detriment of Kalman-based approaches is the covariance matrix representation used in Kalman filters for the uncertainty associated with the "measurements" (e.g., accelerometer and magnetometer readings) involved in the correction phase of the algorithm [17]. The covariance matrix representation of uncertainties makes it difficult to keep those representations updated to exactly reflect the current conditions in which the MARG is operating. If these uncertainty representations are not correct, the Kalman correction phase maybe counterproductive. For example, if magnetometer readings are used for the correction phase of the Kalman filter, with the implicit assumption that the magnetic field is constant in magnitude and direction at all the locations where the MARG will be used, and then the module is used in an area where the magnetic field uniformity assumption is violated, the correction may introduce error in the orientation estimate. Furthermore, since the Kalman filter is a recursive estimator, the injection of error will continue to degrade the orientation estimate for multiple iterations in the future. Some approaches have addressed the correction step of Kalman filters in alternative ways that seek to "alter" the strength of the corrections from accelerometer and magnetometer measurements, factoring instantaneous conditions. One example is the attempted scaling of the measurement covariance matrix, R [37]. These approaches, however, are challenged by the difficulty in modifying each of the multiple (e.g., nineentries in a measurement covariance matrix according to their true meaning (i.e., auto-covariances along the diagonal and cross-covariances elsewhere). Properly estimating the necessary changes to each one of the entries individually is challenging and, commonly, the adjustment is simply implemented as the use of a different scalar factor that multiplies the whole covariance matrix.

A new approach for the correction of an initial orientation estimate that allows for the easier modulation of the influence of the correcting factors has emerged. In this approach, the correction is carried out as an interpolation between the preliminary orientation estimate quaternion and a quaternion representing an orientation that has been corrected with the "full strength" of the information derived from the "measurements" (e.g., the current magnetometer readings). As an example, Valenti et al. explored the use of quaternion interpolation, although they still implemented the concept as auxiliary to a Kalman filter [38]. Further, they find the "full strength" correction quaternions by exclusively algebraic means, as made evident in the name of their approach: the Algebraic Quaternion Algorithm (AQUA).

Our method, which implements the corrections driven by the accelerometer and the magnetometer as quaternion interpolations, was formulated on an intuitive basis, as explained next, and does not involve the structure of a Kalman filter in any way.

## 2. Materials and Methods

In this section, we describe the proposed algorithm for MARG orientation estimation, including the mechanism we are now using to compute of magnetometer trustworthiness, $\mu_K$, without using any source of information that is external to the MARG module itself.

*2.1. GMVDK Algorithm for MARG Orientation Estimation*

In the development of the original "Gravity and Magnetic North Vector correction -Double SLERP" (GMVD) algorithm, we considered the availability of tri-axial gyroscope, accelerometer, and magnetometer readings from the MARG itself every $\Delta t$ seconds and a simultaneous reading of X, Y, and Z MARG position. The position readings were provided by a module containing 3 infrared cameras, which tracked an infrared reflective surface applied on the MARG. (The next subsection describes how this last input of information has now been substituted, such that it is no longer necessary, in the new GMVDK algorithm.)

The goal of our algorithm is to generate, after every new batch of sensor data becomes available, a MARG orientation estimation that takes full advantage of all three sensing modalities present in the MARG, but, simultaneously, protects the estimate from being "contaminated" with invalid influences from the accelerometer and magnetometer, when the preconditions for their corrections are not met. GMVD includes very deliberate provisions to reduce the negative impact that the involvement of the magnetometer could have when the MARG is operating in a region of space where the geomagnetic field is distorted.

We chose to represent the current orientation of the MARG as the rotation that would be necessary to re-align the fixed inertial frame with the current body frame of the MARG.

To simplify the process, we define the initial MARG orientation (at "startup") to coincide with the orientation of the inertial frame. (If a different initial MARG orientation is used, the corresponding rotation from the inertial frame would only need to be known and "added" ("compounded") to the final orientation estimate obtained from each iteration of GMVD). We have chosen to represent the 3D rotation that defines the orientation of the MARG through a unit quaternion (i.e., a quaternion with unit norm). This is a common choice in orientation estimation as this representation is compact, avoids ambiguities such as the "gimbal lock" effect, and is commonly used to drive 3D animation environments (e.g., Unity®) for visualization purposes. Therefore, we seek to find a quaternion, $q_{OUT}$, to represent the orientation of the body frame as the rotation that would modify the orientation of the inertial frame to match the current orientation of the body frame.

There are several excellent books that describe quaternions, their manipulations, and their geometrical interpretations [18–20]. Here, we only summarize critical aspects of quaternions that are necessary for the explanation of our method. A quaternion, $q$, is represented as a four-element array of real numbers, such as $q = [\ q_x,\ q_y,\ q_z,\ q_w\ ]^T$, where 3 of them ($q_x, q_y, q_z$) are called the "vector part" and the fourth number, $q_w$, is called the "scalar part" of the quaternion. We signify quaternions and vectors with boldface. When necessary, we use this operator to form a quaternion from a 3D vector, $q_v$, and a scalar $q_w$:

$$q = H(q_v,\ q_w) \quad (1)$$

The addition of quaternions, the product of a scalar times a quaternion, and the inner product between quaternions are the same as the corresponding operations for four-element vectors, but the product of quaternions is defined in a completely different way [18–20] and is symbolized as "$\otimes$". A quaternion norm can be computed as the square root of the sum of all 4 quaternion elements raised to the second power. Any quaternion can be modified to be a "unit quaternion" by calculating the norm of the quaternion and then dividing each of its components by that norm. The conjugate of quaternion $q$ is simply $q^* = \begin{bmatrix} -q_x, & -q_y, & -q_z, & q_w \end{bmatrix}^T$.

A quaternion of unit norm can be used to represent a 3D rotation, around a 3D vector $u = [u_x, u_y, u_z]$ (which does not need to be parallel to any of the X, Y, or Z axes of the coordinate frame). In that representation, the amount of rotation around the $u$ axis is $\beta$, and the quaternion components will satisfy [39]

$$q_w = \cos\left(\frac{\beta}{2}\right) \quad (2)$$

$$\begin{bmatrix} q_x \\ q_y \\ q_z \end{bmatrix} = u\sin\left(\frac{\beta}{2}\right) = \begin{bmatrix} u_x \\ u_y \\ u_z \end{bmatrix} \sin\left(\frac{\beta}{2}\right) \quad (3)$$

(It should be noted, then, that the quaternion $q_{NOROT} = [0, 0, 0, 1]^T$ indicates "no rotation"). Frequently, the interplay between quaternions and "normal" 3D vectors, such as the ones one would use to indicate a 3D direction, of, for example, the force applied by a rope pulling a mass, takes place by means of "casting" the 3D vector $v = [v_x, v_y, v_z]$ as a "pure" quaternion by adding a scalar component with a value of zero: $q_v = [v_x, v_y, v_z, 0]$. Conversely, if a quaternion has 0 scalar component, it can be "re-interpreted" as just a 3D vector.

There are two critical implications of the use of quaternions to represent the 3D rotations we leverage in our method. First, if a unit quaternion $q$ represents the orientation of the body frame with respect to the inertial frame (i.e., the rotation that would align the inertial frame to the body frame), a 3D vector whose coordinates in the inertial frame are $v_i = v_{ix}, v_{iy}, v_{iz}$ can be "mapped" (from inertial to body frame) with the following quaternion operation to find the coordinates of that same 3D vector in the body frame

$v_b = v_{bx}, v_{by}, v_{bz}$. (For this, $v_i$ must first be "cast as pure quaternion", and the initial quaternion result of the equation must finally be "re-interpreted" as a 3D vector.)

$$v_b = q^* \otimes v_i \otimes q = mapItoB(q, v_i) \tag{4}$$

Because of the practical effect of the above equation, we refer to the process described in it as $v_b = mapItoB(q, v_i)$. Similarly, a 3D vector initially defined in the body frame, $v_b$, can be mapped to the inertial frame with this operation, which we refer to as $v_i = mapBtoI(q, v_b)$.

$$v_i = q \otimes v_b \otimes q^* = mapBtoI(q, v_b) \tag{5}$$

Second, within a single frame of reference (e.g., the body frame), a 3D vector can be rotated from an initial direction 1, where its coordinates are $v_1 = v_{1x}, v_{1y}, v_{1z}$ to a final direction 2, where its coordinates will be $v_2 = v_{2x}, v_{2y}, v_{2z}$, applying the rotation encoded in a unit quaternion $q$, by

$$v_2 = q^* \otimes v_1 \otimes q = qROT(q, v_i) \tag{6}$$

(For this, $v_1$ needs to be "cast as pure quaternion", and $v_2$ must be "re-interpreted" as a vector). Additionally, computing $v_2 = qROT(q_1, v_1)$ and then $v_3 = qROT(q_2, v_2)$ will perform the rotations encoded by $q_1$ and then $q_2$, in sequence, to yield $v_3$ as the final result (i.e., "compounding", where an equivalent quaternion for the compound rotation is $q_{COMPOUND} = q_2 \otimes q_1$).

Using these quaternion manipulations, GMVD will

(a) Create one orientation estimate derived from gyroscope measurements, $q_G$, and two estimates that are independently corrected with information from the accelerometer, $q_{GA}$, and the magnetometer, $q_{GM}$.
(b) Scale down the strength of the accelerometer- and magnetometer-based corrections by interpolating from $q_G$ to $q_{GA}$ and from $q_G$ to $q_{GM}$ using "spherical linear interpolation" (SLERP), operations [40] controlled by the corresponding trustworthiness parameters $\alpha$ and $\mu$. This defines the "scaled" corrected quaternions $q_{SA}$ and $q_{SM}$, respectively.
(c) Finally, fuse $q_{SA}$ and $q_{SM}$, via a "second tier" of SLERP interpolation, to define a final MARG orientation estimation quaternion, $q_{OUT}$, which uses information from the 3 sources available, but would not contain strong corrections directed by the accelerometer or the magnetometer if their preconditions are not met.

The flow of information through one iteration of the GMVD algorithm is diagrammed in Figure 1. The numbers in square brackets indicate the computation of intermediate variables and the overall result, $q_{OUT}$.

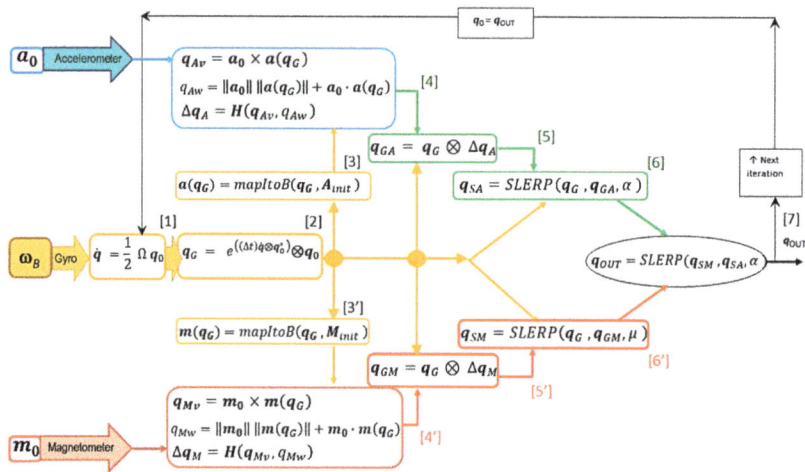

**Figure 1.** Flow of information within a single iteration of the GMVD orientation estimation algorithm. The numbers in square brackets indicate the computation steps.

At "startup", when the orientation of the body frame in the MARG is assumed coincident with the orientation of the inertial frame, several samples of the accelerometer and magnetometer readings are collected and averaged. These initial accelerometer and magnetometer vectors, $A_{init}$ and $M_{init}$, must be obtained with the MARG static and in an area where the geomagnetic field is not distorted. The quaternion $q_{NOROT} = [0, 0, 0, 1]^T$ is assigned as the seed value of the MARG rotation estimate, $q_0 = q_{NOROT}$.

[1] The readings from the gyroscope are "de-biased" by subtracting from each of them the most recent estimate of its bias value, computed every time the MARG is estimated to be static, by linear regression on a window of consecutive samples. This yields the vector of "de-biased" gyroscope readings, $\omega_B = [\omega_{Bx}, \omega_{By}, \omega_{Bz}]$. (Unfortunately, under many circumstances, this "de-biasing" is not 100% successful.) Equation (7) uses the de-biased gyroscope readings to compute the quaternion rate of change:

$$\dot{q} = \frac{1}{2} \begin{bmatrix} 0 & \omega_{Bz} & -\omega_{By} & \omega_{Bx} \\ -\omega_{Bz} & 0 & \omega_{Bx} & \omega_{By} \\ \omega_{By} & -\omega_{Bx} & 0 & \omega_{Bz} \\ -\omega_{Bx} & -\omega_{By} & -\omega_{Bz} & 0 \end{bmatrix} \begin{bmatrix} q_{0x} \\ q_{0y} \\ q_{0z} \\ q_{0w} \end{bmatrix} = \frac{1}{2} \Omega\, q_o \quad (7)$$

[2] The integration of the quaternion rate of change is completed with this quaternion expression to yield the current initial estimate of MARG orientation, which already has incorporated the latest ("de-biased") gyroscope readings:

$$q_G = e^{((\Delta t)\dot{q} \otimes q_0^*)} \otimes q_0 \quad (8)$$

At this point, two pipelines, one for the accelerometer correction and one for the magnetometer correction, are developed "in parallel", as illustrated in Figure 1. As they are functionally equivalent, we outline the steps ([3]–[6]) in the accelerometer (upper) pipeline, with the comments being symmetrically applicable to the magnetometer (lower) pipeline ([3']–[6']).

[3] A "computed body acceleration vector", $a(q_G)$, is calculated as the mapping of the acceleration of gravity represented in the inertial frame, $A_{init}$, to the current orientation of

the body frame, using for this mapping $q_G$, obtained provisionally from just the integration of gyroscope readings:

$$-a(q_G) = mapItoB(q_G, A_{init}) = q_G^* \otimes A_{init} \otimes q_G \qquad (9)$$

The computed body acceleration vector should, ideally, match the current readings of the accelerometer, $a_0$. However, if $q_G$ has already developed drift, the "calculated acceleration vector" will not match the current readings of the accelerometer, $a_0$.

[4] It is possible [41,42] to determine the quaternion $\Delta q_A$, representing the "rotation difference" that must be compounded with the initial $q_G$ into a "fully (accelerometer) corrected" estimate, $q_{GA}$, which would yield a vector matching $a_0$, if used to map $A_{init}$ to the body frame:

$$\Delta q_A = H(q_{Av}, q_{Aw}) \qquad (10)$$

where the "vector" and "scalar" parts being assembled into the quaternion $\Delta q_A$ are

$$q_{Av} = a_0 \times a(q_G) \qquad (11)$$

and

$$q_{Aw} = \|a_0\| \|a(q_G)\| + a_0 \cdot a(q_G) \qquad (12)$$

[5] The compounding is done by this quaternion product:

$$q_{GA} = q_G \otimes \Delta q_A \qquad (13)$$

[6] A key feature of GMVD is that it does not unconditionally accept this "fully" accelerometer-corrected quaternion as a better estimation of the MARG orientation. While $q_{GA}$ is, indeed, infused with information from the instantaneous accelerometer readings, which may correct gyroscopic drift that could have accumulated, it may, itself, be subject to error. That would be the case if the current accelerometer measurements, $a_0$, are not just measuring the acceleration of gravity, which is the precondition assumed to interpret $a_0$ as a body frame observation of the same vector recorded in $A_{init}$ referenced to the initial orientation, i.e., to the inertial frame. Thus, a "scaled" accelerometer-corrected orientation, $q_{SA}$, is found interpolating from $q_G$ to $q_{GA}$ under control of the accelerometer trustworthiness parameter $0 \leq \alpha \leq 1$, using SLERP interpolation. If the accelerometer preconditions are completely absent (i.e., the MARG is moving at variable speed), $\alpha$ will be close to 0, and the interpolated $q_{SA}$ will essentially be $q_G$ (no harm done). If the accelerometer preconditions are met ($\alpha \approx 1$), then $q_{SA}$ will be, essentially, $q_{GA}$. That is, the system will take full advantage of the valid secondary source of MARG orientation information.

$$q_{SA} = SLERP(q_G, q_{GA}, \alpha) \qquad (14)$$

In general, the SLERP interpolation from quaternion $q_1$ to quaternion $q_2$, under control of the parameter $0 \leq h \leq 1$, is computed as [40]

$$cos(\Omega) = q_1 \cdot q_2 \qquad (15)$$

$$SLERP(q_1, q_2, h) = \frac{q_1 sin((1-h)\Omega) + q_2 sin(h\Omega)}{sin(\Omega)} \qquad (16)$$

In the magnetometer pipeline, step [6'] implies the SLERP interpolation from $q_G$ to a "fully" magnetometer-corrected estimate $q_{GM}$, under control of the magnetometer trustworthiness parameter $0 \leq \mu \leq 1$. This step has the same objective as step [6]. In this case, $\mu$ quantifies the level of fulfillment of the magnetometer precondition. The current magnetometer measurements, $m_0$, qualify as an observation of the same vector as $M_{init}$ only if the geomagnetic field at the current location of the MARG is undistorted. The computation of the trustworthiness parameters $\alpha$ and $\mu$ is explained in the following subsection.

[7] Finally, the 2 corrected MARG orientation estimates which potentially took advantage of accelerometer and magnetometer readings, but have curtailed strong corrections if their preconditions were not actually met, are fused with a second-tier SLERP interpolation:

$$q_{OUT} = SLERP(q_{SM}, q_{SA}, \alpha) \qquad (17)$$

2.2. Computation of the Trustworthiness Parameters α and µ

The overall scheme of the GMVD algorithm requires that values of the accelerometer and magnetometer trustworthiness parameters, α and µ, be available for every sampling interval of operation of the MARG. The α parameter must estimate how close to static (or moving without linear acceleration) the MARG might be, so that the current accelerometer readings can be interpreted as a good approximation of gravity components along the body frame axes. The µ parameter must represent how close the magnetic field currently measured by the magnetometer is to the undistorted geomagnetic field first measured at startup.

2.2.1. Original Formulations

Our computation of α was based on a parameter internally calculated by the MARG module we have used (Yost Labs 3-Space Micro USB), called "confidence (factor)", directly readable (along with the accelerometer, gyroscope, and magnetometer values) at every sampling interval [43]. However, we have confirmed that a similar parameter could also be calculated based on the rotational and translational activity of the MARG as a normalized average of the mean squared values of the gyroscope and accelerometer readings. In either case, the original parameter value is processed by a linear equation whose negative slope was set to produce a rapid decrease in the value of α as soon as the MARG begins to depart from a static status. In our intended use of MARG modules for tracking hand-held human–computer interaction devices, it is reasonable to expect that there will be frequent static intervals, in which α will increase to values close to 1. (After the linear function, we also overwrite all negative results to zero, which is the lower bound of the desired values of α, with this conversion: α = (α + |α|)/2.)

The computation of the µ parameter is more involved, as it should represent a property of the environment in which the MARG is currently operating, and not one related to its own motion. Under the assumption that the large ferromagnetic objects around the human body being monitored will not be in constant and rapid motion, we previously sought to model the distortion of the geomagnetic field as a property of three-dimensional space. In a previous implementation of a system for tracking the position and configuration of the hand of a human subject [44,45], the MARG was used in conjunction with a 3-IR camera system (OptiTrack V-120 Trio. OptiTrack is a division of NaturalPoint, Inc., Corvallis, OR, USA), that provided X, Y, and Z position coordinates for the MARG. This previous version of the algorithm, designated as the "Gravity and North Vector Correction with Double Slerp" (GMVD) orientation estimator, used the position information provided by the camera system to manage a 3D map of the space where the MARG is moving, populating small (e.g., 2 cm per side) cubic "voxels", with estimated values of the µ parameter. The computation of µ in the GMVD algorithm details have been documented in [42] and will not be recounted here, as we have now superseded it with the self-contained µ computation detailed next.

2.2.2. New Computation of Magnetometer Trustworthiness, $\mu_K$, without Position Information (GMVDK)

Aware of the existence of application scenarios where the use of a camera system to aid in the storage and retrieval of previously computed µ values may not be practical, we have now developed an alternative magnetometer trustworthiness parameter, $\mu_K$, which is *calculated without involvement of MARG position information*. We seek to encode in $0 \leq \mu_K \leq 1$ the trust we can have that the geomagnetic field at the current location of the MARG is the same as the one initially sensed at startup, $\mathbf{M}_{init}$. If the magnetic vector is the same in magnitude and direction ($\mu_K \approx 1$), then the precondition for the magnetometer-based correction to the orientation estimate is met and it can be applied fully. If the precondition is not fully met, the magnetometer-based correction must be restrained accordingly.

As mentioned previously, the system stores the components of the initial magnetic vector, $\mathbf{M}_{init}$, referenced to the initial orientation of the MARG body frame, which coincided with the inertial body frame at startup. The magnetic vector currently sensed by the MARG, $m_0$, may differ from the initial, $\mathbf{M}_{init}$, with respect to magnitude and direction. Our approach first computes two parameters that focus primarily on each of these possible alterations: $\mu_{KA}$ (based on angle change) and $\mu_{KM}$ (which also uses magnitude change). However, $m_0$ is referenced to the body frame, whereas $\mathbf{M}_{init}$ is referenced to the inertial frame. To compare them properly, we can "map" the current readings of the magnetometer, $m_0$, "back" to the inertial frame. Here, we take advantage of the limited rotational speed expected in the movement of a body segment to approximate the current orientation of the MARG body frame with the final orientation estimate obtained in the previous iteration: $q_{OUT}$. Thus, the magnetometer readings mapped back to the inertial frame will be

$$m_{0i} = mapBtoI(q_{OUT}, m_0) = q_{OUT} \otimes m_0 \otimes q_{OUT}^* \qquad (18)$$

The angle, $\lambda$, between the 2 vectors is

$$\lambda = \cos^{-1}\left[\frac{\mathbf{M}_{init} \cdot m_{0i}}{\|\mathbf{M}_{init}\| \|m_{0i}\|}\right] \qquad (19)$$

The larger the value of $\lambda$, the more distorted the magnetic field is at the current position of the MARG. So, $\mu_{KA}$ is formulated to drop quickly from the value of 1 that it adopts when $\lambda$ is 0 (negative values of $\mu_{KA}$ are overwritten with 0):

$$\mu_{KA} = 1 - 1.5\,\lambda \qquad (20)$$

Changes in magnitude help detect distortion of the magnetic field at the current MARG location but will only be detrimental if there is a simultaneous change in orientation of the magnetic field vector. Thus, $\mu_{KM}$ was set to decay from its ideal value of 1 by a "penalty" value given by

$$penalty = \left[\frac{\|m_{0i}\|}{\|\mathbf{M}_{init}\|}\right](\lambda) \qquad (21)$$

$$\mu_{KM} = 1 - penalty \qquad (22)$$

and negative results for $\mu_{KM}$ are overwritten with a value of 0. The 2 parameters previously described are consolidated as

$$\mu_1 = [\mu_{KA} + \mu_{KM}]/2 \qquad (23)$$

To enforce a continuous suppression of magnetometer corrections when $\mu_1$ is varying rapidly, $\mu_2$ is obtained as the minimum value of $\mu_1$ in the present and past values computed in the last 0.4 s. The final trustworthiness parameter is

$$\mu_K = (\mu_2)(\alpha) \qquad (24)$$

to discourage magnetometer corrections when the MARG might be in rapid rotation (small $\alpha$), as that would violate our approximation of the current MARG orientation by the immediately past $q_{OUT}$.

2.2.3. Pseudocode for Parameter Computation and Parameter Sensitivities

The main flow of information depicted in Figure 1 can be implemented unambiguously by programming Equations (7)–(17). On the other hand, the computation of the trustworthiness parameters $\alpha$ and $\mu_K$, can be characterized more clearly through the corresponding pseudocode programs presented in Appendix B. Further, each of the pseudocode programs (Sections B.1 and B.2) include a parameter. For the computation of $\alpha$, the parameter is "SlopeAlpha", and for the computation of $\mu_K$, the parameter is "SlopeMuk". In each case, the larger the value of the parameter, the more sensitive the corresponding parameter ($\alpha$ or $\mu_K$) will be to departures from the corresponding ideal condition (complete lack of movement for $\alpha$ and complete lack of magnetic field distortion for $\mu_K$). After examination of the behavior of GMVDK in multiple cases, with several values of these parameters, our descriptions in Sections 2.2.1 and 2.2.2 above use the values of these parameters we selected as "nominal values": SlopeAlpha = 1.0 and SlopeMuk = 1.5. Nonetheless, it was of interest to study to what degree the performance of GMVDK might be changed if these parameters are assigned other values—in other words, the sensitivity of GMVDK performance to changes in these parameters. The description of our sensitivity assessment process and the results are detailed in Appendix B. In summary, we found that for a 1% change in SlopeAlpha, the change in GMVDK performance, as defined in our experiment, is of the order of 0.0856%. We also found that for a 1% change in SlopeMuk, the change in GMVDK performance is of the order of 0.1002%. These results seem to indicate that the precise use of the nominal parameter values is not critical to obtain similar levels of performance from GMVDK.

## 3. Results

In this section, we include detailed descriptions of the results obtained from two representative runs of an experimental sequence of manipulations of the 3-Space Micro USB MARG module from Yost Labs (the complete set of specifications offered by the manufacturer for this MARG module are transcribed in Appendix A). In both examples, we show the orientation estimates generated by our new GMVDK algorithm (i.e., GMVD with $\mu_K$), which does not require position information from the system of infrared cameras. In the first example, we compare the results from GMVD (previous version) and GMVDK to orientation estimates generated internally by the MARG according to a Kalman filter algorithm implemented in the chip by the manufacturer. In the second example, we contrast our results with those from two well-known contemporary MEMS MARG orientation estimation algorithms.

### 3.1. Experimental Protocol

We present the orientation estimates obtained by GMVDK from the Yost Labs 3-Space Micro USB MARG module [43], affixed to a small wooden prism, as it was manipulated by a human subject. This was adopted so that the MARG could be placed in repeatable orthogonal orientations by resting the prism on a level surface. During the trial, the MARG was manipulated at three locations. Locations H, A, and B were set at an approximate height of 1 m from the floor. A and B were located on a line approximately oriented west–east. B was located approximately 55 cm west from A. The home location (H) was approximately 30 cm north from A. Whereas all the supporting surfaces used for the experiment were made of wood and placed away from steel or iron furniture, a 0.5 × 3.8 × 37.5 cm steel bar was placed directly under Location B, to have a known distortion of the magnetic field around that location. Figure 2 shows a top view of the plane in which locations H, A, and B were defined and the wooden prism on which the 3-Space Micro USB MARG was mounted.

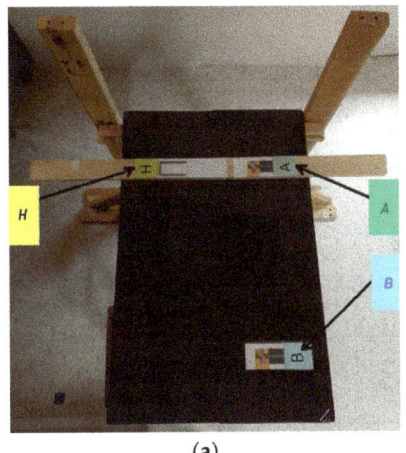 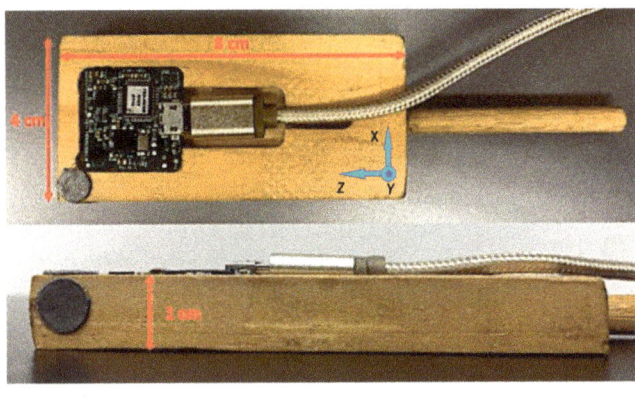

(a)              (b)

**Figure 2.** Recording setup used to perform the rotation sequence described in Table 2. (**a**) MARG recording space (top view); (**b**) 3-Space Micro USB MARG mounted on a wooden prism.

A human subject completed a pre-specified sequence of rotations involving the same orientation changes performed first at Location A, where the geomagnetic field was not distorted, and then at Location B, where the magnetic field was known to be distorted. In between rotations, the subject was asked to hold specific "poses". The pre-defined sequence is indicated in Table 2 (rotations are designated according to the "left-hand rule").

**Table 2.** Sequence of steps in each experimental trial.

| Sequence Step | Location | Rotation | Resulting Pose |
| --- | --- | --- | --- |
| 1 | H | (Initial location and pose for the task) | 1 |
| 2 | A | After translation H to A, yields | 1 |
| 3 | A | +90° Z axis, yields | 2 |
| 4 | A | −90° Z axis, yields | 1 |
| 5 | A | +90° X axis, yields | 3 |
| 6 | A | −90° X axis, yields | 1 |
| 7 | A | +90° Y axis, yields | 4 |
| 8 | A | −90° Y axis, yields | 1 |
| 9 | A | −45° Y axis and + 90° X axis, yields | 5 |
| 10 | A | −90° X axis and + 45° Y axis, yields | 1 |
| 11 | B | Just translation A to B | 6 (same orientation as 1) |
| 12 | B | +90° Z axis, yields | 7 |
| 13 | B | −90° Z axis, yields | 6 |
| 14 | B | +90°X axis, yields | 8 |
| 15 | B | −90° X axis, yields | 6 |
| 16 | B | +90° Y axis, yields | 9 |
| 17 | B | −90° Y axis, yields | 6 |
| 18 | B | −45° Y axis and + 90° X axis, yields | 10 |
| 19 | B | −90° X axis and + 45° Y axis, yields | 6 |
| 20 | H | Just translation back to H | 1 |

The main means of verification for the resilience of GMVDK to magnetic distortion is the similarity of the orientation estimation results obtained for the same sequence of poses (Poses 1–5) adopted first in an environment where the geomagnetic field is not distorted (Location A) and then (Poses 6–10) in a distorted magnetic field (Location B).

## 3.2. Orientation Estimates Obtained from Different Approaches

Figure 3 shows the evolution of the orientation estimates obtained by GMVD (previous version requiring X, Y, and Z position information from the 3-IR camera system, which was recorded for this first experiment) and by our new algorithm, GMVDK. Figure 3 also shows the orientation estimates from the Kalman filter ("KF") implemented within the Yost Labs MARG. This result is included to facilitate the comparison of our method to an orientation estimation algorithm that is well known and has been used extensively for orientation estimation [46–49]. In each case, the orientation estimation is represented by the four quaternion components, which are shown as they evolve through time. For clarification, a vertical green line has been included at the approximate time when the MARG was translated from Location A (not magnetically distorted) to Location B (magnetically distorted). For reference, Figure 3 also includes the two components of the new $\mu_K$ (i.e., $\mu_{KA}$ and $\mu_{KM}$), overlapped, and a plot of the evolution of $\mu_K$ itself, which confirms its significant decrease when the MARG was translated from Location A to Location B and its increase again when the MARG was finally translated back to the home location (H).

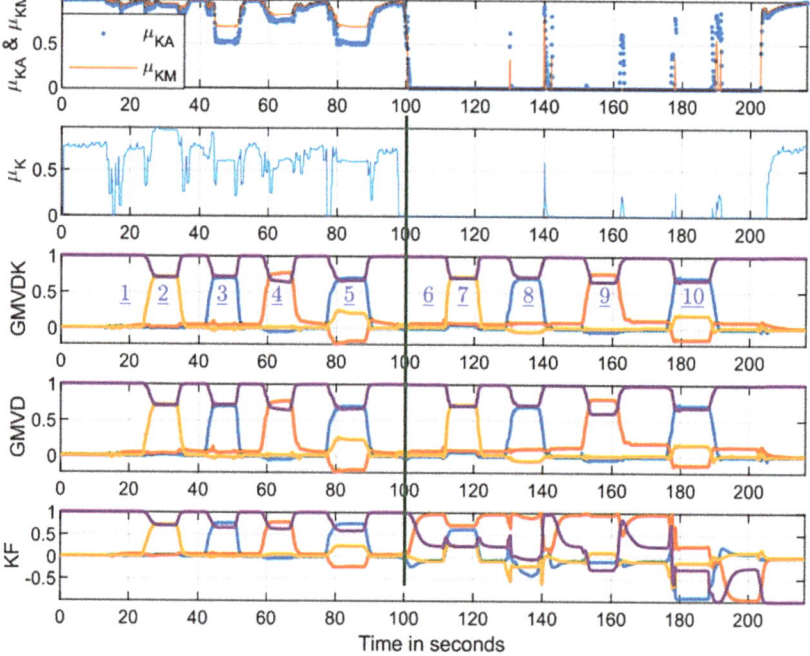

**Figure 3.** New $\mu_K$ trustworthiness parameter and results from three orientation estimation algorithms. All the panels show the evolution of the corresponding values through time. From top to bottom: (1) the two $\mu_K$ components: $\mu_{KA}$ and $\mu_{KM}$; (2) the alternative magnetic trustworthiness parameter, $\mu_K$; (3) the 4 components of the quaternion generated by GMVDK (using $\mu_K$); (4) the 4 components of the quaternion from the previous algorithm version (GMVD); (5) the 4 components generated by the internal Kaman filter (KF) in the MARG module used. The blue underlined numbers are the corresponding pose numbers. The components of all quaternions are color-coded as follows: x in blue; y in red, z in yellow and w in purple.

The plots in Figure 3 show how, while all three approaches, GMVD, GMVDK, and KF, report very similar orientation estimates for all the poses held in Location A (not magnetically distorted), the results generated by KF are markedly different for the same poses at Location B (magnetically distorted).

On the other hand, both GMVD and GMVDK report orientation results that are very similar in Location A and in Location B, confirming their resilience to the distortion of the geomagnetic field.

To assess the orientation differences between results from our proposed GMVDK and the traditional Kalman filter estimator, the quaternion angle difference (QAD) between these quaternions was computed at every sampling instant, using the "dist" command from Matlab® [50], which we verified yields the same results as the formulation to obtain the QAD between quaternions $q_a$ and $q_b$ presented in [34]:

$$QAD(q_a, q_b) = cos^{-1}(2\langle q_a \cdot q_b \rangle - 1) \tag{25}$$

where $\langle q_a \cdot q_b \rangle$ represents the quaternion inner product. We also computed the QAD between orientations from GMVDK and our previous formulation, GMVD. Figure 4 shows the evolution of both those quaternion angle differences. The top panel shows that both our formulations were fairly consistent, displaying minimal QAD throughout. In this top panel, the change in QAD from the first segment of the recording, acquired at Location A, and the second segment of the recording (to the right of the vertical dividing line), acquired at Location B, was very small. The computed RMS values for the segments before and after the dividing line are included in the panel. This reveals that both GMVD and GMVDK performed similarly in the magnetically distorted environment (Location B) and in an environment that was not affected by magnetic distortion (Location A). Since the orientation sequences performed at Location A and at Location B were the same, the top panel of Figure 4 and the traces for GMVD and GMVDK in Figure 3 lead us to the conclusion that both GMVD and GMVDK displayed robustness to the magnetic distortion that existed in Location B.

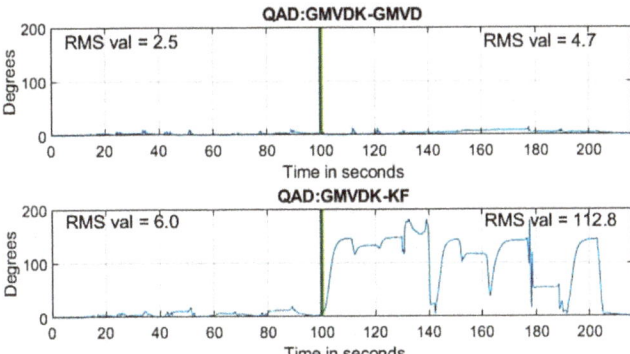

**Figure 4.** Quaternion angle difference (QAD) between GMVDK and GMVD (**top** panel) and quaternion angle difference (QAD) between GMVDK and KF (**bottom** panel). In each case, the RMS value of QAD for the intervals before and after the vertical dividing line is indicated.

On the other hand, the onboard Kalman filter algorithm provided results that are also very close to the ones from GMVDK when the MARG was at Location A (recording a low QAD RMS of 6.0°) but yielded markedly different orientation results when the MARG was at Location B, resulting in large QAD differences with GMVDK in some instants and, overall, reporting a very high QAD RMS of 112.8° after the vertical line (i.e., while the MARG was at Location B).

In Figure 3, the marked change of the four KF traces before and after the dividing line makes it evident that this estimator was impacted by the magnetic distortion at Location B.

But the impact of the errors generated was much better appreciated by 3D visualization of an object oriented according to the estimates. Figure 5 shows renderings of a hand model created in Unity® that has been oriented using the quaternion values yielded by GMVD, GMVDK, and by KF at poses 6, 7, 8, 9, and 10, as identified by the blue underlined numbers inside the GMVDK plot in Figure 3. The leftmost column of renderings shows the hand oriented according with ideal or "reference" orientations that were instructed to the subject (i.e., as the orientations would have been without any error). The error in the orientations estimated by KF ranged in severity, but they did include instances where the reference hand and the KF-oriented hand showed markedly different orientations.

We also had an interest to compare the results of processing the same MARG signals by GMVDK and by two other contemporary MARG orientation estimation approaches. Therefore, we up-sampled data collected in a second representative run (originally collected at 15 samples per second) to match the expected 256 samples per second in the Matlab® (Version R2023a) implementations of Madgwick's [51] and Mahony's [52] MARG orientation estimation algorithms posted by S. Madgwick (at https://x-io.co.uk/downloads/madgwick_algorithm_matlab.zip, accessed on 16 January 2024). We set the adjustable parameters involved in each of the two algorithms to the same values used in the examples provided with the implementation code (Beta was 0.1 for Madgwick's and Kp was 0.5 for Mahony's).

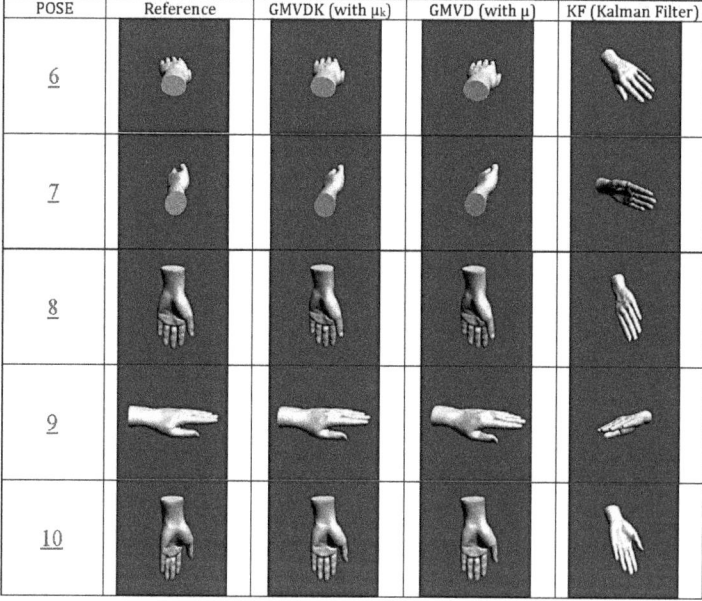

**Figure 5.** Three-dimensional visualization of the orientations obtained from 3 MARG orientation estimation algorithms at the times indicated, with the corresponding blue underlined numbers in the GMVDK panel of Figure 3. From left to right: (1) pose number; (2) visualization of the reference orientations; (3) visualization of the orientations generated by GMVDK; (4) visualization of the orientations generated by GMVD; (5) visualization of the orientations generated by the Kalman filter (KF).

Figure 6 shows the resulting comparison between the algorithm results for this second experimental trial (also following the pre-specified sequence of actions described in Table 2). As it happened for KF, the results from all the algorithms while at Location A (left of the green vertical dividing line) were very similar. However, once the MARG was translated to Location B (right of the green vertical dividing line), GMVDK generated outputs that were very close to the ones obtained for the same sequence in Location A, but both Madgwick's algorithm and Mahony's algorithm generated very different quaternions.

Following the same analysis as we used to create Figure 4, Figure 7 shows the evaluation of the quaternion angle difference between GMVDK and the Madgwick algorithm (top panel) and between GMVDK and the Mahony algorithm (bottom panel). In both cases, the observations were similar to the ones from Figure 4. Both algorithms yielded orientation estimates that were close to those from GMVDK while the MARG was at Location A (to the left of the vertical dividing lines in Figure 7), recording low RMS QAD values of 7.0° and 6.6°. However, when the MARG was manipulated in the magnetically distorted environment at Location B (to the right of the vertical dividing lines in Figure 7), the orientations reported by the Madgwick and Mahony algorithms were significantly different to the orientations reported by GMVDK, recording high RMS QAD vales of 95.0° and 74.9°, respectively.

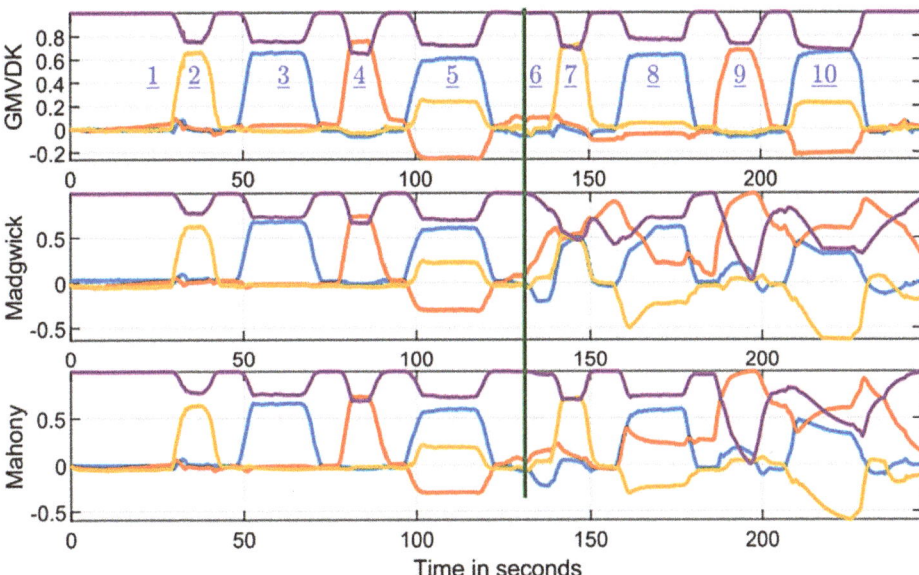

**Figure 6.** Comparison of results from GMVDK and 2 alternative approaches. All three panels display the 4 components of the corresponding orientation quaternion as they evolve through time. From top to bottom: (1) GMVDK; (2) Madgwick's algorithm; (3) Mahony's algorithm. The blue underlined numbers are the corresponding pose numbers. The components of all quaternions are color-coded as follows: x in blue; y in red; z in yellow and w in purple.

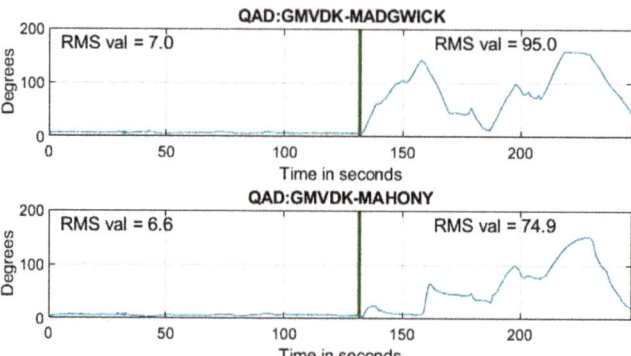

**Figure 7.** Quaternion angle difference (QAD) between GMVDK and Madgwick's algorithm (top panel), and quaternion angle difference (QAD) between GMVDK and Mahoney's algorithm (bottom panel). In each case, the RMS value of QAD for the intervals before and after the vertical dividing line is indicated.

Since the sequence of rotations executed at Location A and at Location B were the same, the markedly different sequences of quaternion components in the second half of the recording generated by the Madgwick and Mahony algorithms led to the conclusion that these two algorithms were much more affected by the magnetic distortion than GMVDK. Further, the QAD differences between GMVDK and each of the other two methods in the right half of Figure 7 provide an approximate quantification of those degradations. Overall, these comparisons support our belief that GMVDK has higher resilience to distortions in the magnetic field.

*3.3. Additional Evaluations of GMVDK Performance*

Beyond the comparisons to alternative orientation estimation methods described in the previous subsection, it was of interest to assess other specific aspects of the performance of the GMVDK algorithm.

Two important aspects to verify in any MARG orientation estimation algorithm are their suitability for real-time implementation in an ordinary computation platform, at a reasonable sampling rate, and their longer-term stability. To determine the execution time required by one iteration of the GMVDK algorithm, on data obtained from one MARG module, we used a real-time C-sharp implementation of GMVDK in which we programmed the recording of a timestamp to a file both just before and immediately after the group of instructions that implement GMVDK. Computing the time difference between 180,000 of those pairs of timestamps, we computed the execution time mean value to be 0.38 milliseconds, with a standard deviation of 0.24 milliseconds. This testing was performed in an ASUS TUF Dash F15 (2021) personal computer, with an 11th Gen Intel Core i7-11370H processor, at a clock rate of 3.30GHz. The mean GMVDK execution time of 0.38 milliseconds indicates that this commodity platform could compute up to 2628 GMVDK iterations per second. Sampling rates commonly used for MARG sensors in the context of human–computer interactions (e.g., 100 Hz) would establish sampling intervals (e.g., 10 milliseconds) that are much longer that the 0.38 millisecond execution time recorded for GMVDK. Accordingly, the execution time of GMVDK seems appropriate for real-time implementation.

To investigate the level of long-term stability that can be expected from the GMVDK algorithm, we performed an overnight, 10 h recording of the quaternion results using the same real-time C-sharp program, running on the same personal computer as described in the previous paragraph. The program logged to file the quaternions from both the onboard Kalman filter and from GMVDK. During this recording, the MARG module was kept static, such that the quaternion expected as the result from both algorithms was quaternion

$q_{NOROT} = [0, 0, 0, 1]^T$. The quaternion values obtained from both algorithms were, indeed, very close to the expected quaternion $q_{NOROT}$. We evaluated the minor deviations from the expected value by computing the QAD between the $q_{NOROT}$ and the GMVDK quaternion for all sampling instants (over the 10 h recording length) and then obtaining the root-mean-squared value of the QAD sequence. We obtained a value of RMSE = 0.4447° for GMVDK. Computing the same metric for the onboard Kalman filter, we obtained RMSE = 2.5291°. This confirmed that GMVDK was, on average, able to keep the orientation errors for this long-term static test to less than half a degree, with the onboard Kalman filter displaying an RMSE value that was bigger.

We also sought to find out if GMVDK would exhibit resilience to different intensities and directions in the magnetic disruption. To study this, the magnetic disruptor was placed 15 cm to the south of Location B (where the second repetition of manipulations resulting in Poses 6 through 10 were still executed), and we verified that GMVDK was less impacted by the magnetic distortion than the Kalman filter under these conditions as well. The details related to this experiment are included in Appendix C, where a second experiment in which the magnetic disruptor was placed 15 cm north from Location B is also described.

Lastly, we studied the ability of GMVDK to provide appropriate estimates of the orientation of the MARG after it had been subjected to a magnetically distorted environment. We did this by translating the MARG from the home location (H) to the magnetically disturbed Location B first and holding the five poses there. After the five poses were completed in Location B, the MARG was translated to the magnetically undisrupted Location A, where the five poses were repeated and then the MARG was translated back to the home location (H). In this so-called H-B-A-H route experiment, we found that GMVDK was able to yield correct orientation estimates for the poses held at Location A immediately, whereas the Kalman filter continued to yield inconsistent orientation estimates for a few milliseconds after the MARG was removed from Location B. A detailed account of this experiment and its results is presented in Appendix D.

## 4. Discussion

Early in this century, the capabilities of the MEMS accelerometer and gyroscopes in MEMS IMUs were found to be insufficient for use in the strapdown configuration for simultaneous position and orientation tracking (e.g., [8]). This led to a narrowing of the goals pursued by using these devices (traying to track orientation only) and to the recruitment of a third kind of MEMS sensor (tri-axial magnetometer) in the effort to provide useful estimates of the orientation of the resulting MARG modules.

Undoubtedly, having the availability of magnetometer signals represents an additional infusion of information to the estimation process, and multiple algorithms have been designed with the implicit aim to fully utilize the information from all three different types of signals (angular rate of change, acceleration, and magnetic field measurements). It may be, however, that too much of an emphasis has been placed in the inclusion of available signals, without enough vigilance regarding the instances in which the information from some of the signals, while available, should better be excluded, or at least restrained.

This began to be realized by the users of MARG orientation estimation algorithms, as noted in the 2009 paper by DeVries et al. [26], where they alerted the biomechanics research community about the possibility of degradation of orientation estimates obtained from some MEMS modules due to geomagnetic disturbances. In that article, they suggested that researchers using this type of MEMS modules for studies tracking the motion of limbs in human subjects should "map" the magnetically distorted areas of their laboratories and avoid using them if MEMS MARG modules were involved. This confirms the severity of the consequences that ignoring magnetic disturbances could have in biomechanical studies.

We have proposed the GMVDK algorithm, described in Section 2, with an emphasis on restraining the involvement of corrections to the initial gyroscopic dead reckoning MARG orientation estimation based on magnetometer or accelerometer signals, if our algorithm finds that the corresponding preconditions are not fully met. While our approach still has the prediction–correction sequence that also exists in Kalman filters, we apply the corrections based on accelerometer and magnetometer information separately, and only to the extent that the corresponding trustworthiness parameter indicates that the correction should be applied. The results presented in Section 3 seem to support our view that GMVDK is more resilient to the degradation of the orientation estimates caused by magnetic disturbances than three other approaches: Kalman filter, Madgwick's algorithm, and Mahony's algorithm.

We designed our experimental protocol so that the same sequence of rotations and poses were performed in both a magnetically undistorted environment (Location A) and a magnetically distorted environment (Location B). Therefore, any significant difference in the output from an algorithm while at Location A in contrast to its output at Location B indicates that the algorithm was impacted by location. By construction, the only important difference between the locations in our setup was the absence/presence of magnetic distortion in Location A/Location B. Therefore, we interpret our observations (Figures 3 and 6) to mean that GMVDK was less affected by magnetic distortion than the Kalman filter, Madgwick's algorithm, and Mahony's algorithm, which is more plainly visualized by the QAD plots contained in Figures 4 and 7. Further, 3D renderings of a virtual object, such as the hand shown in Figure 5, make the degradation of the Kalman filter estimate easier to perceive and gauge intuitively.

The results from both experiments in Figures 4 and 7 seem clear with respect to the strong impact of the conditions at Location B on the orientation estimates produced by the alternative estimation algorithms. In the absence of externally determined ground truth for the orientation in these experiments, we have used the GMVDK estimate as an approximation to the expected correct estimates of orientation. This is because the behavior of the GMVDK estimates at Location A and at Location B, in both Figures 3 and 6, is very similar, as would be expected ideally. Accordingly, we present the QAD comparing the estimates from GMVDK to those from the Kalman filter (Figure 4) and to the estimates from the Madgwick and Mahony algorithms (Figure 5) to quantify the degradation of these alternative methods. We observe very significant changes in the RMS values of those quaternion angular differences before and after the MARG was transitioned from Location A to Location B: Kalman filter: 6° to 112.8°; Madgwick method: 7° to 95°; and Mahony Method: 6.6° to 74.9°.

One additional important observation is that the recurrent nature of Madgwick's and Mahony's algorithms caused them to still yield erroneous orientation estimates even in the last interval of the recording (Figure 6), even when the MARG had already been returned to the home (H) position, where the magnetic field was undistorted. Figure 7 shows that even when the MARG had been translated to a location without magnetic distortion (H), these algorithms were only slowly (progressively) reducing their error, as indicated by the corresponding QAD plots. This further highlights the importance of preventing inappropriate magnetometer-based corrections from taking place in the first place. This is because once such errors enter a recurrent estimator, they are likely to degrade the results for the current iteration and also for future iterations.

## 5. Conclusions

When low-cost miniature MEMS accelerometers and gyroscopes became commercially available, several research communities (biomechanics, human–computer interaction, etc.) wished that they could replicate the capabilities of larger, navigation-grade IMUs. In particular, there was hope that they could be used for tracking the position and orientation of articulated segments of the human body in direct or indirect (e.g., tracking hand-held devices) ways. Micromachining technology had created these miniature, low-power, low-cost devices that were able to produce measurements of accelerations and rotational speeds and the expectations were initially very high.

Unfortunately, even today, the use of these MEMS sensor modules (which now, with the addition of MEMS magnetometers, are consolidated into MARG modules) to confidently track orientation (alone) is challenged by the need to periodically correct orientation estimates initially generated by the gyroscope information. Multiple approaches for efficiently implementing those accelerometer- and magnetometer-based corrections have been proposed through the last few decades, but a definitive answer to this problem remains elusive. In recent years, with the emergence of the IoT, it has become clear that this is also a new and expanding area of application which could benefit from the use of MEMS MARG modules for tracking the orientation of many IoT devices.

In this paper, we presented the "Gravity and Magnetic North Vector, with Double SLERP and $\mu_K$" (GMVDK) algorithm as our proposed approach to the problem, within the operational context of a human–computer interaction application (which may be the same context in which many potential IoT devices would operate).

A key difference in our approach is that GMVDK deliberately implements steps in which the prospective corrections to the initial results of gyroscope signal integration can be "restrained". In fact, the potential accelerometer- or -magnetometer-based corrections are only applied to the extent that the algorithm "trusts" that the preconditions implicitly invoked for the design of the corrections are met.

Therefore, GMVDK also follows the prediction–correction philosophy present in the Kalman filter, but it implements it in a completely different way. In particular, an effort was made in the development of GMVDK to represent the "trustworthiness" of the prospective accelerometer- and magnetometer-based corrections in a single-scalar parameter for each. This was coupled with the consistent representation of orientations (rotations from the initial orientation) as quaternions, leveraging the use of the SLERP interpolations which can partially correct orientations under control of a single scalar parameter.

In this paper, we detailed the new way in which we compute the magnetometer trustworthiness parameter, $\mu_K$, which no longer requires information provided by any external device. Furthermore, we verified the resilience of GMVDK to magnetic distortions that can be found in modern dwellings, offices, and laboratories.

Our verification procedure showed that GMVDK produced the same (quaternion) orientation estimates for a given sequence of MARG orientations ("poses") when it was executed in a magnetically disrupted area (Location B) as when it was executed in a magnetically undisturbed area (Location A). On the other hand, the same accelerometer, gyroscope, and magnetometer data were processed by three other well-known orientation estimators (Kalman filter, Madgwick's and Mahony's algorithms), yielding markedly different results in the magnetically disturbed location, which we interpreted as a degradation of the corresponding estimator performance. We visualized the difference between the results from GMVDK and the alternative estimators by plotting the quaternion angle differences at every sampling instant. Further, we quantified the differences through the RMS aggregate measurements for each QAD plot before and after the MARG was translated from the magnetically undistorted location to the magnetically distorted location.

Overall, the experimental results appear to support our belief that GMVDK provides MARG orientation estimates that are more resilient to the presence of magnetic distortions in the operating space of the MARG than the three alternative orientation estimators.

The degradation in the performance of MARG orientation algorithms due to magnetic distortions is an important challenge that must continue to be studied. We have recently created a freely accessible dataset of MARG signal files with segments that include magnetic distortion, for that purpose [53]. It is our hope that the added robustness exhibited by GMVDK might contribute to a more widespread and successful usage of MEMS MARG modules for a wider range of applications in diverse fields, including human–computer interaction.

**Author Contributions:** Conceptualization, P.S., N.O.-L., N.R. and A.B.; methodology, P.S., N.O.-L., N.R. and A.B.; software (programming of the equations in the manuscript using Matalb®Version 2023a), P.S., N.O.-L., N.R. and A.B.; formal analysis, P.S., N.O.-L., N.R. and A.B.; investigation, P.S., N.O.-L., N.R., M.A. and A.B.; resources, A.B. and M.A.; data curation, P.S. and N.R.; writing—original draft preparation, P.S., N.O.-L. and A.B.; writing—review and editing, N.O.-L., M.A. and

A.B.; visualization, P.S., N.O.-L. and N.R.; supervision, M.A. and A.B.; project administration, M.A.; funding acquisition, M.A. and A.B. All authors have read and agreed to the published version of the manuscript.

**Funding:** This research was funded by The National Science Foundation (NSF), grant number CNS-1920182, and Neeranut Ratchatanantakit and Pontakorn Sonchan were supported by the FIU Dissertation Year Fellowship (DYF) Program.

**Data Availability Statement:** The Matlab® implementations of the Madgwick and Mahoney MARG orientation estimation methods, used to generate two of the traces shown in Figure 6, are available at https://x-io.co.uk/downloads/madgwick_algorithm_matlab.zip, accessed on 16 January 2024.

**Conflicts of Interest:** The authors declare no conflict of interest. The funders had no role in the design of the study; in the collection, analyses, or interpretation of data; in the writing of the manuscript; or in the decision to publish the results.

## Appendix A. Specifications of the 3-Space Micro USB MARG Module from Yost Labs

Below, we transcribe the complete set of specifications offered for the 3-Space Micro USB Attitude and Heading Reference System—AHRS—MARG module from the manufacturer (Yost Labs) at https://yostlabs.com/product/3-space-micro-usb/, accessed on 3 April 2024 ([Specifications] Tab).

**Table A1.** General specifications of the 3-Space Micro USB MARG.

| Parameter | Specification |
|---|---|
| Part number | TSS-MUSB (8G Accelerometer) |
| | TSS-MUSB-HH (24G Accelerometer) |
| | TSS-MUSB-H3 (400G Accelerometer) |
| Dimensions | 23 mm × 23 mm × 2.2 mm (0.9 × 0.9 × 0.086 in.) |
| Weight | 1.3 g (0.0458 oz) |
| Supply voltage | +3.3 v ~ +6.0 v |
| Power consumption | 45 mA @ 5 v |
| Communication interfaces | USB 2.0, asynchronous serial |
| Filter update rate | up to 250 Hz with Kalman AHRS (higher with oversampling) |
| | up to 850 Hz with QCOMP AHRS (higher with oversampling) |
| | up to 1000 Hz in IMU mode |
| Orientation output | absolute and relative quaternion, Euler angles, axis angle, rotation matrix, two vector |
| Other output | raw sensor data, normalized sensor data, calibrated sensor data, temperature |
| Serial baud rate | 1200~921,600 selectable, default: 115,200 |
| Shock survivability | 5000 g |
| Temperature range | −40~85 °C (−40~185 F) |

Table A2. Specifications of the sensors (module, accelerometer, gyroscope, compass).

| Parameter | Specification |
| --- | --- |
| Orientation range | 360° about all axes |
| Orientation accuracy | ±1° for dynamic conditions and all orientations |
| Orientation resolution | <0.08° |
| Orientation repeatability | 0.085° for all orientations |
| Accelerometer scale | ±2 g/±4 g/±8 g selectable for standard models<br>±6 g/±12 g/±24 g selectable for HH models<br>±100 g/±200 g/±400 g selectable for H3 models |
| Accelerometer resolution | 14 bit, 12 bit (HH), 12 bit (H3) |
| Accelerometer noise density | 99 µg/$\sqrt{Hz}$, 650 µg/$\sqrt{Hz}$ (HH), 15 mg/$\sqrt{Hz}$ (H3) |
| Accelerometer sensitivity | 0.00024 g/digit–0.00096 g/digit<br>0.003 g/digit–0.012 g/digit (HH)<br>0.049 g/digit–0.195 g/digit (H3) |
| Accelerometer temperature sensitivity | ±0.008%/°C, ±0.01%/°C (HH, H3) |
| Gyro scale | ±250/±500/±1000/±2000°/s selectable |
| Gyro resolution | 16 bit |
| Gyro noise density | 0.009°/s/$\sqrt{Hz}$ |
| Gyro bias stability @ 25 °C | 2.5°/h average for all axes |
| Gyro sensitivity | 0.00833°/s/digit for ±250°/s<br>0.06667°/s/digit for ±2000°/s |
| Gyro non-linearity | 0.2% full-scale |

Table A2. Cont.

| Parameter | Specification |
| --- | --- |
| Gyro temperature sensitivity | ±0.03%/°C |
| Compass scale | ±0.88 Ga to ±8.1 Ga selectable (±1.3 Ga default) |
| Compass resolution | 12 bit |
| Compass sensitivity | 0.73 mGa/digit |
| Compass non-linearity | 0.1% full-scale |

## Appendix B. Pseudo-Code for Computation of Trustworthiness Parameters and Sensitivity Analysis

*Appendix B.1. Pseudocode for Computation of the Accelerometer Trustworthiness Parameter Alpha*

The goal is to produce a value of alpha (0 < alpha < 1) that represents how much we can trust in accelerometer-based corrections to the MARG orientation estimate (0% to 100%).

At every sampling instant (i.e., when new MARG values are retrieved):

1. Set value of SlopeAlpha (Suggested 1.0)
2. Read (called "Confidence Factor" in the 3-Space module) from MARG
3. alpha1 = (Stillness SlopeAlpha × Stillness) + 1 − SlopeAlpha
4. alpha = [alpha1 + abs(alpha1)]/2

NOTES: Stillness, SlopeAlpha, alpha1, and alpha are all scalar variables.

*Appendix B.2. Pseudocode for Computation of the Magnetometer Trustworthiness Parameter MUK*

The goal is to produce a value of MUK (0 < MUK < 1) that represents how much we can trust in accelerometer-based corrections to the MARG orientation estimate (0% to 100%).

At every sampling instant (i.e., when new MARG values have been retrieved and the final quaternion orientation estimate of the previous iteration, qOUT, is available):

1. Set value of SlopeMuk (Suggested 1.5)
2. Read current magnetometer values, m0, from MARG
3. Compute the mapping of current magnetometer readings, m0, to the inertial frame:

    m0i = qOUT $\otimes$ m0 $\otimes$ qOUT*

4. Compute the cosine of the angle lambda between moi (obtained in 3.) and the initial magnetometer reading, at startup, Minit:

    magnitude_m0i = norm(m0i)
    magnitude_Minit = norm(Minit)
    cos_lambda = dot(m0i, Minit)/(magnitude_m0i $\times$ magnitude_Minit)

5. lambda = acos(cos_lambda)
6. MuKA = 1 − (SlopeMuk $\times$ lambda)
7. penalty_MuKM = (magnitude_moi/magnitude_Minit) $\times$ lambda
8. MuKM = 1 − penalty_MuKM
9. Mu1 = [MuKA + MukM]/2
10. Mu2 = min(current and past Mu1 values in the last 0.4 s)
11. MUK = Mu2 $\times$ alpha

NOTES:
SlopeMuk, magnitude_m0i, magnitude_Minit, cos_lambda, lambda, MuKA, penalty_MuKM, MuKM, Mu1, Mu2, and MUK are all scalar variables.
m0, m0i, and Minit are 3D vectors. qOUT and its conjugate qOUT* are quaternions.
$\otimes$ represents quaternion product; norm( ) is the magnitude of a 3D vector; dot( ) is the dot product of two 3D vectors; abs( ) is the absolute value of a scalar; $\times$ is the product of scalars.

*Appendix B.3. Sensitivity of the GMVDK Performance to Values of SlopeAlpha and SlopeMuk*

The pseudocode listings shown in the previous two subsections indicate that computation of alpha and MUK requires assigning numerical values to the parameters SlopeAlpha and SlopeMuk, respectively. Equation (20) in the body of the article indicates that we used the value SlopeMuk = 1.5 in our implementations. Similarly, in calculating alpha, we used a numerical value of 1.0 for the parameter SlopeAlpha, referenced in Appendix B.1. In both cases, a larger value of the slope parameter would cause the final alpha and MUK values to drop more significantly with a given level of departure from the corresponding ideal conditions (complete absence of movement, for alpha, and complete absence of magnetic distortion, for MUK). In other words, larger SlopeAlpha or SlopeMuk parameters would cause increased caution in accepting orientation corrections based on the corresponding variable (alpha or MUK).

While we set the recommended values, SlopeAlpha = 1.0 and SlopeMuk = 1.5, on the basis of observing the performance of GMVDK in multiple recordings, it is of interest to explore how "sensitive" the performance of GMVDK might be to small shifts from the recommended values for these parameters. We sought to assess these sensitivities (interpreted as the ratio of percentual change in the GMVDK performance to the percentual change in the corresponding parameter) by empirical means, studying a representative case, which is, in fact, the one presented in Figures 6 and 7 of the article. A modified version of Figure 6 is shown below (Figure A1) for the description of our sensitivity assessment procedure.

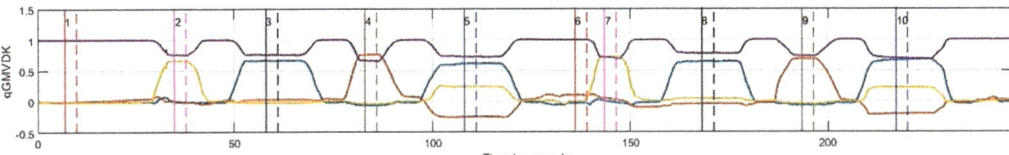

**Figure A1.** Orientation estimations from GMDVK and Kalman filter for the same case as described in Figure 6, but also indicating the time segments where reference quaternions for the poses (numbered) were known (pairs of vertical lines; solid for start and dashed for end of each interval).

Since we are now studying the sensitivity of GMVDK to parameter changes, exclusively, it was important to establish a performance metric that involved GMVDK quaternions only. It was decided to compare the GMVDK quaternions to "ideal" reference quaternions which are known for the time intervals where Poses 1, 2, 3, 4, 5 6, 7, 8, 9, and 10 could be considered to be held by the experimental subject. Segments of the same duration were considered for each one of the poses. Therefore, the QAD values between the quaternion obtained from GMDVK and the corresponding reference quaternion for segments Seg1 to Seg10 were concatenated together into a timeseries All_Segments. Then, the RMSE based on the concatenated time series was computed. This performance value RMSE_from_GMVDK will be a larger positive number when the performance of GMVDK deteriorates, such that the quaternion obtained from GMVDK in the segments studied is more different from the reference quaternions.

Table A3 shows (in its bottom row) the RMSE_from_GMVDK measured as the parameter SlopeAlpha was assigned three values that are lower than 1.0, then it was assigned the recommended value of 1.0 (bolded column), and then three values that were higher than 1.0. The difference in the SlopeAlpha values of any two adjacent columns was 0.25 (25% of the "nominal" value SlopeAlpha = 1.0). All along the value of the other parameter, SlopeMuk was held constant at the "nominal" value of 1.5.

**Table A3.** RMSE of the orientation quaternion from GMVDK (SlopeMuk held at 1.5).

| SlopeAlpha | 0.25 | 0.50 | 0.75 | **1.00** | 1.25 | 1.50 | 1.75 |
|---|---|---|---|---|---|---|---|
| RMSE_from_GMVDK (°) | 9.451 | 9.821 | 9.995 | **9.843** | 9.574 | 9.525 | 9.566 |

The bolded column contains the nominal parameter and resulting RMSE. This column and the two adjacent ones (filled in color) are further explored in the next table.

We observed that RMSE_from_GMVDK did not change much for any of the six non-nominal values tested. Table A4 now shows the values of just the shaded columns from Table A3 normalized by the corresponding nominal value, in order to express them as percentual changes:

**Table A4.** RMSE_from_GMVDK (%) (SlopeMuk held at 1.5).

| SlopeAlpha | 75% | 100% | 125% |
|---|---|---|---|
| RMSE_from_GMVDK | 101.55% | 100% | 97.27% |

The empirical assessment of the sensitivity of the algorithm's performance to variations in this parameter may be approximated as the average of the one-step forward differences before and after the nominal assignment of SlopeAlpha = 1.0, using the percentual values from Table A4:

$$SlopeAlpha_{sensitivity} = \frac{[(100 - 101.55) + (97.27 - 100)]/2}{[(100 - 75) + (125 - 100)]/2} = \frac{-2.14\%}{25\%} = -0.0856 \quad (A1)$$

The sensitivity of GMVDK performance with respect to changes in the value used for the parameter SlopeMuk was investigated in the same way. Table A5 shows (in its bottom row) the RMSE_from_GMVDK measured as the parameter SlopeMuk was assigned three values that were lower than 1.5, then it was assigned the recommended value of 1.5 (bolded column), and then three values that were higher than 1.5. The difference in the SlopeMuk values of any two adjacent columns was 0.25 (16.67% of the "nominal" value SlopeMuk = 1.5). All along the value of the other parameter, SlopeAlpha was held constant at the "nominal" value of 1.0.

**Table A5.** RMSE of the orientation quaternion from GMVDK (SlopeAlpha held at 1.0).

| SlopeMuk | 0.75 | 1.00 | 1.25 | **1.50** | 1.75 | 2.00 | 2.25 |
|---|---|---|---|---|---|---|---|
| RMSE_from_GMVDK (°) | 18.691 | 10.523 | 10.059 | **9.843** | 9.730 | 9.686 | 9.654 |

The bolded column contains the nominal parameter and resulting RMSE. This column and the two adjacent ones (filled in color) are further explored in the next table.

In this case, we observed that the performance of GMVDK remained stable when the value of SlopeMuk was in the neighborhood of the nominal value (e.g., 1.25 < Slope-Muk < 1.75, or even 1.00 < SlopeMuk < 2.00), but if SlopeMuk was decreased too much (e.g., assigning SlopeMuk = 0.75, instead of the suggested SlopeMuk = 1.5), the algorithm did not penalize the presence of magnetic disturbances enough, allowing inappropriate magnetometer-based corrections, which degraded the performance of the algorithm noticeably.

We approximated the sensitivity to changes of SlopeMuk in the neighborhood of the nominal value SlopeMuk = 1.5 with the same procedure as that used for estimating the sensitivity to changes in SlopeAlpha. Table A6 shows the percentual changes:

**Table A6.** RMSE_from_GMVDK (%) (SlopeAlpha held at 1.0).

| SlopeMuk | 83.33% | 100% | 116.66% |
|---|---|---|---|
| RMSE_from_GMVDK | 102.20% | 100% | 98.86% |

Then, proceeding with the calculation of the sensitivity:

$$SlopeMuk_{sensitivity} = \frac{[(100 - 102.20) + (98.86 - 100)]/2}{[(100 - 83.33) + (116.66 - 100)]/2} = \frac{-1.67\%}{16.66\%} = -0.1002 \quad (A2)$$

Overall, the estimated values of these sensitivities (−0.0856 and −0.1002) seemed to indicate that small shifts from the nominal values assigned to both these parameters may not have had a large detrimental effect in the performance of the algorithm. The numbers in Tables A3 and A5 seem to indicate that a large change of the nominal SlopeMuk = 1.5 value (for example to a value of 0.75) will a have noticeable negative impact in the performance of the algorithm.

## Appendix C. GMVDK Performance at Varying Levels and Directions of Magnetic Disruption

In order to evaluate the performance of GMVDK (and the onboard implementation of the Kalman filter included by the manufacturer of the 3-Space Micro USB MARG), we analyzed recordings of two alternative setups:

*Appendix C.1. Magnetic Disruptor Not Placed at Location B, but 15 cm South from Location B*

While the trajectory and sequence of rotations applied to the MARG remained the same as for all the experiments in the body of the article (summarized in Table 2), placing the magnetic disruptor 15 cm to the south of Location B, where the MARG was made to hold Poses 6 to 10, implies that the magnetic disruption challenging the algorithm while these poses were held was different, possibly lower (since there was a longer distance to the disrupter). Figure A2 shows the plot of the orientation quaternions obtained from GMVDK and from the onboard Kalman filter.

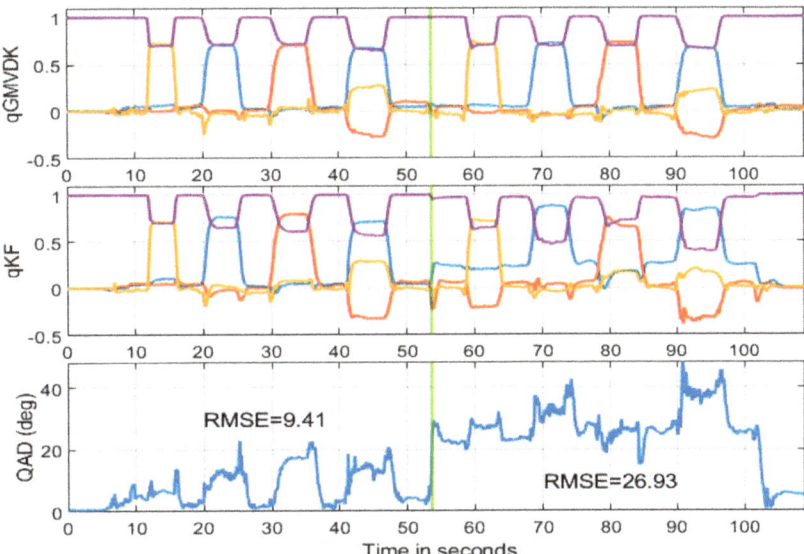

**Figure A2.** Quaternions generated by GMVDK and the onboard Kalman filter when the magnetic disruptor was placed 15 cm south of Location B, and their quaternion angle difference. The RMS values for the QAD before and after the green dividing line are included in the lower panel. The components of all quaternions are color-coded as follows: x in blue; y in red, z in yellow and w in purple.

The similarity of the quaternions obtained for the second set of five poses to the quaternions obtained for the first set of five poses was clearly greater for the output from GMVDK, indicating that it was still more resilient to the magnetic disruption than the Kalman filter. Further, the RMSE values of QAD before and after the vertical dividing line (shown inside the lower panel of the figure) confirmed that the Kalman filter output became more different from the output of GMVDK in the second half of the recording, although in this case, as the magnetic disrupter was farther away, the difference is smaller than the corresponding difference shown in Figure 4 (in the body of the manuscript).

*Appendix C.2. Magnetic Disruptor Not Placed at Location B, but 15 cm North from Location B*

In this second additional experiment, the trajectory and sequence of rotations applied to the MARG remained the same as for all the experiments in the body of the article (summarized in Table 2), but the magnetic disruptor was placed 15 cm to the north of Location B, where the MARG was made to hold Positions 6 to 10. This alternative positioning of the disruptor also implies that the magnetic disruption challenging the algorithm while these poses were held was different, possibly lower (since there was a longer distance to the disrupter). It is worth noting that the alternative position of the magnetic distortion will modify the lines of magnetic field at Location B in yet a third different way. Figure A3 shows the plot of the orientation quaternions obtained from GMVDK and from the onboard Kalman filter.

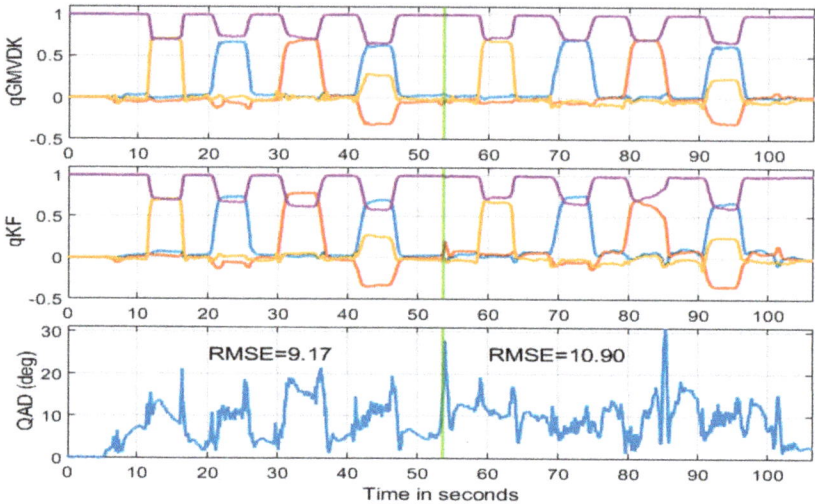

**Figure A3.** Quaternions generated by GMVDK and the onboard Kalman filter when the magnetic disruptor was placed 15 cm north of Location B, and their quaternion angle difference. The RMS values for the QAD before and after the green dividing line are included in the lower panel. The components of all quaternions are color-coded as follows: x in blue; y in red, z in yellow and w in purple.

Interestingly, the Kalman filter in this case seemed to yield orientation estimates that were less affected during the second half of the recording. As in previous cases, the orientation estimates yielded by GMVDK during the second half of the recording are very similar to the ones it yielded for the same 5 poses during the first half of the recording. It seems, therefore, that the level of deterioration of the Kalman filter estimates did not only depend on the closeness of the MARG to the magnetic disruptor but also on the specific directions taken by the lines of magnetic field where the MARG was. The lower level of deterioration of the Kalman filter estimates was confirmed by the similarity of RMSE levels of QAD obtained for the first and second halves of the recording.

In any case, the results of these two additional experiments seem to indicate that GMVDK displays a significance level of resilience to magnetic distortion over a range of magnetic abnormality levels.

**Appendix D. H-B-A-H Route Testing**

We sought to observe if GMVDK has the ability to "recover" from disruptions suffered in its computation by magnetic irregularities once the MARG is translated away from the magnetically distorted area. We were also interested in comparing this kind of ability as it may exist in GMVDK and in other compensation methods (e.g., Kalman filter). Therefore, we present here the results of applying GMVDK and the onboard Kalman filter algorithm to a recording in which the MARG was translated from the home location (H) to the magnetically disturbed Location B first, where the five poses were held. Only then was the MARG translated to the location that did not have a magnetic disruptor under it (Location A). After holding the same five poses in Location A, the MARG was translated back to the home location (H), and the recording ended. Figure A4 shows the quaternions yielded by GMVDK and the Kalman filter, as well as the quaternion angle difference (QAD) between the two series of quaternions, indicating the RMSE QAD values for the first and second halves of the recording, defined by the vertical dividing line in the plots.

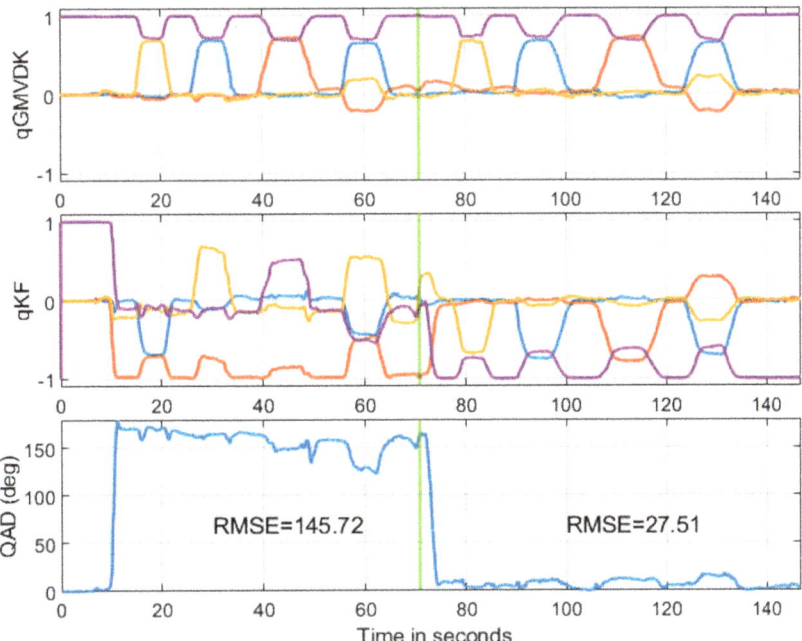

**Figure A4.** Quaternions generated by GMVDK and the onboard Kalman filter when a series of 5 poses was first executed in the magnetically disrupted Location B. The RMS value for the QAD before and after the green dividing line are included in the lower panel. The components of all quaternions are color-coded as follows: x in blue; y in red, z in yellow and w in purple.

Figure A4 shows that the performance of the Kalman filter was significantly deteriorated after translating the MARG from Location H to Location B, yielding a large RMSE QAD value of 145.72° for the first half of the recording. The disruption for GMVDK was much smaller and disappeared very shortly after the vertical line, whereas the Kalman filter took 2 or 3 milliseconds to "recover" until it provided orientation results that were similar to those provided by GMVDK, so that the RMSE QAD of the second half of the recording was much lower. Interestingly, after coming out of the magnetically disrupted Location B, the quaternions generated by the Kalman filter were close to the quaternion generated by GMVDK but with all four of its components multiplied times "−1". In the context of quaternions, one quaternion and another one that was the same as the first with all its four components multiplied by "−1" indicated the same orientation.

## References

1. Lee, I.; Yoon, G.H.; Park, J.; Seok, S.; Chun, K.; Lee, K.-I. Development and analysis of the vertical capacitive accelerometer. *Sens. Actuators A Phys.* **2005**, *119*, 8–18. [CrossRef]
2. Roylance, L.M.; Angell, J.B. A batch-fabricated silicon accelerometer. *IEEE Trans. Electron Devices* **1979**, *26*, 1911–1917. [CrossRef]
3. Johnson, R.C. 3-Axis MEMs gyro chip debuts. *EE Times*, 26 October 2009.
4. Titterton, D.H.; Weston, J.L.; Institution of Electrical Engineers. *Strapdown Inertial Navigation Technology*; Institution of Electrical Engineers: Stevenage, UK, 2004; p. xvii. 558p.
5. Savage, P.G. *Strapdown Analytics*; Strapdown Associates: Maple Plain, MN, USA, 2000.
6. Ignagni, M.B. *Strapdown Navigation Systems: Theory and Application*; Champlain Press: Minneapolis, MN, USA, 2018; 570p.
7. Foxlin, E. Motion Tracking Requirements and Technologies. In *Handbook of Virtual Environments, Design, Implementation, and Applications*; Stanney, K.M., Ed.; Lawrence Earlbaum Associates: Mahwah, NJ, USA, 2002.
8. Woodman, O.J. *An Introduction to Inertial Navigation*; Technical Report No. 696, UCAM-CL-TR-696, ISSN 1476-2980; University of Cambridge: Cambridge, UK, 2007; ISSN 1476-2980.

9. KVHIndustries. *Guide to Comparing Gyro and IMU Technologies—Micro-Electro-Mechanical Systems and Fiber Optic Gyros*; KVHIndustries Inc.: Middletown, RI, USA, 2014; p. 10.
10. Wu, H.; Zheng, X.; Shen, Y.; Wang, X.; Jin, Z.; Ma, Z. A sub-0.1°/h bias-instability MEMS gyroscope using resonant constant-frequency driving technique. In Proceedings of the 2021 IEEE Sensors, Sydney, Australia, 31 October–3 November 2021; pp. 1–4.
11. Wu, H.; Zheng, X.; Wang, X.; Shen, Y.; Ma, Z.; Jin, Z. A 0.09°/h bias-instability MEMS gyroscope working with a fixed resonance frequency. *IEEE Sens. J.* **2021**, *21*, 23787–23798. [CrossRef]
12. Bu, F.; Guo, S.; Fan, B.; Wang, Y. Effect of quadrature control mode on ZRO drift of MEMS gyroscope and online compensation method. *Micromachines* **2022**, *13*, 419. [CrossRef] [PubMed]
13. TRONICS-TDK. GYPRO4300–High Stability Closed-Loop MEMS Gyroscope with Digital Interface Datasheet. 2021. Available online: https://product.tdk.com/system/files/dam/doc/product/sensor/mortion-inertial/gyro/data_sheet/datasheet_gypro3300.pdf (accessed on 6 April 2024).
14. BOSCH. BMI088 6-Axis Motion Tracking for High-Performance Applications Data Sheet. 2018. Available online: https://www.mouser.com/datasheet/2/783/BST_BMI088_DS001-1509549.pdf (accessed on 6 April 2024).
15. InvenSense. *MPU-9150 Product Specification Revision 4.3*; Technical Description of the MPU-9150 Chip; InvenSense Document Number PS-MPU-9150A-00; InvenSense Inc.: Sunnyvale, CA, USA, 2013; 50p.
16. Kalman, R.E. A new approach to linear filtering and prediction problems. *J. Basic Eng.* **1960**, *82*, 35–45. [CrossRef]
17. Barreto, A.; Adjouadi, M.; Ortega, F.R.; O-larnnithipong, N. *Intuitive Understanding of Kalman Filtering with MATLAB*, 1st ed.; CRC Press: Boca Raton, FL, USA, 2020; p. 230.
18. Kuipers, J.B. *Quaternions and Rotation Sequences: A Primer with Applications to Orbits, Aerospace, and Virtual Reality*; Princeton University Press: Princeton, NJ, USA, 1999; 400p.
19. Hanson, A. *Visualizing Quaternions*; Morgan Kaufmann/Elsevier Science Distributor: San Francisco, CA, USA; Amsterdam, The Netherlands; Boston, UK, 2006; 536p.
20. Vince, J. *Quaternions for Computer Graphics*; Springer: London UK; New York, NY, USA, 2011; 154p.
21. Aggarwal, P. *MEMS-Based Integrated Navigation*; Artech House: Boston, MA, USA; London, UK, 2010; 197p.
22. Aggarwal, P.; El-Sheimy, N.; Niu, X.; Syed, Z. A Standard Testing and Calibration Procedure for Low Cost MEMS Inertial Sensors and Units. *J. Navig.* **2008**, *61*, 323–336. [CrossRef]
23. Bar-Itzhack, I.y.; Oshman, Y. Attitude Determination from Vector Observations: Quaternion Estimation. *IEEE Trans. Aerosp. Electron. Syst.* **1985**, *AES-21*, 128–136. [CrossRef]
24. Shuster, M.D.; Oh, S.D. Three-axis attitude determination from vector observations. *J. Guid. Control Dyn.* **1981**, *4*, 70–77. [CrossRef]
25. Ratchatanantakit, N.; O-larnnithipong, N.; Barreto, A.; Tangnimitchok, S. *Consistency Study of 3D Magnetic Vectors in an Office Environment for IMU-Based Hand Tracking Input Development*; Lecture Notes in Computer Science (including subseries Lecture Notes in Artificial Intelligence and Lecture Notes in Bioinformatics); 11567 LNCS; Springer: Cham, Switzerland, 2019; pp. 377–387. [CrossRef]
26. de Vries, W.H.K.; Veeger, H.E.J.; Baten, C.T.M.; van der Helm, F.C.T. Magnetic distortion in motion labs, implications for validating inertial magnetic sensors. *Gait Posture* **2009**, *29*, 535–541. [CrossRef]
27. Coley, B.; Jolles, B.M.; Farron, A.; Bourgeois, A.; Nussbaumer, F.; Pichonnaz, C.; Aminian, K. Outcome evaluation in shoulder surgery using 3D kinematics sensors. *Gait Posture* **2007**, *25*, 523–532. [CrossRef] [PubMed]

28. Zijlstra, W.; Bisseling, R. Estimation of hip abduction moment based on body fixed sensors. *Clin. Biomech.* **2004**, *19*, 819–827. [CrossRef] [PubMed]
29. Kortier, H.; Schepers, H.M.; Sluiter, V.I.; Veltink, P.H. Ambulatory Assesment of Hand Kinematics, using an instrumented glove. In Proceedings of the 12th International Symposium on 3-D Analysis of Human Movement, 3DMA 2012: Technology & Treatment, Bologna, Italy, 18–20 July 2012; pp. 15–18.
30. Kortier, H.G.; Sluiter, V.I.; Roetenberg, D.; Veltink, P.H. Assessment of hand kinematics using inertial and magnetic sensors. *J. Neuroeng. Rehabil.* **2014**, *11*, 70. [CrossRef] [PubMed]
31. Noort, J.V.D.; Dijk, K.V.; Kortier, H.; Beek, N.V.; Verhagen, R.; Bour, L.; Veltink, P. Applications of the PowerGlove for Measurement of Finger Kinematics. In Proceedings of the 2014 11th International Conference on Wearable and Implantable Body Sensor Networks Workshops, Zurich, Switzerland, 16–19 June 2014; pp. 6–10.
32. Lin, B.S.; Lee, I.J.; Chen, J.L. Novel Assembled Sensorized Glove Platform for Comprehensive Hand Function Assessment by Using Inertial Sensors and Force Sensing Resistors. *IEEE Sens. J.* **2020**, *20*, 3379–3389. [CrossRef]
33. Nazarahari, M.; Rouhani, H. 40 years of sensor fusion for orientation tracking via magnetic and inertial measurement units: Methods, lessons learned, and future challenges. *Inf. Fusion* **2021**, *68*, 67–84. [CrossRef]
34. Nazarahari, M.; Rouhani, H. Sensor fusion algorithms for orientation tracking via magnetic and inertial measurement units: An experimental comparison survey. *Inf. Fusion* **2021**, *76*, 8–23. [CrossRef]
35. McGee, L.A. *Discovery of the Kalman Filter as a Practical Tool for Aerospace and Industry*; National Aeronautics and Space Administration: Washington, DC, USA, 1985; Volume 86847.
36. Grewal, M.S.; Andrews, A.P. Applications of Kalman Filtering in Aerospace 1960 to the Present [Historical Perspectives]. *IEEE Control Syst. Mag.* **2010**, *30*, 69–78. [CrossRef]
37. Himberg, H.; Motai, Y.; Barrios, C. R-adaptive kalman filtering approach to estimate head orientation for driving simulator. In *Proceedings of the 2006 IEEE Intelligent Transportation Systems Conference, Toronto, ON, Canada, 17–20 September 2006*; IEEE: New York, NY, USA, 2006.
38. Valenti, R.G.; Dryanovski, I.; Xiao, J. A linear Kalman filter for MARG orientation estimation using the algebraic quaternion algorithm. *IEEE Trans. Instrum. Meas.* **2015**, *65*, 467–481. [CrossRef]
39. Xiaoping, Y.; Bachmann, E.R.; McGhee, R. A Simplified Quaternion-Based Algorithm for Orientation Estimation From Earth Gravity and Magnetic Field Measurements. *Instrum. Meas. IEEE Trans.* **2008**, *57*, 638–650. [CrossRef]
40. Shoemake, K. Animating rotation with quaternion curves. *SIGGRAPH Comput. Graph.* **1985**, *19*, 245–254. [CrossRef]
41. larnnithipong, N. Hand Motion Tracking System Using Inertial Measurement Units and Infrared Cameras. Ph.D. Thesis, Florida International University, Miami, FL, USA, 2018.
42. Ratchatanantakit, N.; O-larnnithipong, N.; Sonchan, P.; Adjouadi, M.; Barreto, A. A sensor fusion approach to MARG module orientation estimation for a real-time hand tracking application. *Inf. Fusion* **2023**, *90*, 298–315. [CrossRef]
43. YostLabs. *3-Space Sensor Miniature Attitude & Heading Reference System with Pedestrian Tracking User's Manual*; Manual for the 3Space Family of Modules; Yost Labs: Portsmouth, OH, USA, 2017; 81p. Available online: https://yostlabs.com/wp-content/uploads/pdf/3-Space-Sensor-Users-Manual-3.pdf (accessed on 27 March 2024).
44. Ratchatanantakit, N.; O-larnnithipong, N.; Sonchan, P.; Adjouadi, M.; Barreto, A. Live Demonstration: Double SLERP Gravity-Magnetic Vector (GMV-D) orientation correction in a MARG sensor. In Proceedings of the 2021 IEEE Sensors, Sydney, Australia, 31 October–3 November 2021.
45. Ratchatanantakit, N.; O-larnnithipong, N.; Sonchan, P.; Adjouadi, M.; Barreto, A. *Statistical Evaluation of Orientation Correction Algorithms in a Real-Time Hand Tracking Application for Computer Interaction*; Springer: Cham, Switzerland, 2022; pp. 92–108.
46. Grewal, M.S.; Andrews, A.P.; Bartone, C. *Global Navigation Satellite Systems, Inertial Navigation, and Integration*, 4th ed.; John Wiley & Sons, Inc.: Hoboken, NJ, USA, 2020; 608p.
47. Xiaoping, Y.; Aparicio, C.; Bachmann, E.R.; McGhee, R.B. Implementation and Experimental Results of a Quaternion-Based Kalman Filter for Human Body Motion Tracking. In Proceedings of the 2005 IEEE International Conference on Robotics and Automation, Barcelona, Spain, 18–22 April 2005; pp. 317–322.
48. Yun, X.; Lizárraga, M.I.; Bachmann, E.R.; McGhee, R.B. An improved quaternion-based Kalman filter for real-time tracking of rigid body orientation. In Proceedings of the 2003 IEEE/RSJ International Conference on Intelligent Robots and Systems (IROS 2003) (Cat. No.03CH37453), Las Vegas, NV, USA, 27–31 October 2003; Volume 2, pp. 1074–1079.
49. Marins, J.L.; Xiaoping, Y.; Bachmann, E.R.; McGhee, R.B.; Zyda, M.J. An extended Kalman filter for quaternion-based orientation estimation using MARG sensors. In Proceedings of the 2001 IEEE/RSJ International Conference on Intelligent Robots and Systems. Expanding the Societal Role of Robotics in the the Next Millennium (Cat. No.01CH37180), Maui, HI, USA, 29 October–3 November 2001; Volume 2004, pp. 2003–2011.
50. Mathworks. Dist: Angular Distance in Radians. Available online: https://www.mathworks.com/help/nav/ref/quaternion.dist.html (accessed on 15 February 2023).
51. Madgwick, S. An efficient orientation filter for inertial and inertial/magnetic sensor arrays. *Rep. x-Io Univ. Bristol* **2010**, *25*, 113–118.

52. Mahony, R.; Hamel, T.; Pflimlin, J.M. Complementary filter design on the special orthogonal group SO(3). In Proceedings of the 44th IEEE Conference on Decision and Control, Seville, Spain, 15 December 2005; pp. 1477–1484.
53. Sonchan, P.; Ratchatanantakit, N.; O-larnnithipong, N.; Adjouadi, M.; Barreto, A. Benchmarking Dataset of Signals from a Commercial MEMS Magnetic-Angular Rate-Gravity (MARG) Sensor Manipulated in Regions with and without Geomagnetic Distortion. *Sensors* **2023**, *23*, 3786. [CrossRef]

**Disclaimer/Publisher's Note:** The statements, opinions and data contained in all publications are solely those of the individual author(s) and contributor(s) and not of MDPI and/or the editor(s). MDPI and/or the editor(s) disclaim responsibility for any injury to people or property resulting from any ideas, methods, instructions or products referred to in the content.

Article

# Geometry Scaling for Externally Balanced Cascade Deterministic Lateral Displacement Microfluidic Separation of Multi-Size Particles †

Heyu Yin [1], Sylmarie Dávila-Montero [2] and Andrew J. Mason [3,*]

[1] Department of Electrical and Computer Engineering, Columbia University, New York, NY 10027, USA; hy2693@columbia.edu
[2] Department of Electrical and Computer Engineering, The Citadel College, Charleston, SC 29409, USA; sdavilam@citadel.edu
[3] Department of Electrical and Computer Engineering, Michigan State University, East Lansing, MI 48824, USA
* Correspondence: mason@msu.edu
† This paper is an extended version of our paper published in the 2020 IEEE International Symposium on Circuits and Systems (ISCAS), Seville, Spain, 10–21 October 2020.

**Abstract:** To non-invasively monitor personal biological and environmental samples in Internet of Things (IoT)-based wearable microfluidic sensing applications, the particle size could be key to sensing, which emphasizes the need for particle size fractionation. Deterministic lateral displacement (DLD) is a microfluidic structure that has shown great potential for the size fractionation of micro- and nano-sized particles. This paper introduces a new externally balanced multi-section cascade DLD approach with a section-scaling technique aimed at expanding the dynamic range of particle size separation. To analyze the design tradeoffs of this new approach, a robust model that also accounts for practical fabrication limits is presented, enabling designers to visualize compromises between the overall device size and the achievement of various performance goals. Furthermore, results show that a wide variety of size fractionation ranges and size separation resolutions can be achieved by cascading multiple sections of an increasingly smaller gap size and critical separation dimension. Model results based on DLD theoretical equations are first presented, followed by model results that apply the scaling restrictions associated with the second order of effects, including practical fabrication limits, the gap/pillar size ratio, and pillar shape.

**Keywords:** deterministic lateral displacement (DLD); particle separation; externally balanced cascade multi-size separation; I-shaped pillar

## 1. Introduction

Point-of-care (PoC) technology for personal health monitoring and environmental monitoring enables wellness management capabilities, including the early detection of diseases, lifestyle supervision, and the reduction of exposure risks [1]. As Internet of Things (IoT) technologies advance, a growing number of IoT-based biosensing and chemical sensing platforms are being developed, all toward the goal of miniaturized, wearable, and personalized health analysis [2,3]. To support IoT-based personalized sensing platforms, the explosion of micro- and nano-scale technologies has accelerated the need for new methods to analyze sample particles at the micron to nanometer scale. Particles of interest span from natural biological particles to foreign and synthetic particles, and particle analysis applications range from scientific exploration in healthcare and environmental studies to the commercialization of instruments. Particle analysis devices must infer the information of real-world significance from particle parameters including size, shape, concentration, and chemical composition. Because the nature and reactivity of particles often vary significantly with their size [4,5], an increasingly important capability is the separation of particles into size-specific bins, commonly referred to as size fractionation. For example, it is well known that the aerodynamic diameter of foreign particles such as particulate matter (PM) greatly determines their penetration into the human respiratory system and their subsequent related impacts on human health. Due to the critical role PM size plays in health impacts, multi-size particle separation, for example, through microfluidic technologies, is strongly desirable, providing valuable analytical capability across many/all size fractions within real-world PM samples [6]. Especially valuable in PM monitoring are technologies that permit the realization of compact instruments allowing susceptible individuals to regularly monitor the air quality within their everyday microenvironments.

Microfluidic devices are well known for their miniaturization potential and high throughput. IoT-based wearable microfluidic sensing can non-invasively process and analyze biological and environmental components that could affect personal health, enabling smart personalized wellness monitoring [7–9]. Several microfluidic separation technologies have been introduced to continuously sort or separate sample particles by size fractions. Active separation technologies, such as dielectrophoresis, electrophoresis, acoustophoresis, immunomagnetic force, or optical force [10–15], must incorporate an external force, and thus are complex and not readily miniaturizable. However, passive methods, such as those of cyclone separator, impactor, and deterministic lateral displacement (DLD) devices permit miniaturization because they only rely on the internal forces within specifically designed microchannel structures [14,16–18]. DLD has shown great results for the separation of micron-scale particles, such as bacteria and blood components, in centimeter-scale devices [19,20]. Utilizing the asymmetric bifurcation of laminar flow around pillar arrays, DLD was first introduced to perform rapid particle separation (40 s) at a high resolution (10 nanometers). DLD is a passive and label-free technique that is widely used because of its simplicity, predictability, and high separation resolution [20]. Several key advances beyond the basic DLD design have been reported, including the use of I-shaped pillars that induce particle rotation and enhance separation efficiency [21,22]. Additionally, nanoscale DLD arrays of uniform gap sizes ranging from 25 to 235 nm have demonstrated the potential for the on-chip sorting and quantification of nanometer-scale biocolloids [23]. Traditional DLD devices inherently separate particles at only one "critical dimension" related to particle size. However, recent work has shown that DLD can be used to fractionate particles into several different size bins by internally cascading hydrodynamically balanced sections of different geometric parameters [24,25]. These internally balanced DLD devices have different sections that separate at a different critical size by varying pillar parameters, but they must continue to process the larger particles separated in prior sections, and thus the pillar gap, which is a vital parameter in size selection, can never be set to be smaller than the largest particle size. As a result, the range of particle sizes that can be separated is very limited or overall channel lengths become inappropriately long.

Working toward a goal of expanding the dynamic range over which particle sizes can effectively be separated into multiple size bins, this paper introduces a new cascade design, called an externally balanced cascade multi-size gap-scaled DLD. In this new approach, each DLD section separates particles at an increasingly smaller critical dimension and then the separated particles are extracted at the end of that section, permitting the next section's pillar geometries, including its pillar gap, to be optimized for separating smaller particles. By introducing external components to balance the hydraulic resistance in each section of a multi-section DLD system, this device maintains a laminar flow throughout all sections and achieves a combination of important capabilities that no prior device has reported. Namely it can simultaneously provide a wide dynamic range of size fractionation, separate samples into multiple size bins, and separate particles down to a nanometer in size, all while optimizing the total channel length. Expanding on preliminary results [26], this paper introduces an in-depth description of the multi-section design and more thoroughly defines and evaluates the mathematical model that permits the determination of the available design space and the analysis of interactions between design parameters and device dimensions. Results from this model are presented to help designers better understand the design considerations of externally balanced cascade multi-section DLD separators. The presented approach and model enable a new generation of DLD multi-size separators that push the barriers of dynamic range and minimum particle separation size in many IoT and POC applications.

## 2. DLD Theory

The theoretical operation of DLD separation by particle size is well established and relies on the laminar flow through a periodic array of micrometer-scale obstacles, which are typically pillars within a microfluidic flow channel. As shown in Figure 1a, when each row of pillars is offset by a slight angle or gradient, $\theta$, the fluid emerging from a gap, $g$, between two pillars will encounter a pillar in the next row and will bifurcate as it moves around the pillar. Within this fluid flow in the axial or flow direction, the $x$ direction, particles that are smaller than a critical separation diameter, $D_c$, which is determined by pillar geometry, will follow the streamlines and go back to their original lanes after a specific number of rows, $N$. In Figure 1a, this "zigzag mode" can be observed for particles with sizes of $D_{p3}$ and $D_{p4}$. In contrast, particles of size $D_{p2}$, which is larger than the critical size $D_c$, are bumped off their initial flow path and displaced laterally to follow the pillar gradient. Based on theoretical analysis and experimental verification, the critical separation diameter can be calculated by [21,27]

$$Dc = 1.4gN^{-0.48} \qquad (1)$$

where $g$ is the gap between pillars and $N$ is the number of rows needed for particles larger than $Dc$ to be shifted by one column from their original position. The pillar gradient is given by

$$\tan(\theta) = 1/N \qquad (2)$$

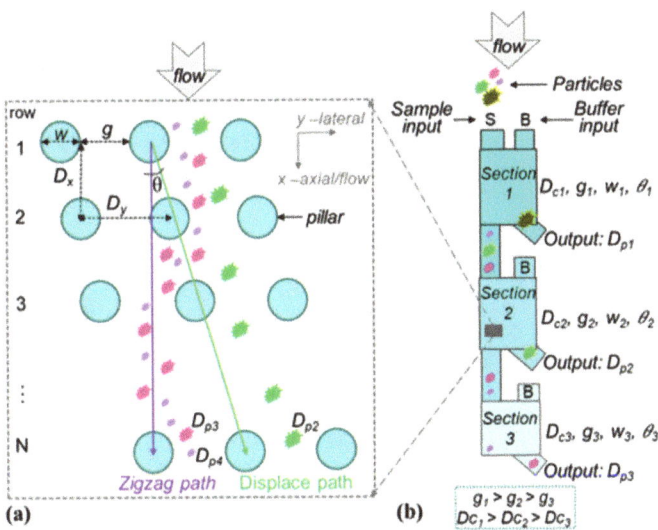

**Figure 1.** (a) An illustration of the pillar array for DLD separation showing key model parameters, and (b) a concept diagram for the externally balanced multi-size DLD separator with multiple cascade gap-scaled sections.

The length of a DLD device can be calculated by

$$L = mND_x \qquad (3)$$

where $D_x$ is the center-to-center distance of the pillars in the flow direction and $m$ represents the number of row-shifts chosen to ensure an adequate separation at the output.

## 3. Cascade Multi-Section DLD Separator

### 3.1. Externally Balanced Cascade Multi-Section DLD Approach

Many applications can benefit from separating at multiple size thresholds. Although DLD has generally been utilized to separate particles at a single critical dimension, a few efforts toward multi-size DLD separation have been reported. These prior efforts implement multi-section DLD devices where each section separates at a different critical size by varying only the pillar gradient or geometry parameters, resulting in either the separation of only a small dynamic range of a particle size or a higher dynamic range at the cost of very long devices [24,25,27–34]. Importantly, these devices can separate only a small range of particle sizes because the pillar gap can never be set to be smaller than the largest particles in the sample to avoid clogging, and the pillar gap, $g_k$, is the most effective parameter for changing the separation sizes. In contrast, the externally balanced approach introduced in this paper cascades multiple sections in a way that the particles separated by each section will be collected at the end of that section. By removing the largest particles after each section, the following section can optimize its geometries, including the pillar gap, for a smaller size range of particle sizes, without concern that larger particles, which have been removed, could clog the channel. Laminar flow is maintained through all sections by introducing external components to balance the hydraulic resistance in each section of a multi-section DLD system. As results below will demonstrate, by eliminating the largest particles before each subsequent section, the design of each section can be optimized for an increasingly smaller range of particle sizes, which greatly expands the design space of each section and allows for a high dynamic range of separation sizes within a reasonable overall channel length.

Figure 1b illustrates an externally balanced multi-section cascade DLD with scaled design geometries in each section permitting multi-size particle separation with a wide dynamic range ($D_{p\_max}/D_{p\_min}$). This illustration shows three cascaded adjacent sections where the critical diameter of particle separation has been gradually decreased in each section due to the gap scaling design. Taking Section 2 as an example, a particle of size $D_{p2}$, which is larger than $D_{c2}$, will be displaced and collected from the Section 2 output while all particles with a diameter smaller than $D_{c2}$ will be conducted into Section 3 for further size fractionation.

### 3.2. Fluidic Mechanisms and Design Rules for the Multi-Section Cascade DLD

Due to the limited understanding of flow fields and the particle dynamics in inertial DLD flows with a Reynolds number well above unity, most current DLD devices focus on a Reynolds number smaller than unity where the fluid inside the microchannel will follow a laminar or even creeping flow model [35]. Although fluid-only predictions are insufficient for explaining experimentally observed critical size behavior, simulated measures agree well with the analytical prediction that a finite size limit on experimentally achievable particle separation often uses a row shift range between 0 and 0.1 ($0 \leq \varepsilon \leq 0.1$) from an assumed parabolic velocity profile [35,36]. Similarly, this paper targets applications that work in laminar flow or creeping flow, so the Reynolds number was set as smaller than 1 and the row shift was set within the range ($0 \leq \varepsilon \leq 0.1$). Although the full understanding of particle mode behaviors remains elusive, decades of work on DLD have concluded that to design a multi-section cascade DLD device, there are three critical parameters that have to be considered: the hydraulic resistance, gap/pillar ratio, and anisotropic affect.

Firstly, the balance of hydraulic resistance across the lateral direction is critical for all DLD designs. Unbalanced hydraulic resistance would contradict the predictable intrinsic particle movement due to the great lateral pressure induced by the resistance difference across the interface. In addition, an awareness of the hydraulic resistance of each component is vital in determining the applied pressure for the successful operation of the DLD system.

The second figure of merit that is essential in the design is the gap/pillar ratio, which not only affects the pressure drop but also determines the surface area-to-volume ratio for a given overall channel geometry [37]. Moreover, varying the aspect ratio also correlates to hydraulic resistance modification. Additionally, pillar shape not only affects how particles will be displaced, but it will also determine the achievable pillar and gap size range due to fabrication limitations.

Thirdly, the anisotropic effect, which will reduce the in situ critical diameter, has to be carefully considered. The most apparent example of the anisotropic effect is the fluidic resistance of an array of obstacles along the channel wall that will introduce a lateral pressure difference and thus induce the anisotropic effect. Thus, the design of pillars on the boundary must be carefully considered to reduce the anisotropic effect [38,39].

Another parameter that affects the device length or critical size behavior is the $D_x/D_y$ ratio. A modification to row shift fraction can be used to replace the traditional model if one chooses that $D_y$ is not equal to $D_x$ [40]. However, the relationship between the $D_x/D_y$ ratio with $Dc$ size has not been fully determined. In this paper, we assume $D_x$ to be equal to $D_y$.

Channel depth is another important design parameter. Increasing the channel depth will certainly decrease the array's flow resistance and reduce clogging. However, deeper channels will limit the pillar and/or gap feature size due to practical fabrication limitations associated with the surface aspect ratio. In this paper, to support the goal of a high dynamic range, channel depth was set to be just higher than the largest particle, $D_{p\_max}$ ($D_p$ is defined as the hydrodynamic diameter of the particle) to avoid particle clogging.

To start a DLD device design, the pillar array's parameters must be set based on the fluidic mechanisms and fabrication design rule. After choosing the design parameter limits for a specific application, the hydraulic resistance of each section can be calculated. Then, a suitable balancing methodology, such as a long serpentine microchannel design or commercially available valves, can be used to balance the hydraulic resistance between sections. Once the pillar array design parameters are defined, the length of each section and then the whole DLD device can be geometrically calculated. Finally, to compare different design parameter choices, the total device length can be set as a figure of merit to analyze the trade-offs. The design parameters used in this paper are summarized in Table 1.

Table 1. Definitions of important variables for the design of multi-section DLD devices.

| Variables | Definition |
| --- | --- |
| $D_{ck}$ | Critical diameter of the particle |
| $D_{p\_max}$ | Biggest particle that will be separated |
| $D_{p\_min}$ | Smallest particle that will be separated |
| $w$ | Diameter of the pillar |
| $g$ | Gap between the pillar (in the lateral direction) |
| $D_x$ | Center-to-center distance in the flow direction |
| $D_y$ | Center-to-center distance in the lateral direction |
| $L$ | Total length of the device |
| $NoS$ | Number of sections |
| $SSF$ | Section-scaling factor |
| $N$ | Number of rows required for one column shift |
| $\gamma$ | Pillar diameter to gap ratio ($\gamma = w/g$) |
| $\beta$ | 1.1—design tolerance |
| $\theta$ | Gradient angle ($\tan(\theta) = 1/N$) |
| $m$ | 1 (Number of columns to be displaced) |

## 4. Multi-Section Mathematical Model

Since shorter devices not only have less flow resistance but also require less space, to develop a model for our gap-scaled multi-size DLD separation device, the overall goal was defined as minimizing the device length while maintaining a high separation performance at each desired size fraction. As shown in Figure 2, our model assumes that each cascaded section of the device separates at an increasingly smaller critical size, and a section scaling factor (SSF) was established to define the critical size ratio between adjacent sections. Thus, for any section, $k$,

$$Dc_{k+1} = Dc_k / SSF \qquad (4)$$

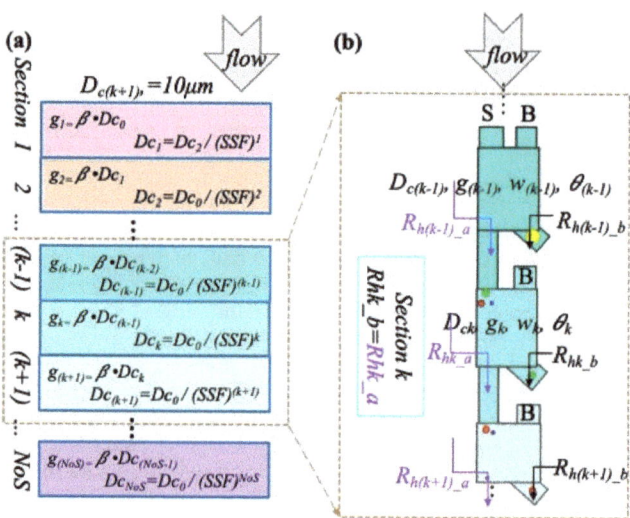

**Figure 2.** Mathematical model that permits the analysis of design tradeoffs: (**a**) the mathematical model shows the relationship between g and Dc with the SSF and NoS set, and (**b**) section $k-1$, $k$, and $k+1$, as examples to illustrate how the hydraulic resistance should be externally balanced.

Notice that, for any total number of sections (NoS), the separator will generate an NoS+1 unique size output. Assuming from Figure 2b that the left output of each section is the input to the next cascade section, the gap in any section needs to be larger than the largest incoming particle to avoid clogging, which is determined by the $Dc$ of the prior section. Thus,

$$g_{k+1} = Dc_k \cdot \beta \qquad (5)$$

where $\beta$ is a design tolerance variable to avoid particle clogging, which is set to 1.1 in this work. If we further assume that the ratio, $\gamma$, between the pillar size and the pillar gap is a constant across all cascading sections, then we have

$$w_k = g_k \cdot \gamma_k \qquad (6)$$

where the optimum choice of $\gamma_k$ will be studied in Section 5. In different applications, $\gamma_k$ can be set as >1 to reduce fabrication complexity [37], or as <1 if a migration angle adjustment needs to be considered [41].

Defining $Dc_0$ as the largest particle size at the input of Section 1 ($Dc_0 = D_{p\_max}$), (4) and (5) can be reformed into general expressions for any $k^{th}$ section as

$$Dc_k = \frac{Dc_0}{SSF^k}, \quad g_k = \beta \cdot \frac{Dc_0}{SSF^{(k-1)}} \qquad (7)$$

The expression in (7) allows us to define the maximum resolution (smallest separated particle size), $D_{p\_min}$, for a device with NoS sections as

$$D_{p\_min} = Dc_{Nos} = \frac{Dc_0}{SSF^{Nos}} \quad (8)$$

Due to the known tradeoff in DLD separation between device length and particle resolution, we can complete the model by expressing the length, $L$, of a multi-section device as the sum of the lengths of all $k^{th}$ sections using (3) and (6), thus

$$L = \sum_{k=1}^{NoS}[m \cdot N_k \cdot g_k(1 + \gamma_k)] \quad (9)$$

where $N_k$ is related to $Dc_k$ and $g_k$ by (1).

To explore the design space for multi-size DLD separation with gap-scaled cascade sections, these modeling equations were implemented in MATLAB, and the simulation results across several parameters of interest are presented in Sections 5 and 6. In these simulations, $m$ was set to 1, $D_{p\_max}$ was set to 10 µm, and the other design parameters are defined above.

## 5. Preliminary Analysis to Define the Relationship between Design Parameters and Device Length

The externally balanced cascade multi-size gap-scaled DLD model was first simulated to see relationships and design tradeoffs without any practical fabrication limits. The device length, $L$, and the separation device resolution, $D_{p\_min}$, were explored to define the optimal range of the total number of sections, NoS, and the section scale factor, SSF. In this preliminary analysis, because the pillar and gap size has no minimum limit, we set $\gamma_k$ to 3 as a starting point, based on our previous study [26].

### 5.1. L vs. NoS Relationship for Different SSFs

Figure 3 shows the relationship between the total length of the multi-size DLD device and the NoS for different values of the SSF ranging from 1.5 to 3.4 with 0.2 steps. This plot shows that the $L$ increases with both the NoS and SSF, as expected. However, the $L$ saturates for an NoS larger than ~5–10. This reflects the fact that, due to scaling the gap size down, the length of each subsequent section is smaller and smaller, allowing the total length to reach a saturation point that varies with the SSF.

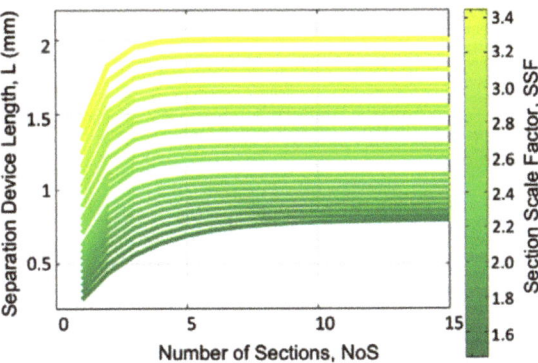

**Figure 3.** The total device length (L) as a function of the number of sections (NoS) and section scale factor (SSF).

## 5.2. L vs. SSF Relationship for Different NoS

Figure 4 plots the device length as a function of the *SSF* for different values of the *NoS* ranging from 1 to 15. This plot shows an interesting behavior of the *SSF*, with a minimum around an SSF = 1.4 and no increase in the *L* for a *NoS* greater than ~5.

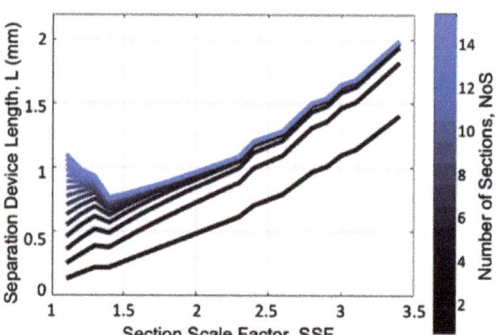

**Figure 4.** The total device length (L) as a function of the section scale factor (SSF) for different values of the number of sections (NoS).

The behaviors from Figures 3 and 4 are further clarified by the 3D plot in Figure 5, which shows the *L* as a function of both the *NoS* and the *SSF*. Here, we can see the valley of the lowest *L* values and observe the tradeoff between the *NoS* and the *SSF*.

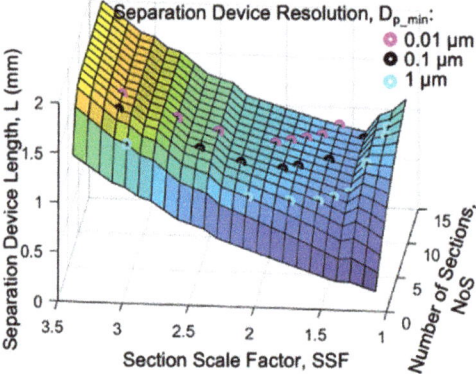

**Figure 5.** 3D plot of the L as a function of both the SSF and the NoS. Colored dots show where various values of device resolution ($D_{p\_min}$) can be achieved.

## 5.3. The Dp_min vs. NoS Relationship for Different SSFs

An important design parameter not considered in Figures 3 and 4 is the device resolution, $D_{p\_min}$. Figure 6 plots the $D_{p\_min}$ (in log scale) as a function of the *NoS* for various values of *SFFs* ranging from 1.5 to 3.4. This plot is helpful to determine which values of the *NoS* and the *SSF* can achieve the desired final particle size separation resolution. For example, very small values of *SSF* would be unable to achieve a resolution of less than 0.1 µm. This helps to put some bounds on the useful values of the *NoS* and the *SSF*. To better highlight this, $D_{p\_min}$ values of 0.01 µm, 0.1 µm, and 1 µm (error margin ± 15%) were extracted from data and added to the 3D plot in Figure 5. This allows us to see which values of the *NoS* and the *SSF* can achieve the desired separation resolution.

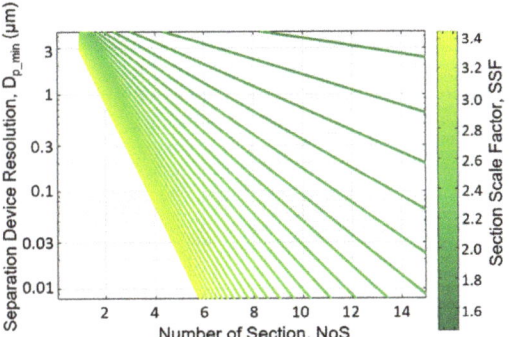

**Figure 6.** Theory analysis to study how the section numbers (NoS) and section scale factor (SSF) affect the device resolution ($D_{p\_min}$).

## 6. Analysis of Secondary Design Considerations

In practical implementations of a DLD device, multiple factors we define as "secondary design considerations" can influence and limit the ideal model defined above. In this section, we analyze several of them, including fabrication limits, gamma variations, and pillar shapes.

### 6.1. Fabrication Limit

Both the pillar size and the gap between pillars will experience limits due to resolutions in the fabrication capability. A DLD device is typically fabricated using deep-RIE with a traditional soft-lithography process [18]. Deep-RIE has feature size limitations depending on the etching depth, which will directly restrict the gap/pillar ratio, $\gamma_k$. For example, if we design a 15 μm depth channel, the smallest achievable $\gamma_k$ is about 0.2 ($w_k/(w_k + g_k) > 0.2$). E-Beam lithography could provide a higher resolution (sub-micron) but is not suitable for centimeter-sized devices. For some applications that do not require extremely high resolutions, soft lithography or even 3D printing-based DLD devices with over 10 micro resolutions are used. To demonstrate the impact of fabrication limits on multi-size gap scaling, we define parameters $g_{k\_fablim}$ and $w_{k\_fablim}$ as the minimum feature size for the gap and the pillar size that can be achieved by device fabrication, respectively. For example, if we define the fabrication resolution as $F_R$ for a simple pillar shape design, such as circular, triangle and square shapes, $w_{k\_fablim}$ will be equal to $F_R$. Thus, the $w_{k\_fablim}$ for circular, triangle and square shapes will be the diameter, side size and width that can be fabricated in different fabrication methods. In contrast, for a specific complicated non-spherical pillar shape that is designed for a specific purpose, the $w_{k\_fablim}$ needs to be designed to be several times larger than the $F_R$, accordingly. Taking an I-shape as an example, in order to introduce particle rotation efficiently, the dimension of the two protrusions that is set as $F_R$ will determine the smallest pillar size ($w_{k\_fablim}$) which has to be at least three times larger than $F_R$. The parameters $g_{k\_fablim}$ and $w_{k\_fablim}$ were added to the simulation model as follows to effectively disable scaling once the pillar and/or gap size reaches this fabrication limit.

$$g_k < g_{k\_fablim}, \text{ else, } g_k = g_{k\_fablim}$$

$$w_k < w_{k\_fablim}, \text{ else, } w_k = w_{k\_fablim} \qquad (10)$$

In our model, the parameters $g_{k\_fablim}$ and $w_{k\_fablim}$ can be chosen by the DLD device designer to match the limitations of a given fabrication facility and process flow. In our fabrication facilities (Lurie Nanofabrication Facility and W.M. Keck Microfabrication Facility), according to the MEMS process we chose, for a circle-shaped pillar design, both the smallest gap size and the smallest circle pillar size were set to 1 μm. Figure 7a plots the simulation results after accounting for fabrication limits and shows that these practical limits significantly change the relationship between the $L$, $NoS$, and $SSF$ compared to Figure 3. We no longer see the $L$ saturate for large values of the $NoS$, and in fact, the $L$ grows to 100 s per meter for a large $NoS$ and $SSF$. With the modification in (10), once section scaling reaches the fabrication limit, the gap can no longer be scaled down and, according to (1), the number of rows, $N$, grows increasingly larger with each subsequent section. Because the $L$ values in meters are impossibly large, the available design space is significantly constrained. To better show what the model predicts for reasonable values of $L$, Figure 7b shows an example where a 0.01 µm resolution is achieved around $L = 200$ mm (depending on the $NoS$) when we zoom into a specific $D_{p\_min}$ value. Note that $NoS = 1$ defines an exceptional, single-section, device that is not practical for achieving high dynamic range goals, so an amount of $NoS \geq 2$ was used in this model.

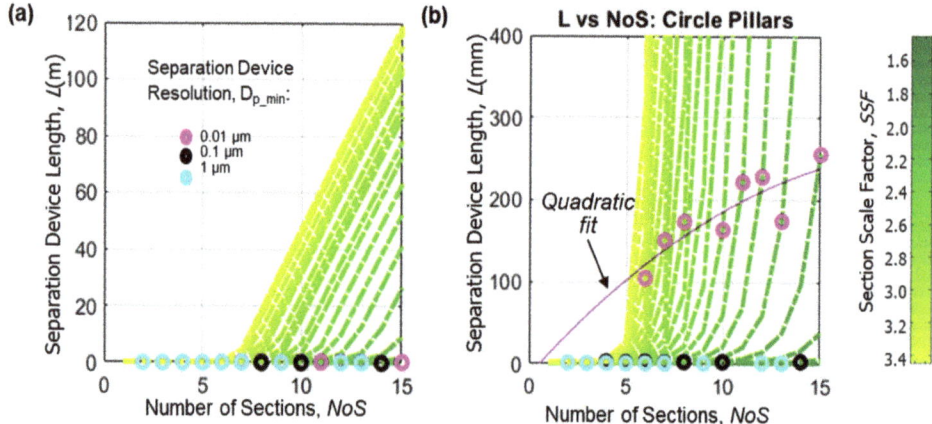

**Figure 7.** (**a**) The L as a function the $NoS$ and $SSF$ after implementing practical fabrication limits for circle-shaped pillars. Zooming in (**b**) helps to illustrate the design space suitable for achieving a 10 nm resolution, $D_{p\_min}$.

### 6.2. Gamma Variation ($\gamma_k$)

In Section 4, $\gamma$ was set to a constant that was used across all sections in order to simplify the relationship among the critical design parameters of a multi-section cascaded DLD device. However, the value of $\gamma$ strongly influences device dimensions, and in light of the impact of fabrication limits, to reduce the total length of the device, the model was modified to permit the $\gamma$ value to be varied in each section.

The gamma value for each section $k$ that achieves the minimum total device length is defined as $\gamma_k$, the ideal gamma for each section. To investigate the optimal values of $\gamma_k$ for each of the sections, a range of values from 0.3 to 3 was used by the model. For any given device resolution, $D_{p\_min}$, the fabrication limit parameters, and the NoS of interest, the model returns the length of each section for each value of $\gamma$ evaluated. For example, for applications such as the size fractionation of PM samples, which range from coarse particles ($PM_{10}$) to ultrafine particles ($PM_{0.1}$) requiring a 0.01 µm particle separation resolution, if the NoS is set to 10, it would result in 11 size fractions of particles with a size dynamic range of 1000. Figure 8 shows the simulation results for this example case ($D_{p\_min} = 0.01$ µm and NoS = 10). Figure 8 presents the section lengths in millimeters for different values of $\gamma$, with a close-up of Sections 1–7 shown in the inset. For a goal of the smallest overall device length, the smallest length for each section, $L_{k\_min}$, can be achieved by selecting the ideal $\gamma_k$ value. In the case presented in Figure 8, if $k > 2$, the ideal $\gamma_k$ is 1, while if $k \leq 2$, the ideal $\gamma_k$ value will be less than 1.

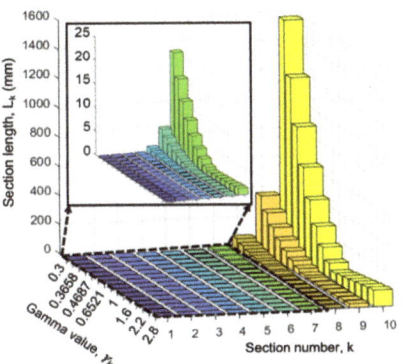

**Figure 8.** Length of each section, $L_k$, as a function of $\gamma_k$ with implementing practical fabrication limits for circle-shaped pillars when the required $D_{p\_min}$ is 10 nm and the NoS is 10. Details for Sections 1–7 were zoomed in and presented in the inset.

This analysis was repeated for applications such as blood separation, to analyze a variety of biological particles that require a 1 µm separation resolution while maintaining the 11 size fractions. Figure 9a shows the simulation results for this case ($D_{p\_min}$ = 1 µm and NoS = 10), where the specific section that reaches the fabrication limit is indicated. In Figure 9a, the smallest section length, $L_{k\_min}$, marked by a star, can be achieved by selecting the $\gamma_k$ equal to 1 for a $k > 6$ and a $\gamma_k$ less than 1 for a $k \leq 6$. Comparing these results to the example case in Figure 8, we see that changing the device resolution, $D_{p\_min}$, will change the ideal sectional $\gamma_k$. The ideal sectional $\gamma_k$ was also observed to be a function of the NoS, as shown in Figure 9b where the NoS was decreased to five while keeping the $D_{p\_min}$ at 1 µm.

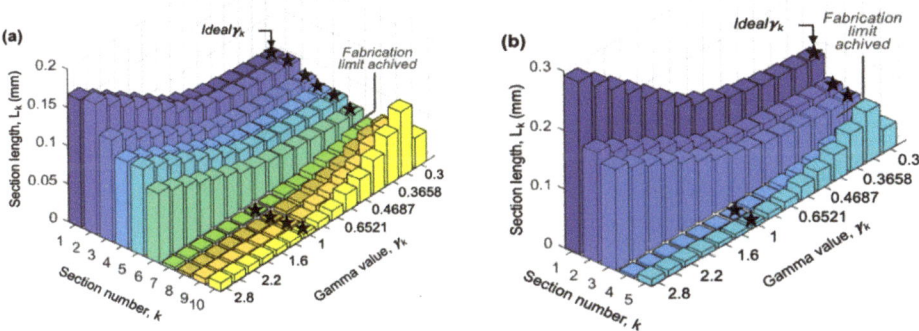

**Figure 9.** Length of each section, $L_k$, as a function of $\gamma_k$ with implementing practical fabrication limits for circle-shaped pillars when the required $D_{p\_min}$ is 1 µm and the NoS is 10, (**a**), and 5, (**b**).

Because the ideal sectional $\gamma_k$ can vary depending on the number of sections and device resolution, we set our model to look for the shortest device length when the ideal sectional $\gamma_k$ is selected for different values of the NoS and $D_{p\_min}$. Figure 10a shows the results of the shortest device length by adding the shortest section lengths together as a function of the NoS. In general, the total device length can be dramatically decreased by the proper choice of the NoS. From the inset in Figure 10a, results show that increasing the device resolution, $D_{p\_min}$, will not necessarily increase the device length significantly, if the ideal sectional $\gamma_k$ is selected for each section separately.

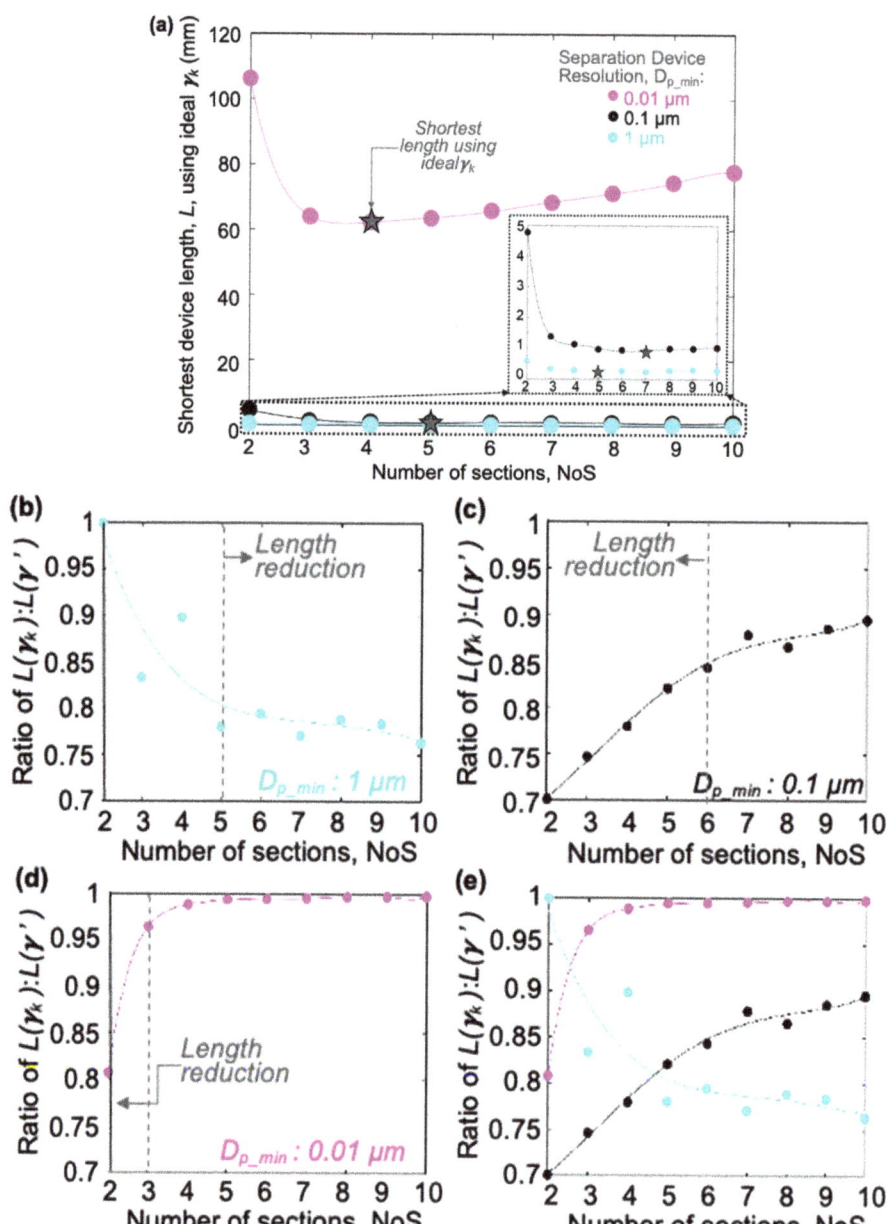

**Figure 10.** (a) Plots the of L vs. NoS for circle-shaped pillars with fabrication limits; different $D_{p\_min}$ values highlight important regions of consideration; (b–e) Plots of the $L(\gamma_k)/L(\gamma')$ ratio vs. the NoS for 1 μm fabrication limit ($g_{k\_fablim} = w_{k\_fablim} = 1$ μm) at different device resolutions, where $L(\gamma_k)$ is the minimum length using the ideal gamma per section, while $L(\gamma')$ is the minimum length using the same gamma across all sections.

It is worth noting that, in some cases, setting the same $\gamma$ value for each section will not necessarily extend the device length too much but can significantly simplify the design process. Figure 10b–e present the ratio of device length for different device resolution conditions. As a comparison, the shortest device length using ideal sectional $\gamma_k$ for each section, $L(\gamma_k)$, is divided by the shortest device length using the same $\gamma$, defined as $\gamma'$, for all sections, $L(\gamma')$. The $L(\gamma_k)/L(\gamma')$ ratio can predict if it is worthwhile to choose different $\gamma$ values for different sections, which will increase the design complexity. For example, as shown in Figure 10d, for the $D_{p\_min}$ = 0.01 μm case, choosing an ideal gamma per section, $\gamma_k$, could reduce the total device length when the NoS is small; however, when the NoS is greater than five, using an ideal gamma per section no longer reduces the total device length but will increase design complexity. In contrast, as shown in Figure 10b, for the $D_{p\_min}$ = 1 μm case, when the NoS is larger than five, utilizing an ideal gamma strategy can save ~25% of the device length. Due to fabrication limits being reached at larger NoS values for the $D_{p\_min}$ = 1 μm case, there is greater variation in optimal gamma values depending on the NoS. Therefore, the $L(\gamma_k)/L(\gamma')$ ratio shown in Figure 10b does not follow a smooth trendline compared to Figure c, d. In summary, as shown in Figure 10e, for small $D_{p\_min}$ cases, where the gap and pillar width will reach the fab limitation faster, very little length reduction is achieved by utilizing the ideal sectional gamma strategy; however, for lower resolution cases, adding more sections with the ideal sectional gamma strategy not only provides more size bins but also achieves an up to ~25% length reduction.

*6.3. I-Shaped Pillar (More Complex Pillar Geometry)*

Different pillar shapes, such as circle shape, triangular shape, T-shape, L-shape, and I-shape, were studied to identify different advantages for specific applications [22,42]. Different pillar shapes require different smallest feature size fabrication capabilities. In our model, $w_{k\_fablim}$ was used to represent the pillar shape and further study how the shape affects the total device length. In real applications, particles will have varied shapes and can be categorized as non-spherical particles. Circle-shaped pillars are the most common in the DLD literature. However, I-shaped pillars can systematically enhance separation efficiency for non-spherical particles. Although, I-shaped pillars require forming a complex groove shape that restricts fabrication capability. Here, the I-shape is analyzed to show how the pillar shape can also be included in the model.

For the I-shaped pillar DLD design, the model was altered to set the smallest pillar size to 6 μm and the smallest pillar gap to 2 μm. Results for the shortest device length using an ideal sectional $\gamma_k$ for each section are shown in Figure 11a. Here it can be observed that a 0.01 μm resolution can be achieved around $L(\gamma_k)$ = 1500 mm in an I-shaped design, whereas the circle-shaped design only requires ~100 mm. Thus, in comparison to the circle-shape, the total DLD device length is about 15 times longer for I-shaped pillars due to the smaller fabrication limit when $D_{p\_min}$ is chosen as 0.01 μm. This suggests a 0.01 μm resolution would be very difficult to achieve with I-shaped pillars. However, a 0.1 μm resolution is achievable in ~14 mm using a *NoS* = 5 and an *SSF* around 2.6. In addition, as shown in the Figure 11a inset, the total device length can be reduced dramatically if we sacrifice resolution. Furthermore, Figure 11b shows the $L(\gamma_k)/L(\gamma')$ ratio has similar trends for a I-shape as it did for a circle-shape in Figure 10. Interestingly, as shown in Figure 11b, choosing the ideal sectional $\gamma_k$ value for each section in the I-shaped design can always reduce the total device length. However, as the NoS increases, the length reduction gradually decreases.

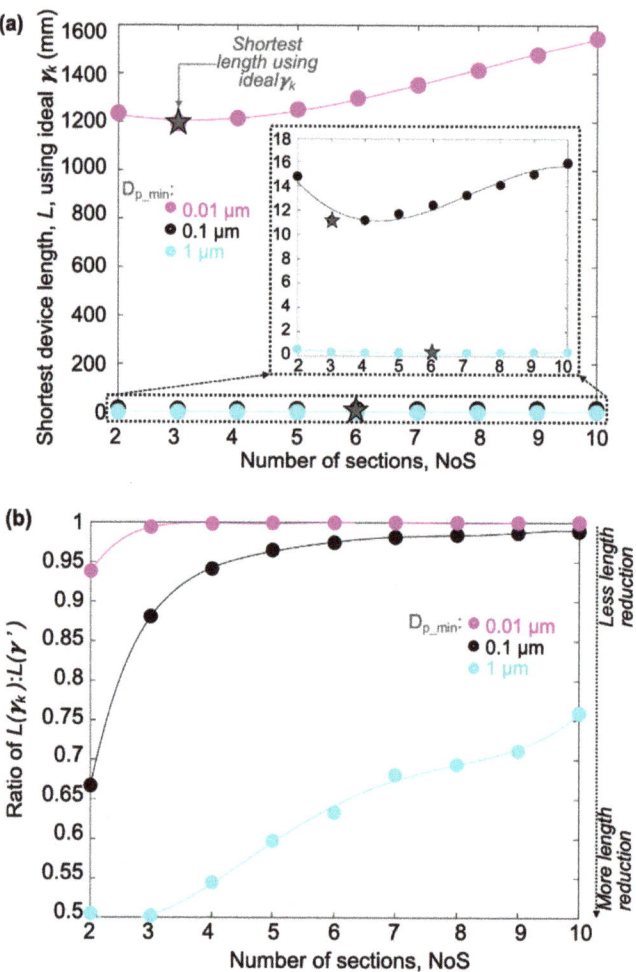

**Figure 11.** (a) Plots of the L vs. NoS for I-shaped pillars with fabrication limits; different $D_{p\_min}$ values highlight important regions of consideration; (b) Plots of the $L(\gamma_k)/L(\gamma')$ ratio vs. the NoS for I-shaped pillar consideration ($g_{k\_fablim}$ = 2 µm, $w_{k\_fablim}$ = 6 µm), where $L(\gamma_k)$ is the minimum length using the ideal gamma per section, while $L(\gamma')$ is the minimum length using the same gamma across all sections.

## 7. Case Study

With refinements to the model presented in Section 5, the mathematical model can be utilized to estimate the total device length for specific study cases. In this section, three case studies, as shown in Table 2, are presented to better illustrate how the final device length is impacted by choosing specific design parameters. The MATLAB Code is provided as Supplementary Material.

Table 2. Comparison for different fabrication limit cases.

| Cases | | DLD Device Length |
|---|---|---|
| Case 1:<br>I-shaped pillar;<br>$g_{k\_fablim} = 2$ μm; $w_{k\_fablim} = 6$ μm; | $D_{p\_min} = 0.01$ μm; $D_{p\_max} = 10$ μm | $L = 60$ m<br>Dynamic range = 1000 |
| Case 2:<br>Circle-shaped pillar;<br>$g_{k\_fablim} = w_{k\_fablim} = 0.1$ μm; | $D_{p\_min} = 1$ μm; $D_{p\_max} = 10$ μm; | $L \sim 0.3$ mm<br>Dynamic range = 10 |
| | $D_{p\_min} = 0.01$ μm; $D_{p\_max} = 10$ μm; | $L \sim 41$ mm<br>Dynamic range = 1000 |
| Case 3:<br>Circle-shaped pillar;<br>$g_{k\_fablim} = w_{k\_fablim} = 10$ μm; | $D_{p\_min} = 1$ μm; $D_{p\_max} = 100$ μm; | $L \sim 8$ mm<br>Dynamic range = 100 |

## 7.1. Case 1

The first case is the design of an I-shaped pillar-based DLD device for an application that requires a size separation dynamic range of 1000 and a $D_{p\_min}$ of 0.01 μm using only one section. This design would separate particles having diameters between 0.01 μm and 10 μm into one bin (output) and those less than 0.01 μm into a different output. As calculated by our model, the total device length is around 60 m. Because a 60 m device length is impossible to fabricate, Case 1 demonstrates that it is not practical to achieve both a high separation resolution (0.01 μm) and a very wide dynamic range (1000) in a device with only one section. However, as Case 2 and Case 3 will show, the externally balanced cascade DLD concept introduced in this paper can meet these goals.

## 7.2. Case 2

The second case aims to again achieve a dynamic range of 1000 but now using a multi-section externally balanced cascade DLD design. Assuming a 100 nm fabrication limit (achievable with EB lithography and Stepper lithography) and circle-shaped pillars, our model was applied to estimate the results for two different $D_{p\_min}$ targets. For a $D_{p\_min} = 1$ μm, as would be common for biological particles, our model calculates that a cascade DLD device with five sections can achieve these goals with a total length of only ~0.3 mm. Similarly, for a $D_{p\_min} = 0.01$ μm, which would be reasonable for PM fractionation, our model calculates that a cascade DLD device can achieve this with four sections and a total length of ~41 mm, which is suitable for many applications. Overall, this case study demonstrates that, to expand the dynamic range, the externally balanced cascade DLD design provides solutions without increasing the device length to an impractical size.

## 7.3. Case 3

The use of a high-resolution fabrication process increases both the cost and fabrication complexity. For some applications, such as mineral processing, a 1 μm separation resolution is sufficient, and low-resolution fabrication methods, such as 3D-printing or PDMS-based soft-lithography, with 10 μm fabrication limits can be utilized. Because these low-resolution cases can use polymer or soft materials and do not need to be formed in silicon wafers, restrictions to the total device length are relaxed, and lengths up to many centimeters can be tolerated. Here, the designer can simply choose a suitable *NoS* and *SSF* and set the $\gamma_k$ to be constant for each section to achieve design goals with a low fabrication complexity. For an example application case requiring particles to be separated into five size bins with a $D_{p\_min} = 1$ μm and a dynamic range of 100, our model shows that the DLD device can be implemented in a total device length of ~8 mm.

## 8. Conclusions

This paper introduced a new approach for multi-size DLD separation using gap-scaled cascaded sections, and it defined a detailed model of design parameter interactions for this approach. Simulations of the model illuminate informative relationships between design variables that aid in analyzing design tradeoffs. Applying practical fabrication dimension limits to our model significantly impacts the simulation results of the device lengths and narrows down the available design space. Results were also shown for circle and I-shaped pillars, which further highlight the impacts of design choices. The model and results in this paper can enable the achievement of desired separation size resolutions within tolerable device lengths, expediting the development of IoT-based wearable microfluidic sensing platforms.

**Supplementary Materials:** The following supporting information can be downloaded at: https://www.mdpi.com/article/10.3390/mi15030405/s1.

**Author Contributions:** Conceptualization, H.Y. and A.J.M.; methodology, H.Y.; software, S.D.-M.; validation, H.Y.; formal analysis, H.Y.; investigation, A.J.M.; resources, A.J.M.; data curation, H.Y. and S.D.-M.; writing—original draft preparation, H.Y.; writing—review and editing, H.Y., S.D.-M. and A.J.M.; visualization, H.Y.; supervision, A.J.M.; project administration, A.J.M.; funding acquisition, A.J.M.. All authors have read and agreed to the published version of the manuscript.

**Funding:** This work was funded by the National Institutes of Health under grants R01ES022302, R01AI113257, and R01ES033515.

**Data Availability Statement:** Data is contained within the article (and Supplementary Materials).

**Acknowledgments:** The authors would like to thank the staff at the Lurie Nanofabrication Facility (LNF) and W.M. Keck Microfabrication Facility for fabrication assistance.

**Conflicts of Interest:** The authors declare no conflict of interest.

## References

1. Mejía-Salazar, J.R.; Rodrigues Cruz, K.; Materon Vasques, E.M. Microfluidic Point-of-Care Devices: New Trends and Future Prospects for eHealth Diagnostics. *Sensors* **2020**, *20*, 1951. [CrossRef] [PubMed]
2. Ardalan, S.; Hosseinifard, M.; Vosough, M.; Golmohammadi, H. Towards smart personalized perspiration analysis: An IoT-integrated cellulose-based microfluidic wearable patch for smartphone fluorimetric multi-sensing of sweat biomarkers. *Biosens. Bioelectron.* **2020**, *168*, 112450. [CrossRef] [PubMed]
3. Blachowicz, T.; Ehrmann, A. 3D printed MEMS technology—Recent developments and applications. *Micromachines* **2020**, *11*, 434. [CrossRef] [PubMed]
4. Dai, Q.; Bertleff-Zieschang, N.; Braunger, J.A.; Björnmalm, M.; Cortez-Jugo, C.; Caruso, F. Particle Targeting in Complex Biological Media. *Adv. Heal. Mater.* **2017**, *7*, 1700575. [CrossRef] [PubMed]
5. Chikaura, H.; Nakashima, Y.; Fujiwara, Y.; Komohara, Y.; Takeya, M.; Nakanishi, Y. Effect of particle size on biological response by human monocyte-derived macrophages. *Biosurface Biotribol.* **2016**, *2*, 18–25. [CrossRef]
6. Kim, K.-H.; Kabir, E.; Kabir, S. A review on the human health impact of airborne particulate matter. *Environ. Int.* **2015**, *74*, 136–143. [CrossRef]
7. Chen, S.; Qiao, Z.; Niu, Y.; Yeo, J.C.; Liu, Y.; Qi, J.; Fan, S.; Liu, X.; Lee, J.Y.; Lim, C.T. Wearable flexible microfluidic sensing technologies. *Nat. Rev. Bioeng.* **2023**, *1*, 950–971. [CrossRef]
8. Palekar, S.; Kalambe, J.; Patrikar, R.M. IoT enabled microfluidics-based biochemistry analyzer based on colorimetric detection techniques. *Chem. Pap.* **2023**, *77*, 2935–2945. [CrossRef]
9. Pechlivani, E.M.; Papadimitriou, A.; Pemas, S.; Ntinas, G.; Tzovaras, D. IoT-Based Agro-Toolbox for Soil Analysis and Environmental Monitoring. *Micromachines* **2023**, *14*, 1698. [CrossRef]
10. Blom, M.T.; Chmela, E.; Oosterbroek, R.E.; Tijssen, R.; Berg, A.v.D. On-Chip Hydrodynamic Chromatography Separation and Detection of Nanoparticles and Biomolecules. *Anal. Chem.* **2003**, *75*, 6761–6768. [CrossRef] [PubMed]
11. Janča, J.; Berneron, J.-F.; Boutin, R. Micro-thermal field-flow fractionation: New high-performance method for particle size distribution analysis. *J. Colloid Interface Sci.* **2003**, *260*, 317–323. [CrossRef]
12. Hartley, L.; Kaler, K.V.I.S.; Yadid-Pecht, O. Hybrid integration of an active pixel sensor and microfluidics for cytometry on a chip. *IEEE Trans. Circuits Syst. I Regul. Pap.* **2007**, *54*, 99–110. [CrossRef]
13. Chakrabarty, K. Design Automation and Test Solutions for Digital Microfluidic Biochips. *IEEE Trans. Circuits Syst. I Regul. Pap.* **2009**, *57*, 4–17. [CrossRef]

14. Antfolk, M.; Laurell, T. Continuous flow microfluidic separation and processing of rare cells and bioparticles found in blood—A review. *Anal. Chim. Acta* **2017**, *965*, 9–35. [CrossRef] [PubMed]
15. Bhagat, A.A.S.; Bow, H.; Hou, H.W.; Tan, S.J.; Han, J.; Lim, C.T. Microfluidics for cell separation. *Med. Biol. Eng. Comput.* **2010**, *48*, 999–1014. [CrossRef]
16. Doering, F.L.; Paprotny, I.; White, R.M. MEMS AirMicrofluidic Sensor for Portable Monitoring of Airborne Particulates. *Sens. Actuators A Phys.* **2013**, *201*, 505–516.
17. Marple, V.A.; Olson, B.A. Sampling and Measurement Using Inertial, Gravitational, Centrifugal, and Thermal Techniques. In *Aerosol Measurement: Principles, Techniques, and Applications*; John Wiley & Sons, Inc.: Hoboken, NJ, USA, 2011.
18. Yin, H.; Wan, H.; Mason, A.J. Separation and Electrochemical Detection Platform for Portable Individual PM2.5 Monitoring. In Proceedings of the IEEE International Symposium on Circuits and Systems, Baltimore, MD, USA, 28–31 May 2017.
19. Poenar, D.P. Microfluidic and micromachined/MEMS devices for separation, discrimination and detection of airborne particles for pollution monitoring. *Micromachines* **2019**, *10*, 483. [CrossRef] [PubMed]
20. McGrath, J.; Jimenez, M.; Bridle, H. Deterministic lateral displacement for particle separation: A review. *Lab A Chip* **2014**, *14*, 4139–4158. [CrossRef]
21. Zeming, K.K.; Ranjan, S.; Zhang, Y. Rotational separation of non-spherical bioparticles using I-shaped pillar arrays in a microfluidic device. *Nat. Commun.* **2013**, *4*, 1625. [CrossRef]
22. Ranjan, S.; Zeming, K.K.; Jureen, R.; Fisher, D.; Zhang, Y. DLD pillar shape design for efficient separation of spherical and non-spherical bioparticles. *Lab A Chip* **2014**, *14*, 4250–4262. [CrossRef]
23. Wunsch, B.H.; Smith, J.T.; Gifford, S.M.; Wang, C.; Brink, M.; Bruce, R.L.; Austin, R.H.; Stolovitzky, G.; Astier, Y. Nanoscale lateral displacement arrays for the separation of exosomes and colloids down to 20 nm. *Nat. Nanotechnol.* **2016**, *11*, 936–940. [CrossRef] [PubMed]
24. Pariset, E.; Pudda, C.; Boizot, F.; Verplanck, N.; Revol-Cavalier, F.; Berthier, J.; Thuaire, A.; Agache, V. Purification of complex samples: Implementation of a modular and reconfigurable droplet-based microfluidic platform with cascaded deterministic lateral displacement separation modules. *PLoS ONE* **2018**, *13*, e0197629. [CrossRef] [PubMed]
25. Tottori, N.; Hatsuzawa, T.; Nisisako, T. Separation of main and satellite droplets in a deterministic lateral displacement microfluidic device. *RSC Adv.* **2017**, *7*, 35516–35524. [CrossRef]
26. Yin, H.; Dávila-Montero, S.; Mason, A.J. Analysis of Section Scaling for Multiple-Size DLD Microfluidic Particle Separation. In Proceedings of the ISCAS 2020 IEEE International Symposium on Circuits and Systems, Sevilla, Spain, 10–21 October 2020; pp. 1–5. [CrossRef]
27. Davis, J.A.; Inglis, D.W.; Morton, K.J.; Lawrence, D.A.; Huang, L.R.; Chou, S.Y.; Sturm, J.C.; Austin, R.H. Deterministic hydrodynamics: Taking blood apart. *Proc. Natl. Acad. Sci. USA* **2006**, *103*, 14779–14784. [CrossRef] [PubMed]
28. Karabacak, N.M.; Spuhler, P.S.; Fachin, F.; Lim, E.J.; Pai, V.; Ozkumur, E.; Martel, J.M.; Kojic, N.; Smith, K.; Chen, P.-I.; et al. Microfluidic, marker-free isolation of circulating tumor cells from blood samples. *Nat. Protoc.* **2014**, *9*, 694–710. [CrossRef] [PubMed]
29. Holmes, D.; Whyte, G.; Bailey, J.; Vergara-Irigaray, N.; Ekpenyong, A.; Guck, J.; Duke, T.; David, H.; Graeme, W.; Joe, B.; et al. Separation of blood cells with differing deformability using deterministic lateral displacement. *Interface Focus* **2014**, *4*, 20140011. [CrossRef]
30. Inglis, D.W.; Morton, K.J.; Davis, J.A.; Zieziulewicz, T.J.; Lawrence, D.A.; Austin, R.H.; Sturm, J.C. Microfluidic device for label-free measurement of platelet activation. *Lab A Chip* **2008**, *8*, 925–931. [CrossRef]
31. Holm, S.H.; Beech, J.P.; Barrett, M.P.; Tegenfeldt, J.O.; Holm, S.H.; Beech, J.P.; Barrett, M.P.; Tegenfeldt, J.O. Separation of parasites from human blood using deterministic lateral displacement. *Lab A Chip* **2011**, *11*, 1326–1332. [CrossRef]
32. Zeming, K.K.; Thakor, N.V.; Zhang, Y.; Chen, C.-H. Real-time modulated nanoparticle separation with an ultra-large dynamic range. *Lab A Chip* **2015**, *16*, 75–85. [CrossRef]
33. Laki, A.J.; Botzheim, L.; Iván, K.; Tamási, V.; Civera, P. Separation of Microvesicles from Serological Samples Using Deterministic Lateral Displacement Effect. *BioNanoScience* **2014**, *5*, 48–54. [CrossRef]
34. Au, S.H.; Edd, J.; Stoddard, A.E.; Wong, K.H.K.; Fachin, F.; Maheswaran, S.; Haber, D.A.; Stott, S.L.; Kapur, R.; Toner, M. Microfluidic isolation of circulating tumor cell clusters by size and asymmetry. *Sci. Rep.* **2017**, *7*, 2433. [CrossRef] [PubMed]
35. Hochstetter, A.; Vernekar, R.; Austin, R.H.; Becker, H.; Beech, J.P.; Fedosov, D.A.; Gompper, G.; Kim, S.-C.; Smith, J.T.; Stolovitzky, G.; et al. Deterministic Lateral Displacement: Challenges and Perspectives. *ACS Nano* **2020**, *14*, 10784–10795. [CrossRef] [PubMed]
36. Inglis, D.W.; Davis, J.A.; Austin, R.H.; Sturm, J.C. Critical particle size for fractionation by deterministic lateral displacement. *Lab A Chip* **2006**, *6*, 655–658. [CrossRef] [PubMed]
37. Yeom, J.; Agonafer, D.D.; Han, J.-H.; A Shannon, M. Low Reynolds number flow across an array of cylindrical microposts in a microchannel and figure-of-merit analysis of micropost-filled microreactors. *J. Micromechanics Microengineering* **2009**, *19*, 065025. [CrossRef]
38. Vernekar, R.; Krüger, T.; Loutherback, K.; Morton, K.; Inglis, D.W. Anisotropic permeability in deterministic lateral displacement arrays. *Lab A Chip* **2017**, *17*, 3318–3330. [CrossRef] [PubMed]
39. Inglis, D.; Vernekar, R.; Krüger, T.; Feng, S. The fluidic resistance of an array of obstacles and a method for improving boundaries in deterministic lateral displacement arrays. *Microfluid. Nanofluidics* **2020**, *24*, 18. [CrossRef]

40. Zeming, K.K.; Salafi, T.; Chen, C.-H.; Zhang, Y. Asymmetrical Deterministic Lateral Displacement Gaps for Dual Functions of Enhanced Separation and Throughput of Red Blood Cells. *Sci. Rep.* **2016**, *6*, 22934. [CrossRef]
41. Kim, S.-C.; Wunsch, B.H.; Hu, H.; Smith, J.T.; Austin, R.H.; Stolovitzky, G. Broken flow symmetry explains the dynamics of small particles in deterministic lateral displacement arrays. *Proc. Natl. Acad. Sci. USA* **2017**, *114*, E5034–E5041. [CrossRef]
42. Loutherback, K.; Chou, K.S.; Newman, J.; Puchalla, J.; Austin, R.H.; Sturm, J.C. Improved performance of deterministic lateral displacement arrays with triangular posts. *Microfluid. Nanofluidics* **2010**, *9*, 1143–1149. [CrossRef]

**Disclaimer/Publisher's Note:** The statements, opinions and data contained in all publications are solely those of the individual author(s) and contributor(s) and not of MDPI and/or the editor(s). MDPI and/or the editor(s) disclaim responsibility for any injury to people or property resulting from any ideas, methods, instructions or products referred to in the content.

Article

# Application of Braided Piezoelectric Poly-l-Lactic Acid Cord Sensor to Sleep Bruxism Detection System with Less Physical or Mental Stress

Yoshiro Tajitsu [1,*], Saki Shimda [2], Takuto Nonomura [2], Hiroki Yanagimoto [1], Shun Nakamura [1], Ryoma Ueshima [1], Miyu Kawanobe [1], Takuo Nakiri [1], Jun Takarada [1], Osamu Takeuchi [3], Rei Nisho [4], Koji Takeshita [4], Mitsuru Takahashi [5] and Kazuki Sugiyama [5]

1. Electrical Engineering Department, Graduate School of Science and Engineering, Kansai University, Osaka 564-8680, Japan; k99454567@kansa-u.ac.jp (H.Y.); k99484565@kansa-u.ac.jp (S.N.); k98554060@kansa-u.ac.jp (R.U.); k99554173@kansa-u.ac.jp (M.K.); k93440043@kansa-u.ac.jp (T.N.); k03456433@kansa-u.ac.jp (J.T.)
2. Nishikawa Co., Ltd., Chuo, Tokyo 103-0006, Japan; sshimada@nishikawa1566.com (S.S.); ttnonomura@nishikawa1566.com (T.N.)
3. Faculty of Foreign Language Studies, Kansai University, Osaka 564-8680, Japan; k95480112@kansa-u.ac.jp
4. Teijin Frontier Co., Ltd., Kita, Osaka 530-8605, Japan; nishior7@teijin-frontier.com (R.N.); takeshita-ko34@teijin-frontier.com (K.T.)
5. Revoneo LLC, Fushimi, Kyoto 600-8086, Japan; contactmt@revoneo.com (M.T.); contactks@revoneo.com (K.S.)
* Correspondence: tajitsu@kansai-u.ac.jp; Tel.: +81-6-6368-1121

**Abstract:** For many years, we have been developing flexible sensors made of braided piezoelectric poly-l-lactic acid (PLLA) fibers that can be tied and untied for practical applications in society. To ensure good quality of sleep, the occurrence of bruxism has been attracting attention in recent years. Currently, there is a need for a system that can easily and accurately measure the frequency of bruxism at home. Therefore, taking advantage of the braided piezoelectric PLLA cord sensor's unique characteristic of being sewable, we aimed to provide a system that can measure the frequency of bruxism using the braided piezoelectric PLLA cord sensor simply sewn onto a bed sheet on which the subject lies down. After many tests using trial and error, the sheet sensor was completed with zigzag stitching. Twenty subjects slept overnight in a hospital room on sheets integrated with a braided piezoelectric PLLA cord. Polysomnography (PSG) was simultaneously performed on these subjects. The results showed that their bruxism could be detected with an accuracy of more than 95% compared with PSG measurements, which can only be performed in a hospital by a physician and are more burdensome for the subjects, with the subjects simply lying on the bed sheet with a braided piezoelectric PLLA cord sensor sewn into it.

**Keywords:** poly-l-lactic acid; piezoelectricity; braided cord; sensing; PLLA

## 1. Introduction

Sleep affects humans in many ways; the lack of it causes fatigue, affects immunity, memory and learning, performance, and mental health, and most recently, causes dementia [1–8]. For mental health care, it is important to record daily sleep conditions and maintain and improve sleep duration and quality [9]. Polysomnography (PSG) is the gold standard for the objective assessment of sleep status [10,11]. However, PSG is a sophisticated diagnostic method, and the collection and accurate interpretation of results require specialized knowledge and skills in PSG measurement. The measurement system is also complex and expensive. Only a limited number of hospitals and laboratories offer PSG. Measurement during sleep requires the use of numerous sensors. Furthermore, the data obtained are visually analyzed by doctors [9–11]. Thus, the burden on the subject during measurement is great, and the greatest drawback is that measurement cannot be performed routinely at home. Therefore, with the growing interest in sleep in recent years, there is a need for a sleep tracker that can simply and routinely measure sleep status [9–11]. However, current sleep trackers are less accurate than PSG. With these as a background, our goal is to develop a device that can measure health status during sleep at home with an accuracy comparable to that of PSG without causing any burden or discomfort to the subject.

We previously conducted research using a braided piezoelectric poly-l-lactic acid (PLLA) cord [12,13], which has been attracting attention as a wearable sensor [14–20]. The braided piezoelectric PLLA cord sensor we have developed to date has many unique features. The following is a brief summary of the important features we have reported thus far [12,13]. First, plant-derived piezoelectric PLLA fibers are used as a motion-sensing material, and compared with other practical piezoelectric materials such as lead zirconate titanate (PZT), they do not contain heavy metals, such as lead or fluorine, and have less environmental impact [21–28]. Fibers are braided into a coaxial cable-like structure, making it resistant to electrical disturbances, as shown in Figure 1a. This structure of the piezoelectric PLLA braided cord was already reported [24,27]. The core of the piezoelectric PLLA braided cord is a conductive fiber bundle, and PLLA and PET fibers are wound around it. Furthermore, the conductive fibers cover them to realize a coaxial cable structure. The core was wrapped with PET fiber to form a braided piezoelectric PLLA cord. The cord is as mechanically strong as packing cords and is water-resistant. It can also be tied and untied due to its braided structure. On the other hand, PLLA fibers are monofunctional sensors that respond to bending motions and basically do not respond to stretching. However, if they are formed into, for example, a decorative knot to make an accessory-type sensor as shown in Figure 1b, they can respond to various motions [16,17,27,28]. This is a very significant feature of this braided piezoelectric PLLA cord sensor that PZT and other sensors do not have. The braided piezoelectric PLLA cord sensor can also be formed into various stitches with an embroidery needle. This is not only a design feature, but selectivity in sensing motion also can be achieved by embroidering a decorative knot or fabric [27]. For example, a choker with a lucky knot charm can detect only pulsation without being affected by the body's motion even when the body is making a large motion. When chain stitches are embroidered on a denim fabric, only specific movements of each body part can be detected [12,13,19,20,27]. These results are supported by the findings of analysis with the finite element method (FEM), which identifies the bending displacement of the stitched braided piezoelectric PLLA cord sensor [12,13]. Therefore, taking advantage of the braided piezoelectric PLLA cord sensor's unique characteristic of being sewable, we aimed to provide a system that can measure the frequency of bruxism by simply sewing the braided piezoelectric PLLA cord sensor onto a bed sheet.

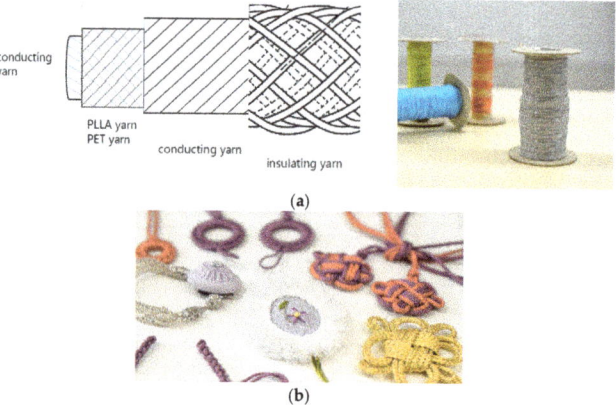

**Figure 1.** (**a**) Braided piezoelectric PLLA cord and (**b**) its decorative knots.

In this study, we constructed and improved a system using a braided piezoelectric PLLA cord as a sensor. Then, the frequencies of bruxism in many subjects during one night of sleep were acquired using the sensor simultaneously with PSG measurement. The data thus obtained were compared to verify the accuracy of our system. As a result, we obtained a comparable accuracy to PSG. The results are reported below.

## 2. Difficulty in Measuring Bruxism

Bruxism is defined by the American Sleep Society as a "repetitive jaw muscle activity characterized by the clenching or grinding of teeth and/or the fixation or thrusting of the mandible" [29,30]. When teeth grinding occurs, the teeth are clenched hard and rub against each other repeatedly, which can aggravate gum sensitivity and periodontal disease [29–38]. It is also considered to cause temporomandibular joint disorder, facial pain, headaches, and stiff shoulders [31]. For those who sleep in the same room with others, bruxism can generate noise and also deteriorate the sleep quality of those in the same room. Stress and anxiety have been suggested as causes of bruxism [32], but a clear cause is not yet known. Treatment options are limited and include dental treatment and the use of a mouthpiece to prevent tooth wear [33,34]. There are two methods to diagnose bruxism: one is by interviewing subjects with abnormal dental conditions such as tooth wear caused by teeth grinding during sleep [9] and the other is using sensors such as those in PSG to detect bruxism [9–11,34–38]. In the former method, the only option is treatment because the diagnosis is made in the advanced state of symptoms. On the other hand, the method of directly detecting teeth grinding requires a device to be worn on the jaw, which is burdensome and cannot be used for daily measurement. These hurdles make it difficult to conduct research. When this diagnosis is conducted in the hospital, the subject wears the testing device and sleeps in a hospital bed overnight, and data are collected. Figure 2 below shows an illustration of this process. Sleeping in this state is stressful both physically and mentally due to the burden imposed by the testing equipment. In addition, since the examination equipment can only be used in a hospital, it is not possible to monitor the daily sleep status of a subject at home. Therefore, there is a need for a technology that can routinely monitor the condition of bruxism during sleep in a noncontact, nonburdensome manner. This would be useful in elucidating the causes, treatment, and prevention of bruxism.

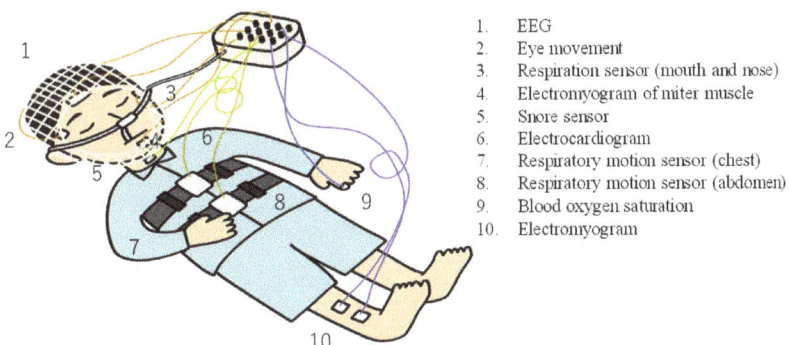

1. EEG
2. Eye movement
3. Respiration sensor (mouth and nose)
4. Electromyogram of miter muscle
5. Snore sensor
6. Electrocardiogram
7. Respiratory motion sensor (chest)
8. Respiratory motion sensor (abdomen)
9. Blood oxygen saturation
10. Electromyogram

**Figure 2.** Illustration of PSG measurement during sleep.

## 3. Braided Piezoelectric PLLA Cord Sensor

As a system for detecting the occurrence of bruxism that allows subjects to sleep soundly overnight without any psychological or physical stress, we considered integrating a braided piezoelectric PLLA cord sensor into a bed sheet. Changes in the subject's sleeping posture cause major problems when detecting signals indicating the occurrence of bruxism for the following reasons. Originally, the braided piezoelectric PLLA cord sensor was based on the piezoelectricity of PLLA fibers. Piezoelectricity is a phenomenon that generates an electric charge in response to strain or stress applied to a material [39,40]. Therefore, if a bed sheet is subjected to a large amount of strain or stress due to body movement or tossing and turning during sleep, a large signal is generated on the basis of the piezoelectricity of the PLLA fibers. In other words, if a signal larger than the piezoelectric signal that would be generated by teeth grinding is generated by tossing and turning, it is superimposed on the signal generated by teeth grinding. The separation of these signals is expected to be difficult. In addition, there are various postures such as lying on one's back or on one's side. Piezoelectric sensors such as those constructed using PZT [39,40] are now in practical use. However, since the size of a PZT sensor is usually 3–5 cm, considering that the subject changes their lying position while sleeping, many PZT sensors must be spread over the entire bed and wired to each other. This is not a practical way when considering the time and effort required to do this. In contrast, a single braided piezoelectric PLLA cord can be easily sewn into a bed sheet over a large area that is responsive to changes in the posture of a subject lying on the bed sheet. The major problem here is the detection of the vibration generated by bruxism. The site of bruxism generation is considered to be around the jaw and mouth. However, the braided piezoelectric PLLA cord is sewn into the bed sheet. The braided piezoelectric PLLA cord is not directly in contact with the site of bruxism generation, but is rather in contact with the subject's back and other parts below the neck. Common sense suggests that it would be difficult to detect the occurrence of bruxism with the braided PLLA piezoelectric cord sewn into the bed sheet under this condition. In previous studies, when such sensing was not possible, FEM was conducted to search for conditions under which sensing was possible [41,42], and a prototype sensor was successfully fabricated on the basis of FEM results. In this study, we followed the same approach and first conducted FEM to search for conditions under which the braided piezoelectric PLLA cord can be sewn into bed sheets to sense the vibration generated by bruxism.

*3.1. FEM*

The posture of the subject on the bed during sleep should be in a way such that the subject does not move away from the sensing area with the braided piezoelectric PLLA cord. Furthermore, considering that the way of contact with the braided piezoelectric PLLA cord changes depending on the subject's posture, it is necessary to consider the method of sewing the braided piezoelectric PLLA cord. There are two main patterns of embroidering the braided piezoelectric PLLA cord on sheets. One is straight stitching, in which the braided piezoelectric PLLA cord is stitched perpendicularly to the fabric as if it were sewn with a regular sewing machine, and the other is zigzag stitching, in which the braided piezoelectric PLLA cord is placed on the fabric surface and fastened with a different thread. Since the site at which the subject comes in contact with the braided piezoelectric PLLA cord varies depending on the subject's posture and sleeping position, it is important that the signal does not change at that time, which translates into system accuracy and simplicity. Therefore, we first investigated via FEM whether there is a difference in response between zigzag and straight stitching. Figure 3 shows the piezoelectric response of a model with the braided piezoelectric PLLA cord zigzag-stitched in a circle and applied with a stress of 10 N perpendicularly to the entire fabric. The color of the piezoelectric response indicates the magnitude of the response. The model with zigzag stitching shows almost the same piezoelectric response throughout the circumference. In other words, the piezoelectric response is the same regardless of the point of stress application on the circle. In contrast, as shown in Figure 4, the model with straight stitching shows a large piezoelectric response at the point where it touches the fabric and at the point of stress application on the fabric where the curvature of the folded braided piezoelectric PLLA cord changes. During sleep, the posture and position of the subject's body vary from subject to subject, and even for the same subject, it varies from time to time. In other words, it is impossible to predict how the braided piezoelectric PLLA cord will come in contact with the subject's body in this study.

That is, it is strongly suggested that zigzag stitches, which generate the same piezoelectric response no matter where the braided piezoelectric PLLA cord comes in contact with the subject's body, are suitable for the purpose of this study.

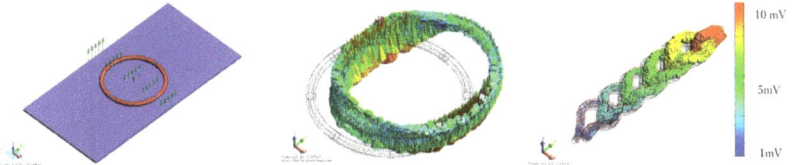

**Figure 3.** FEM calculation results of piezoelectric response of a model with braided piezoelectric PLLA cord zigzag-stitched in a circle.

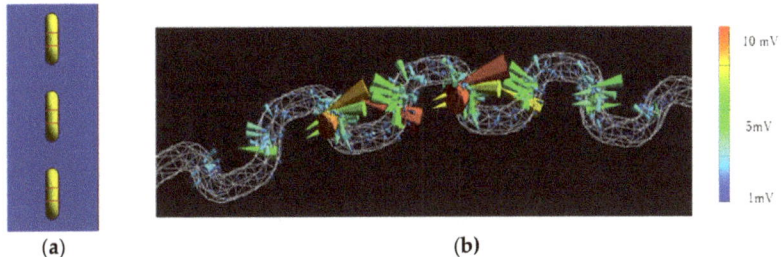

**Figure 4.** FEM calculation results of piezoelectric response of a model with braided piezoelectric PLLA cord stitched straight: (**a**) top view; (**b**) bird's-eye view.

If we adopt zigzag stitching, the braided piezoelectric PLLA cord must be designed to have an inflection point that covers the entire bed sheet. To determine the effect of this design, the piezoelectric response was calculated for the braided piezoelectric PLLA cord having a curvature as shown in Figure 5. In particular, we paid attention to whether the piezoelectric response at the inflection point is much larger than that at other locations, as observed in straight stitching. As shown in the figure, a very detailed analysis of the calculation results shows that the piezoelectric response at the inflection point is indeed larger than those at other locations, but the rate of increase is less than 10%. From the calculation results, the final configuration of a single braided piezoelectric PLLA cord to be sewn onto a bed sheet was designed as shown in Figures 6 and 7. Figure 6 shows that the cord covers a relatively large area of curvature where a constant piezoelectric response can be expected. On the other hand, the stitch pattern in Figure 7 has a longer-period curve than that in Figure 6. For Figure 6, a constant piezoelectric response is obtained. For Figure 7, the magnitude of the piezoelectric response is not affected by the addition of shorter-period curves. From these calculations, we decided to sew a single braided piezoelectric PLLA cord in a zigzag pattern to achieve a bed sheet with a configuration that provides a long-period curve.

**Figure 5.** FEM calculation results of piezoelectric response of a model with braided piezoelectric PLLA cord stitched straight: (**a**) model with piezoelectric PPLA braided cord sewn onto fabric; (**b**) calculated values of response signal.

**Figure 6.** FEM calculation results of the configuration of the braided piezoelectric PLLA cord sewn onto the sheet (I): (**a**) model with piezoelectric PPLA braided cord sewn onto fabric; (**b**) calculated values of response signal.

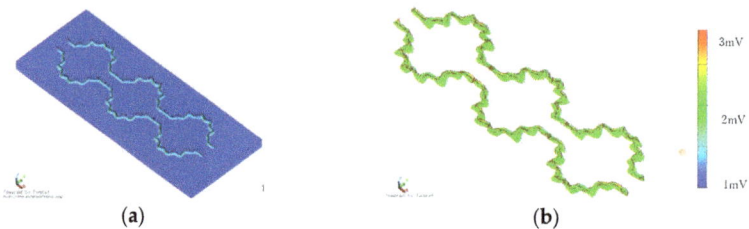

**Figure 7.** FEM calculation results of the configuration of the braided piezoelectric PLLA cord sewn onto the sheet (II): (**a**) model with piezoelectric PPLA braided cord sewn onto fabric; (**b**) calculated values of response signal.

*3.2. Bed-Sheet Sensor Blueprint*

The shape of the braided piezoelectric PLLA cord sewn in a zigzag pattern was determined from the results of FEM calculations. The specific size of the actual sheet sensor was determined from this shape, with particular consideration given to ensuring that the braided piezoelectric PLLA cord would always be in contact with the body in any position of the subject while lying in bed, even for a petite woman. To determine the size of zigzag stitching, human body dimensions were considered. Table 1 shows such data published by the Ministry of Economy, Trade and Industry, Japan. From this table, it can be determined that for the braided piezoelectric PLLA cord to be always in contact with the convexity of the body, even when a petite woman lies on her back, side, or at an angle in bed, the embroidery spacing of the braided piezoelectric PLLA cord must be 20 cm or less. Figure 8 shows a bed-sheet blueprint with the braided piezoelectric PLLA cord stitched in a zigzag pattern determined via FEM (hereafter, bed-sheet-type sensor).

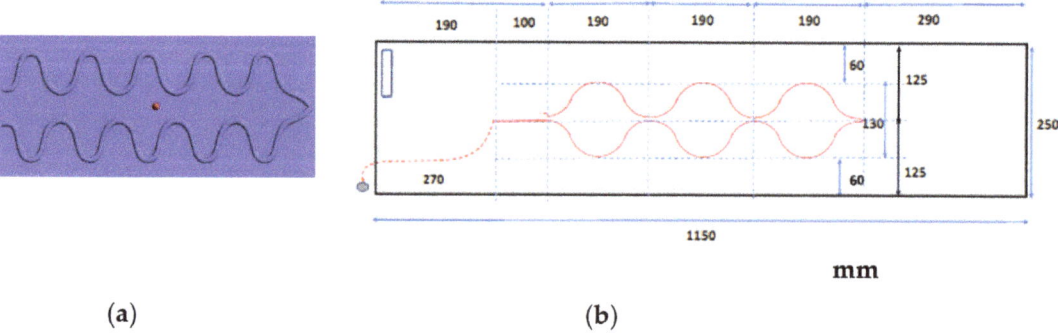

**Figure 8.** Illustration of the actual shape and dimensions of the sheet sensor prepared. (**a**) FEM model for bed-sheet type sensor. (**b**) Actual blueprint of bed-sheet-type sensor.

Table 1. Human body dimension measurement results (H19-10-1).

| Distance between Right and Left Acromion (Not Necessarily a Straight Line) | | Vertical Distance from the Acromion to the Lower End of the Elbow Bone Bent at a Right Angle | | Horizontal Linear Distance between the Anterior and Posterior Surfaces of the Chest at the Nipple Point | |
|---|---|---|---|---|---|
| male | female | male | female | male | female |
| mm | | mm | | mm | |
| 403 | 358 | 340 | 338 | 201 | 201 |
| 404 | 360 | 315 | 309 | 212 | 200 |
| 406 | 358 | 341 | 335 | 222 | 200 |
| 406 | 359 | 316 | 308 | 225 | 204 |
| 404 | 360 | 342 | 329 | 228 | 209 |
| 403 | 359 | 314 | 307 | 230 | 210 |
| 399 | 358 | 341 | 329 | 231 | 214 |
| 395 | 359 | 314 | 306 | 229 | 219 |
| 391 | 356 | 339 | 330 | 225 | 222 |
| 388 | 352 | 312 | 305 | 228 | 229 |
| 385 | 350 | 339 | 328 | 230 | 231 |
| 380 | 347 | 309 | 300 | 228 | 233 |

From the Ministry of Economy, Trade and Industry Japan

## 4. Bed-Sheet Sensor

From the FEM results, a design that fits the Japanese body shape was created, and a single braided piezoelectric PLLA cord was stitched in a zigzag pattern using a computerized sewing machine to make the stitches firm, as shown in Figure 9 (bed-sheet-type sensor). The sewn bed-sheet-type sensor was placed on the bed and covered with a mattress pad. In this system, which is the same as the previously reported system for the signal detection circuit, the sensed signal is received by a preamplifier and then amplified 400 times by an amplifier [41,42]. Basic measurements were conducted to confirm the responsiveness based on the piezoelectricity of the bed-sheet-type sensor. The sheet with a braided piezoelectric PLLA cord as the sensor sewn onto it was clamped at positions 10 cm to the left and right from the center of the bed sheet, and a static tension of 1 N was applied so that the sheet would not sag. An AC tensile strain with a frequency of 1 Hz and a distortion of 0.1% was applied, and a 700-fold amplified response signal was received. An example of the response signal is shown in Figure 10. The response waveform shows good reproducibility and continuity. Next, sheets were placed on actual beds used in sleep experiments. The sheet cover used in a sleep experiment was placed over another sheet to determine the pressure response. Deformation was applied by pulsing a 0.5 mm diameter circular brass rod pushed 0.5 mm into the bed-sheet-type sensor at the center. An example of the response is shown in Figure 11. The bed-sheet-type sensor was found to respond well to sharp pulses.

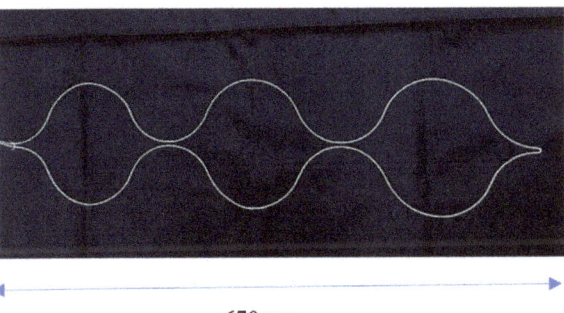

670 mm

**Figure 9.** Photo of the completed bed-sheet-type sensor to be used in the experiment.

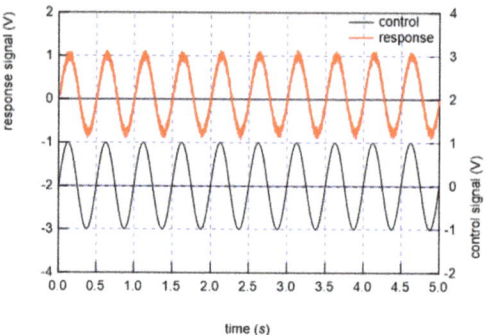

**Figure 10.** Example of response signal from the bed-sheet-type sensor when an AC tensile strain of 1 Hz frequency and a 0.1% strain rate were applied.

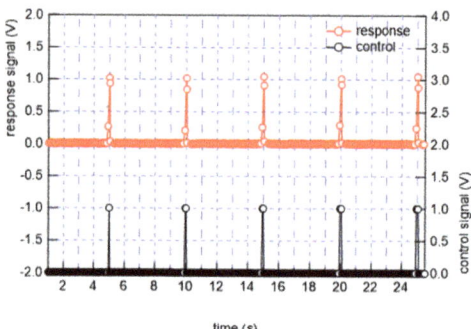

**Figure 11.** Response of bed-sheet-type sensor to the push-in displacement of the sensor.

## 5. Detection of Bruxism during Sleep

Subjects who agreed to participate in our experiment were fitted with PSG equipment during an overnight sleep at a sleep clinic. The experiment was conducted as follows. In this measurement, the bed-sheet-type sensor was placed on the bed where the subject slept, and PSG measurements were conducted simultaneously throughout the night. We emphasize here that the results obtained from the bed-sheet-type sensor in this experiment can be accurately contrasted with PSG results obtained under the supervision of a physician.

*5.1. Medical Diagnostic Measurements*

A brief description of the medical equipment used in the sleep clinic is given in [9–11] that focuses on features related to this experiment. PSG measurements (Figure 2) include electroencephalography (EEG) at Fp2-A1, F4-A1, C4-A1, and O2-A1; jaw and leg electromyography (EMG); bilateral electrooculography (BEOG); nasal airflow measurement; chest and abdominal respiratory movement detection; fingertip oxygen saturation measurement; electrocardiography (ECG); and a positional detection sensor fixed on the skin at the center of the sternum. Electrodes for EEG were placed on the head surface according to the international 10–20 method [9–11,34–38]. In addition, the body position during sleep was confirmed with an infrared camera. Specifically, EEG, BEOG, EMG of the jaw, ECG, abdominal movement detection, nasal airflow measurement, and oxygen saturation and snoring sound measurements were conducted. Video images and activity levels were also recorded. EEG and EMG of the jaw were conducted to determine tooth grinding [34–38]. Here, rhythmic masticatory muscle activity (RMMA) was determined by the technician on the basis of sleep stages, arousal, the visual assessment of movements, and EMG of the masseter muscle according to the AASM criteria [9,37]. The reason why the diagnosis of teeth grinding is so precise in such a sleep clinic is that it is based on the International Classification of Sleep Disorders [29,34–38], and sleep bruxism is considered to be one of the most common sleep disorders [9,37,38]. The appropriate processing of the data from the all-night EMG measurements in the hospital room was conducted by a clinical laboratory technician, and sleep bruxism was determined by a sleep specialist on the basis of the following criteria [9].

1) Mean amplitude of the electromyogram: More than 10% of the maximum occlusal force (masseter muscle) at waking time.
2) Muscle contraction pattern during sleep bruxism episodes:
    (a) Phasic episode: 3 or more bursts (duration of 0.25 s to 2.0 s for each burst).
    (b) Tonic episode: one burst lasting more than 2 s.
    (c) Mixed episodes: bursts of both phasic and tonic episodes are present.

*5.2. Bed-Sheet-Type Sensor Measurements*

A mattress pad was placed on a sleep clinic bed for measurement. The bed-sheet-type sensor was placed directly on the mattress pad and fixed with pins to prevent it from shifting. The pad and the bed-sheet-type sensor were covered with a quick-drying, water-absorbent box sheet; thus, the bed-sheet-type sensor was not visible to the subject. The subject can sleep in any position, and our bed-sheet-type sensor does not restrict the subject's sleeping posture. The subject lies down naturally with a pillow in the desired position, covers themselves with a blanket, and goes to sleep. The response signal from the bed-sheet-type sensor was amplified 400 times through a preamplifier and an amplifier and then stored in a data logger (NR-600B, Keyence corporation, Osaka, Japan; settings: 1 kHz and 12 bits) placed under the bed, as shown in Figure 12. The purpose of the circuit configuration is briefly explained below. First, since the impedance of the braided piezoelectric PLLA cord is very large, a preamplifier is used for impedance matching with the circuit. Furthermore, since this signal contains noise related to the power supply, Twin-T CR is used to remove the noise and amplify the weak signal for detection. Since the current measurement target is a human and the frequency bands of respiration, pulse, and body motion are 0.1 Hz to 10 Hz, the band-pass filter is used to detect these signals with high accuracy.The data obtained during sleep were stored overnight along with the PSG data described in the previous section.

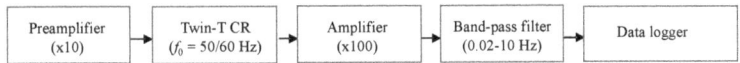

**Figure 12.** Circuit system for bed-sheet-type sensor.

**6. Results**

A demonstration experiment to show the effectiveness of the bed-sheet-type sensor sewn with a braided piezoelectric PLLA cord was conducted at a medical institution using a sleep test to diagnose sleep apnea syndrome. Note that the experiment at such a medical institution was conducted only when the physician had confirmed that the bed sheet embroidered with the braided piezoelectric PLLA cord as the sensor did not interfere with the diagnosis of sleep apnea syndrome and the subject consented to participating in the demonstration experiment.

*6.1. Medical Judgment*

In the experiment, subjects slept overnight in a clinic; they were fitted with PSG and other medical devices necessary to diagnose apnea, and they slept lying down on a bed. The bed-sheet-type sensor was placed under the bed cover. Before the measurement began, the subjects were first asked to lie down on the bed, and the medical staff examined them for any possible sleep disturbances caused by the bed-sheet-type sensor. All potential subjects responded that they felt no discomfort at all. Fifteen subjects participated in the demonstration experiment. During the overnight examination, some subjects experienced bruxism and others did not. An example of an actual all-night measurement of the waveforms during sleep obtained from the medical devices is shown. Figure 13 shows the ECG waveform, EMG waveform, activity levels, and postural changes. Note that EMG can detect weak signals at rest, but when body movements detected by the sleeping posture and activity meter (such as turning over) occur, EMG also detects a large response signal. On the other hand, ECG measures heartbeats continuously, and although the effect of body movements on ECG should not be large in terms of the measurement principle, body movements actually generate a large signal because the electrodes attached to the body change their state of adhesion. The amount of activity here refers to the total amount of activity per minute as measured by the physician with a small accelerometer attached to the subject's waist. Since the amount of activity is constantly changing, the amount of activity in the figure is averaged over a 10 min period in order to show an overall trend. In the all-night measurement, a wide variety of signals are generated, which should not be the case in principle. It is very labor-intensive for technicians to individually determine sleep stages from the data obtained in such a complex environment. Overall, it can be seen in Figure 13 that the activity meter signal increases when the sleeping posture changes. In other words, the activity level increases at the timing of sleep turning. Sleep levels also change at this timing. At this time, a large signal is also generated in the ECG, which in principle should not be affected by turning over. Furthermore, determining the occurrence of teeth grinding is even more difficult. As mentioned earlier, the onset of bruxism is determined from changes in electromyographic signals in accordance with the aforementioned diagnostic rules [9,34–39]. The difficulty is that when the sleep duration is 6 h (21,600 s), a characteristic waveform lasting only 10–20 s is found in the data, from which the occurrence of bruxism is identified. For a subject who grinds their teeth, more than 50 episodes of bruxism occur in a single night. This indicates that even if such a complex PSG device could be fitted at home (which is not possible), it would be impractical to continue to observe the frequency of bruxism over a long period, even with current diagnostic methods, where a specialist must make the diagnosis.

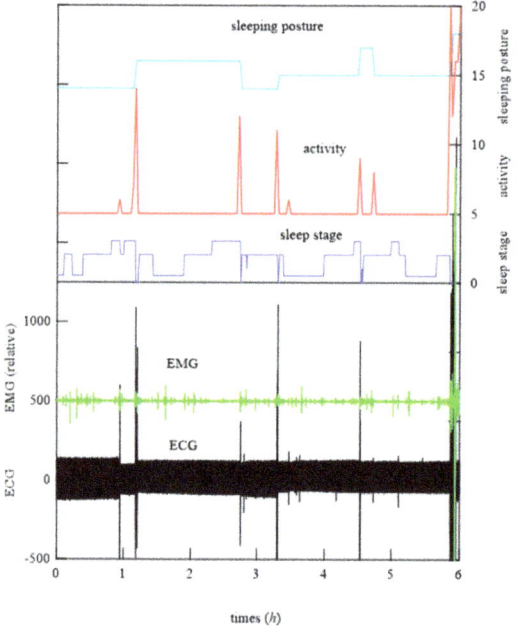

**Figure 13.** Example of ECG waveform, EMG waveform, activity levels, and postural changes during overnight sleep.

## 6.2. Demonstration of Bed-Sheet-Type Sensor

To begin the analysis, the PSG signals when the subject was at rest and turned over and at all points where teeth grinding occurred were compared with the signals from the bed-sheet-type sensor. Representative results are shown in Figures 14–16. In these figures, the magnitudes of the time and response signal axes are the same; thus, the magnitude and period of signals can be intuitively understood. As shown in Figure 14, when the subject was at rest, the signals from the bed-sheet-type sensor synchronized with the ECG signals. When the subject turned over in the bed, both the ECG and EMG signals were large, as shown in Figure 15. The signals from the bed-sheet-type sensor were also large. In the case of teeth grinding, the ECG signal was almost unchanged from the resting state, as shown in Figure 16. In the case of EMG, a high-frequency signal can be seen, although it is difficult to see on this scale (an enlarged image will be shown later). The high-frequency signal from the bed-sheet-type sensor also appears to be observed for a short period. From the above, a fast Fourier transform (FFT) process was applied to the overnight sleep signals to characterize the signal data obtained from the bed-sheet-type sensor during overnight sleep. Figure 17 shows a representative example of the signals obtained with the bed-sheet-type sensor for one night and the results of the FFT. In other words, these measurement data are unprocessed measurement signals that include all small signals, such as the vital signals of the subject and signals from body movements; the FFT results show that the detected signals are within a wide frequency range. In particular, the FFT results suggest that the absolute magnitudes of the signals are separated by frequency bands such as 0.1 Hz to 1 Hz, 1 Hz to 2 Hz, and 3 Hz to 7 Hz.

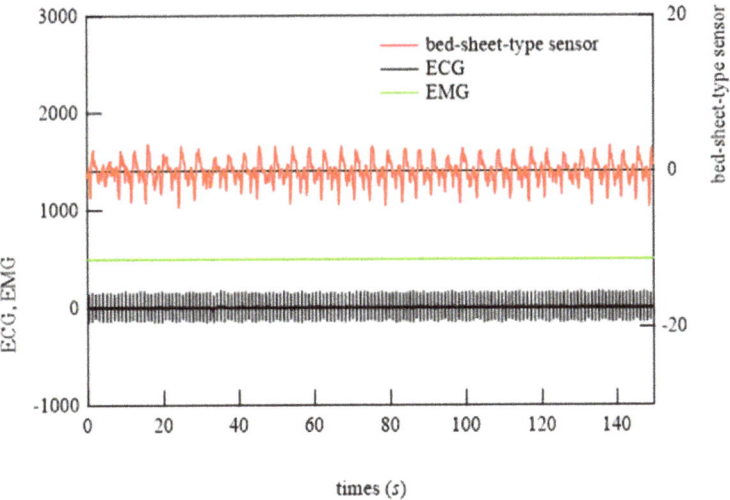

**Figure 14.** Typical examples of ECG, EMG, and bed-sheet-type sensor signals when the subject was at rest.

Using these results, primary filtering processing of the signals, low pass filter (LPF) processing (cut-off frequency: 0.5 Hz), band pass filter (BPF) processing (cutoff frequency: 0.8–1.5 Hz), and high pass filter (HPF) processing (cutoff frequency: 8 Hz) were performed to determine the characteristics of the signals from the bed-sheet-type sensor during the resting state, turning, and teeth grinding as determined by the physician. The results are shown below. First, representative results of the resting state are shown in Figure 18. LPF processing shows that the signal precisely synchronized with the respiratory signal, as shown at the bottom of Figure 18. The HPF-processed signal shows sharp pulses, indicating that it synchronized with the ECG signal, as shown at the top of Figure 18. Next, Figure 19 shows the data when the subject turned from lying on their belly to lying on their back. The top of Figure 19 shows the sleeping posture as determined by the physician. The activity level shows a peak when the sleeping posture changes. Here, the activity level is the sum of the amount of activity per minute. Thus, it can be seen that the activity level captures the change in sleeping posture well; the EMG signal is also larger, indicating that it is responding to the turning over. It also shows that the ECG which should not be affected by the measurement principle is also affected by a large amount. On the other hand, the bed-sheet type sensor responds from the beginning to the end of the turning over. In PSG of teeth grinding, ECG shows that the heart rate is unchanged as usual, as shown in Figure 20. Only EMG shows a characteristic signal. We sought to determine whether the bed-sheet-type sensor detects a characteristic signal. A typical example is shown below.

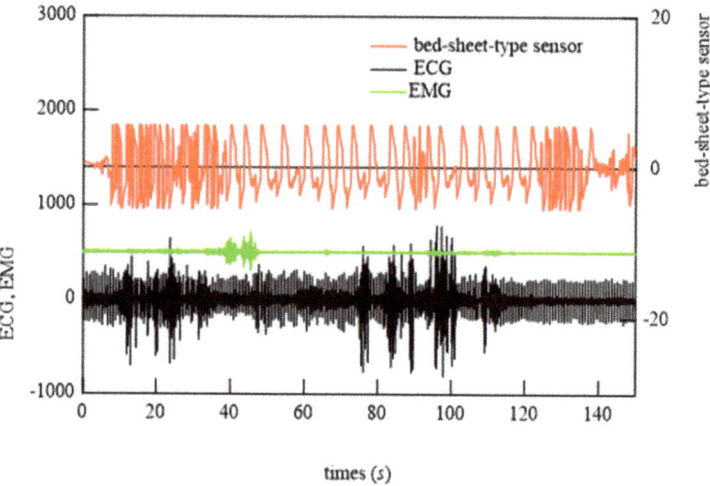

**Figure 15.** Typical examples of ECG, EMG, and bed-sheet-type sensor signals during turning over.

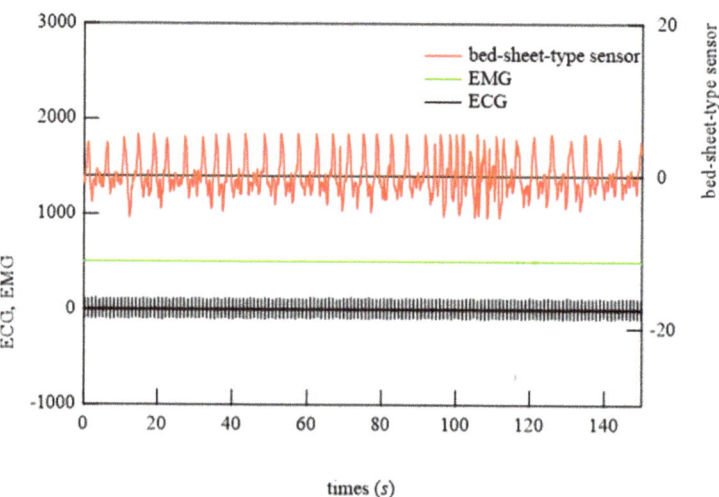

**Figure 16.** Typical examples of ECG, EMG, and bed-sheet-type sensor signals at the onset of bruxism.

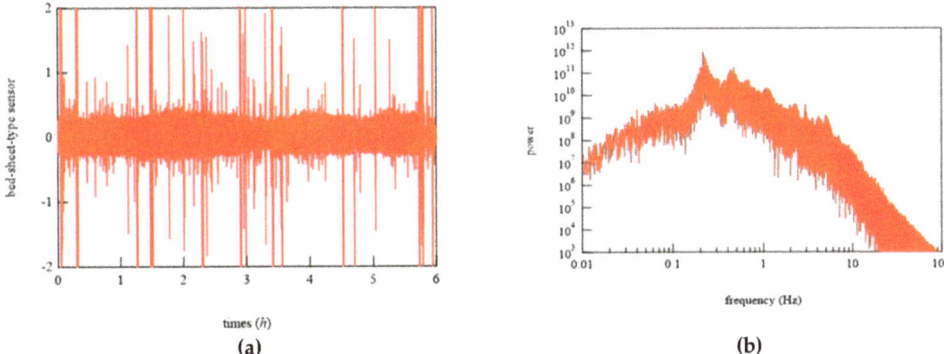

**Figure 17.** (**a**) Representative example of the signals obtained by the bed-sheet-type sensor for one night and (**b**) the results of the FFT.

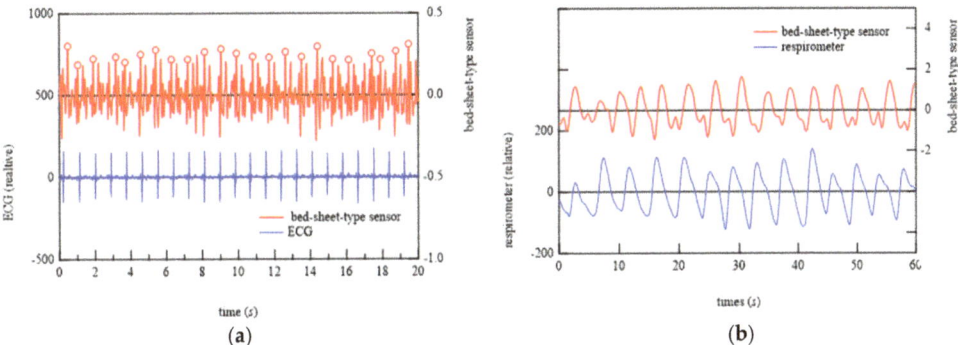

**Figure 18.** Response signals obtained from the bed-sheet-type sensor in this experiment obtained for (**a**) pulsation and (**b**) respiration.

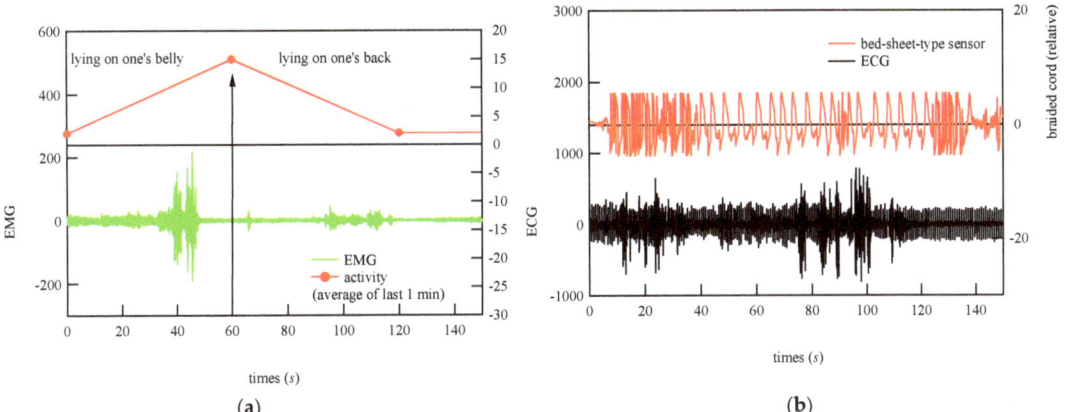

**Figure 19.** Typical example of simultaneous measurements with medical equipment (sleeping posture, activity, (**a**) EMG, and (**b**) ECG) in a hospital and with the bed-sheet-type sensor on a bed during a period that includes turning over in bed.

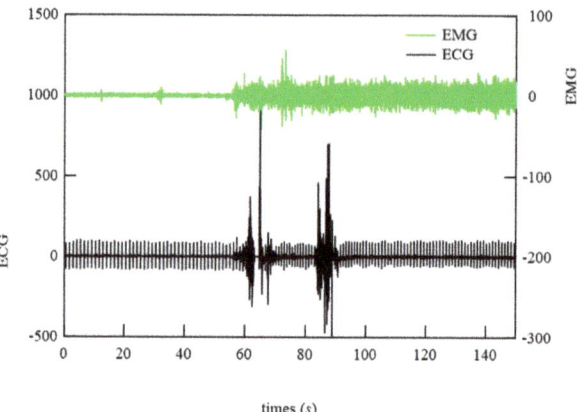

**Figure 20.** Two representative examples of EMG and ECG signals at the time of the physician-confirmed occurrence of bruxism.

Figure 21 shows EMG and HPF-filtered bed-sheet-type sensor signals at time when the doctor identified the occurrence of bruxism. It can be seen that the HPF-filtered bed-sheet-type sensor signal was different from the pulsation-based signal observed in the resting state shown in Figure 18a.

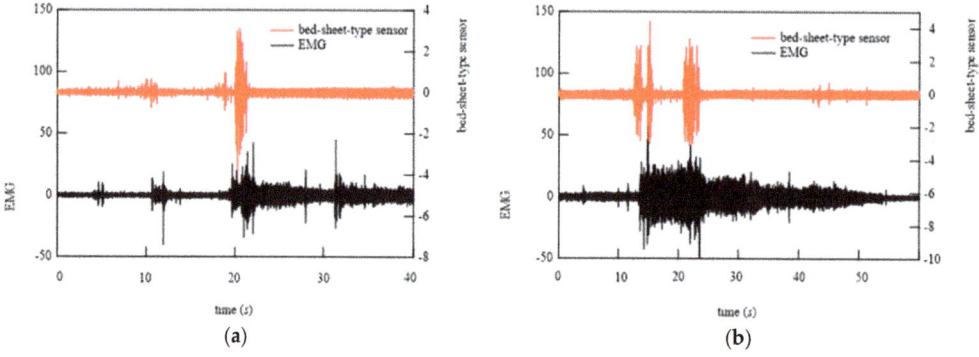

**Figure 21.** (a,b) Two representative examples of EMG signals and signals from the bed-sheet-type sensor at the time of the physician-confirmed occurrence of bruxism.

*6.3. Algorithm for Bruxism Detection*

As described above, it was found that the bed-sheet-type sensor can detect characteristic signals during teeth grinding; therefore, we constructed an algorithm for detecting the occurrence of teeth grinding using the bed-sheet-type sensor. To understand the characteristics of the waveforms obtained from the bed-sheet-type sensor, FFT processing was performed. The results are shown in Figure 22, which also shows the results in the resting state and when the subject turned over in bed. As can be seen, a large signal is generated below 1 Hz during turning over and other body movements. In contrast, when only grinding is occurring, the signal at frequencies below 1 Hz is small, as in the resting state. However, the signal at frequencies higher than 3 Hz is large only when teeth grinding occurs. This trend is compared with that of the results of EMG, which is used by physicians to determine teeth grinding. Figure 23 shows that the EMG and the bed-sheet-type sensor results show good agreement. Based on these results, the following processing flow was established to detect the occurrence of teeth grinding using only signals from the bed-sheet type sensor: (1) Real-time sensing is carried out by a bed-sheet-type sensor. (2) First-order

HPF processing (cutoff frequency: 3 Hz) is applied to the sensor signals. (3) If the amplitude of the signal from the bed-sheet sensor is 5 times the average value over the past 20 min and continues for 2 s, a start flag is set. If the amplitude is less than 5 times the average value, an end flag shall be applied. (4) The FFT of the signal in the interval between the flags in (3) is calculated. The sum of the calculated amplitudes between 3 Hz and 7 Hz is obtained. When this sum is 100 times or more than the normal value (the average of the data from the time of falling asleep to the present), the occurrence of bruxism is judged.

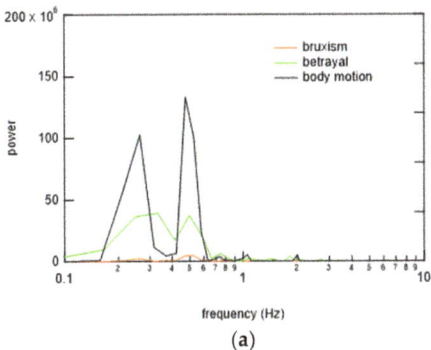
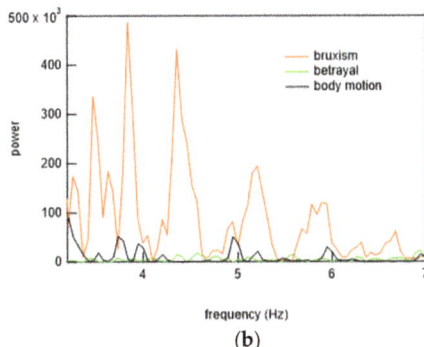

**Figure 22.** (a,b) Comparison of FFT signals of bed-sheet-type sensor during bruxism, turning over, and body movements.

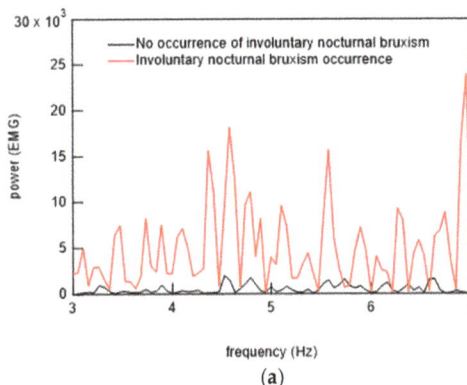
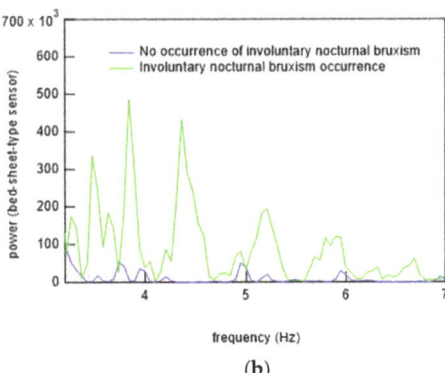

**Figure 23.** Comparison of FFT signals of (a) EMG and (b) bed-sheet-type sensor during bruxism and restful sleep.

In the physician's evaluation, 3 of the 15 subjects were found to have no incidence of bruxism, whereas the other 12 subjects were found to have incidence of bruxism. For these 12 subjects, the occurrence of teeth grinding was independently determined from the bed-sheet-type sensor data using the algorithm described above. The results are summarized in Table 2. The following is an explanation of the table. For example, for subject No. 1, the physician identified the number of times teeth grinding occurred from the EMG waveform during one night of sleep, which was 150 times. On the other hand, the number of times teeth grinding occurred was independently identified by the above-mentioned process using the data of the bed-sheet-type sensor, which was 158 times. The number of times that the doctor's judgment was consistent with the bed-sheet-type sensor was 150 times and the number of times of misjudgment was 8. In other words, every one of the 150 instances of teeth grinding identified by the physicians was precisely confirmed by the data from the bed-sheet-type sensors. It should be emphasized that the occurrence of teeth grinding was not missed at all by our bed-sheet-type sensor. However, the bed-sheet-type sensor results included eight false positives. We have carefully examined all the cases of misjudgment, and in all of them, a small body movement occurred for a moment. Table 2 shows that the average success rate for each subject when using the bed-sheet-type sensor in this study was over 90%.

Table 2. Comparison of diagnostic and bed-sheet-type sensor judgment results of bruxism.

| Subject | Gender | Age | Height (cm) | Weight (kg) | BMI | Medical Diagnosis | Bed-Sheet-Type Sensor | | | |
|---|---|---|---|---|---|---|---|---|---|---|
| | | | | | | | Determination by Bed-sheet-Type Sensor | Number of Times of Correct Judgment | Number of Times of Misjudgment | Success Rate (%) |
| 1 | male | 34 | 163 | 96 | 36 | 150 | 158 | 150 | 8 | 94.9 |
| 2 | female | 81 | 159 | 58 | 23 | 30 | 35 | 30 | 5 | 85.7 |
| 3 | male | 39 | 173 | 67 | 22 | 51 | 53 | 51 | 2 | 96.2 |
| 4 | male | 35 | 155 | 60 | 25 | 34 | 38 | 34 | 4 | 89.5 |
| 5 | male | 15 | 164 | 49 | 18 | 46 | 54 | 46 | 8 | 85.2 |
| 6 | male | 47 | 181 | 68 | 21 | 88 | 94 | 88 | 6 | 93.6 |
| 7 | male | 46 | 170 | 90 | 31 | 91 | 97 | 91 | 6 | 93.8 |
| 8 | male | 58 | 170 | 64 | 22 | 93 | 99 | 93 | 6 | 93.9 |
| 9 | female | 38 | 170 | 82 | 28 | 41 | 47 | 41 | 6 | 87.2 |
| 10 | male | 61 | 167 | 69 | 25 | 74 | 82 | 74 | 8 | 90.2 |
| 11 | male | 64 | 175 | 98 | 32 | 47 | 49 | 47 | 2 | 95.9 |
| 12 | female | 58 | 163 | 66 | 25 | 47 | 49 | 47 | 2 | 95.9 |
| | | | | | | | | | average of success rate (%) | 91.9 |

Teeth grinding is a phenomenon that occurs in many people without their being aware of it, and its presence is usually only recognized when serious dental damage or sleep disturbances occur. Currently, the standard treatment for sleep disorders is to wear a mouthpiece, which many patients find physically and mentally stressful, thereby affecting their sleep quality. The widespread application of bed-sheet-type sensor systems for routine data collection has the potential to contribute significantly to advancements in research in this field. That is, the data routinely accumulated by bed-sheet-type sensor systems could facilitate the development of innovative therapies that are currently considered unfeasible. For example, it could lead to the development of specialized pillows and household products designed to facilitate sleep positions that prevent teeth grinding. The experimental results of the bed-sheet-type sensor system obtained in this study strongly suggest that the system may lead to the development of new treatments.

## 7. Conclusions

Bruxism is attracting attention as one of the factors that interfere with the maintenance and improvement of sleep quality, as it grinds and cracks teeth, aggravates gum sensitivity and periodontal disease, and causes temporomandibular joint disorder, facial pain, headaches, and stiff shoulders. In addition, for those sleeping together with others in the same room, the noise generated by bruxism worsens the quality of sleep of the other people. Thus, bruxism has a negative impact on health, and stress and anxiety have been suggested as some of the causes. However, a method of measuring it continuously in daily life has not been established. In this study, we developed a bed-sheet-type sensor consisting of a braided piezoelectric PLLA cord as a device that can measure bruxism without causing any burden or discomfort. This bed-sheet-type sensor offers a nonintrusive method of measuring bruxism, eliminating the need for direct body contact, unlike PSG. We have developed a device and an algorithm to identify and detect the unique waveform of bruxism. We used the device on subjects in a sleep clinic and determined the consistency of its results with judgments made by physicians on the basis of PSG results as a demonstration experiment. As a result, we obtained surprising results showing consistency with all the judgments made by the physicians. However, there were several cases in which the bed-sheet-type sensor system showed the occurrence of teeth grinding, but the physician judged that no teeth grinding occurred. In the future, we aim to fully implement the bed-sheet-type sensor system in society by enhancing collaboration with medical specialists, gathering more data from a diverse range of subjects, and refining the accuracy of its waveform analysis.

**Author Contributions:** Conceptualization, Y.T.; methodology, S.S. and M.T.; validation, S.S.; formal analysis, Y.T., M.T. and K.S.; investigation, H.Y., S.N., T.N.(Takuo Nakiri), T.N.(Takuto Nonomura) and J.T.; resources, R.N., K.T., S.S. and O.T.; ethics review planning, O.T.; data curation, R.U., M.K. and J.T.; writing—original draft preparation, Y.T.; writing—review and editing, Y.T.; project administration, Y.T.; funding acquisition, Y.T. All authors have read and agreed to the published version of the manuscript.

**Funding:** The development of the experimental measurement system for this work was supported in part by a Grant-in-Aid for Scientific Research (No. 21K05188) from the Ministry of Education, Culture, Sports, Science and Technology of Japan.

**Institutional Review Board Statement:** The experiments in this paper were approved by Kansai University-FFLS, which reviewed the ethical aspects of the project (approval number 23–47. 2023).

**Data Availability Statement:** Data are contained within the article.

**Acknowledgments:** The authors would like to thank Tachibana Eletech Co., Ltd.; Shinsei Electronics Co., Ltd.; and Alpha Media Technology Corporation for their kind assistance in circuit production. The authors would also like to thank NGK Insulators, Ltd. for their kind assistance with the use of a thin rechargeable battery. We would also like to express our deepest gratitude to H. Shinjyo of Tsukasa Kogyo Co., Ltd.; Y. Kimura; and K. Ikeda for their dedicated assistance in the production of the embroidery.

**Conflicts of Interest:** The authors declare no conflict of interest. KT and RN are employees of Teijin Frontier Co., Ltd., SS and TN are employees of Nishikawa Co., Ltd., and MT ad KS are employees of Revoneo LLC. The paper reflects the views of the scientists, and not all companies.

# References

1. Hobson, J.A. Sleep is of the brain, by the brain and for the brain. *Nature* **2005**, *437*, 1254–1256. [CrossRef] [PubMed]
2. Imeri, L.; Opp, M. How (and why) the immune system makes us sleep. *Nat. Rev. Neurosci.* **2009**, *10*, 199–210. [CrossRef] [PubMed]
3. Hanlon, E.C.; Van Cauter, E. Quantification of sleep behavior and of its impact on the cross-talk between the brain and peripheral metabolism. *Proc. Natl. Acad. Sci. USA* **2011**, *108*, 15609–15616. [CrossRef] [PubMed]
4. Faraut, B.; Boudjeltia, K.Z.; Vanhamme, L.; Kerkhofs, M. Immune, inflammatory and cardiovascular consequences of sleep restriction and recovery. *Sleep Med. Rev.* **2012**, *16*, 137–149. [CrossRef] [PubMed]
5. Schmidt, M.H. The energy allocation function of sleep: A unifying theory of sleep, torpor, and continuous wakefulness. *Neurosci. Bowbearer.* **2014**, *47*, 122–153. [CrossRef] [PubMed]
6. Zielinski, M.R.; McKenna, J.T.; McCarley, R.W. Functions and mechanisms of sleep. *AIMS Neurosci.* **2016**, *3*, 67–104. [CrossRef] [PubMed]
7. Rockstrom, M.D.; Chen, L.; Taishi, P.; Nguyen, J.T.; Gibbons, C.M.; Veasey, S.C.; Krueger, J.M. Tumor necrosis factor alpha in sleep regulation. *Sleep Med. Rev.* **2018**, *40*, 69–78. [CrossRef]
8. Harding, E.C.; Franks, N.P.; Wisden, W. Sleep and thermoregulation. *Curr. Opin. Physiol.* **2020**, *15*, 7–13. [CrossRef]
9. Berry, R.B.; Brooks, R.; Gamaldo, S.E.; Harding, M.; Lloyd, R.; Quan, S.; Troester, M.; Vaughn, B. *The AASM Manual for the Scoring of Sleep and Associated Events Rules, Terminology and Technical Specifications Version 2.4*; American Academy of Sleep Medicine: Darien, IL, USA, 2017.
10. Geyer, J.; Carney, P. *Atlas of Polysomnography*; Wolters Kluwer Health: Alphen aan den Rijn, The Netherlands, 2018.
11. Mattice, C.; Brooks, R.; Chiong, L. *Fundamentals of Sleep Technology*; Wolters Kluwer Health: Alphen aan den Rijn, The Netherlands, 2020.
12. Tajitsu, Y.; Kawase, Y.; Katsuya, K.; Tamura, M.; Sakamoto, K.; Kawahara, K.; Harada, Y.; Kondo, T.; Imada, Y. New wearable sensor in the shape of a braided cord (Kumihimo). *IEEE Trans. Dielectr. Electr. Insul.* **2018**, *25*, 772–777. [CrossRef]
13. Tajitsu, Y. Development of E-textile Sewn Together with Embroidered Fabric Having Motion-Sensing Function Using Piezoelectric Braided Cord for Embroidery. *IEEE Trans. Dielectr. Electr. Insul.* **2020**, *27*, 1644–1649. [CrossRef]
14. Mukhopadhyay, S. *Wearable Electronics Sensors: For Safe and Healthy Living (Smart Sensors, Measurement and Instrumentation, 15)*; Springer: Berlin, Germany, 2015.
15. King, C. Application of Data Fusion Techniques and Technologies for Wearable Health Monitoring. *Med. Eng. Phys.* **2017**, *42*, 1–12. [CrossRef] [PubMed]
16. Witt, R.; Kellogg, A.; Snyder, P.; Dunn, J. Windows into Human Health through Wearables Data Analytics. *Curr. Opin. Biomed. Eng.* **2019**, *9*, 28. [CrossRef] [PubMed]
17. Iqbal, A.; Mahgoub, E.; Leavitt, A.; Asghar, W. Advances in Healthcare Wearable Devices. *Flex. Electron.* **2021**, *5*, 9. [CrossRef]

18. Chakraborty, T.; Ghosh, I. Real-time Forecasts and Risk Assessment of Novel Coronavirus (COVID-19) Cases: A Data-Driven Analysis. *Chaos Solitons Fractals* **2020**, *135*, 109850. [CrossRef] [PubMed]
19. Ates, H.C.; Nguyen, P.Q.; Gonzalez-Macia, L.; Morales-Narváez, E.; Güder, F.; Collins, J.J.; Dincer, C. End-to-end Design of Wearable Sensors. *Nat. Rev. Mater.* **2022**, *7*, 887. [CrossRef] [PubMed]
20. Demrozi, F.; Borzì, L.; Olmo, G. Wearable Sensors for Supporting Diagnosis, Prognosis, and Monitoring of Neurodegenerative Diseases. *Electronics* **2023**, *12*, 1269. [CrossRef]
21. Galetti, P.; DeRossi, D.; DeReggi, A. *Medical Applications of Piezoelectric Polymers*; Wiley: New York, NY, USA, 1988.
22. Nalwa, H. *Ferroelectric Polymers*; Marcel Dekker: New York, NY, USA, 1995.
23. Fukada, E. History and Recent Progress in Piezoelectric Polymers. *IEEE Trans. Ultrason. Ferroelectr. Freq. Control* **2000**, *47*, 1277–1290. [CrossRef]
24. Tajitsu, Y. Piezoelectricity of Chiral Polymeric Fiber and Its Application in Biomedical Engineering. *IEEE Trans. Ultrason. Ferroelectr. Freq. Control* **2008**, *55*, 1000–1008. [CrossRef]
25. Carpi, F.; Smela, E. *Biomedical Applications of Electroactive Polymer Actuators*; Wiley: Chichester, UK, 2009.
26. Okuzaki, H.; Asaka, K. *Soft Actuators*; Springer: Cham, Switzerland, 2014.
27. Ohki, Y. Development of a Braided Piezoelectric Cord for Wearable Sensors. *Electr. Insul. Mag.* **2020**, *36*, 59–64. [CrossRef]
28. Yin, L. A Self-sustainable Wearable Multi-modular E-textile Bioenergy Microgrid System. *Nat. Commun.* **2021**, *12*, 1542–1555. [CrossRef]
29. Sateia, M. *International Classification of Sleep Disorders*, 3rd ed.; The American College of Chest Physicians: Glenview, IL, USA, 2014; Volume 146, pp. 1387–1394.
30. Thomas, D.; Patel, J.; Kumar, S.; Dakshinamoorthy, J.; Greenstein, Y.; Ravindran, H.; Pitchumani, P. Sleep Related Bruxism— Comprehensive Review of the Literature Based on a Rare Case presentation. *Front. Oral Maxillofac. Med.* **2024**, *6*, 3–23. [CrossRef]
31. Kato, T.; Yamaguchi, T.; Okura, K.; Abe, S.; Lavigne, G. Sleep less and bite more: Sleep disorders associated with occlusal loads during sleep. *J. Prosthodont. Res.* **2013**, *57*, 69–81. [CrossRef] [PubMed]
32. Clark, T.; Rugh, D.; Handelman, L. Nocturnal Masseter Muscle Activity and Urinary Catecholamine Levels in Bruxers. *J. Dent Res.* **1980**, *59*, 1571–1576. [CrossRef] [PubMed]
33. Macedo, R.; Silva, A.; Machado, A.; Saconato, H.; Prado, F. Occlusal splints for Treating Sleep Bruxism (tooth grinding). *Cochrane Database Syst Rev.* **2007**, *17*, CD005514. [CrossRef] [PubMed]
34. Seraidarian, P.; Seraidarian, I.; Cavalcanti, B.; Marchini, L.; Neves, C. Urinary Levels of Catecholamines among Individuals with and without Sleep bruxism. *Sleep Breath* **2009**, *13*, 85–88. [CrossRef] [PubMed]
35. Furihata, R.; Saitoha, K.; Suzuki, M.; Jike, M.; Kaneita, Y.; Ohida, T.; Buysse, D.; Uchiyama, M. A Composite Measure of Sleep Health Is Associated with Symptoms of Depression among Japanese Female Hospital Nurses. *Compr. Psychiatry* **2012**, *97*, 152151. [CrossRef]
36. Nonoue, S.; Mashita, M.; Haraki, S.; Mikami, A.; Adachi, H.; Yatani, H.; Yoshida, A.; Taniike, M.; Kato, T. Inter-scorer Reliability of Sleep Assessment Using EEG and EOG Recording System in Comparison to Polysomnography. *Sleep Biol. Rhythm.* **2017**, *15*, 39–48. [CrossRef]
37. Borbely, A.; Achermann, P. Sleep homeostasis and models of sleep regulation. In *Principles and Practice of Sleep Medicine*, 5th ed.; Kryger, M.H.; Roth, T., Dement, W.C., Eds.; Elsevier Saunders: Philadelphia, PA, USA, 2010; pp. 431–444.
38. Urade, Y.; Hayaishi, O. Prostaglandin D2 and sleep/wake regulation. *Sleep Med. Rev.* **2011**, *15*, 411–418. [CrossRef]
39. Nye, J. *Physical Properties of Crystals*; Clarendon Press: Oxford, UK, 1985.
40. Lang, S.B. Guide to the Literature of Piezoelectricity and Pyroelectricity. *Ferroelectrics* **2005**, *321*, 91–204. [CrossRef]
41. Tajitsu, Y.; Takarada, J.; Takatani, K.; Nakanishi, R.; Yanagimoto, H.; Shiomi, S.; Nakagawa, I.; Kawahara, I.; Nakiri, T.; Shimda, S.; et al. Prototype Sensor System Using Fabricated Piezoelectric Braided Cord for Work-Environment Measurement during Work from Home. *Micromachines* **2021**, *12*, 966–978. [CrossRef]
42. Tajitsu, Y.; Takarada, J.; Tokiya, H.; Sugii, R.; Takatani, K.; Yanagimoto, H.; Riku Nakanishi, R.; Seita Shiomi, S.; Daiki Kitamoto, D.; Nakiri, T.; et al. Application of Piezoelectric PLLA Braided Cord as Wearable Sensor to Realize Monitoring System for Indoor Dogs withLess Physical or Mental Stress. *Micromachines* **2023**, *14*, 143–163. [CrossRef] [PubMed]

**Disclaimer/Publisher's Note:** The statements, opinions and data contained in all publications are solely those of the individual author(s) and contributor(s) and not of MDPI and/or the editor(s). MDPI and/or the editor(s) disclaim responsibility for any injury to people or property resulting from any ideas, methods, instructions or products referred to in the content.

Article

# Low-Power Consumption IGZO Memristor-Based Gas Sensor Embedded in an Internet of Things Monitoring System for Isopropanol Alcohol Gas

Myoungsu Chae [1,†], Doowon Lee [1,2,†] and Hee-Dong Kim [1,*]

1. Department of Semiconductor Systems Engineering, Convergence Engineering for Intelligent Drone, Institute of Semiconductor and System IC, Sejong University, 209, Neungdong-ro, Gwangjin-gu, Seoul 05006, Republic of Korea
2. IHP GmbH—Leibniz Institute for Innovative Microelectronics, Im Technologiepark 25, 15236 Frankfurt (Oder), Germany
* Correspondence: khd0708@sejong.ac.kr
† These authors contributed equally to this work.

**Abstract:** Low-power-consumption gas sensors are crucial for diverse applications, including environmental monitoring and portable Internet of Things (IoT) systems. However, the desorption and adsorption characteristics of conventional metal oxide-based gas sensors require supplementary equipment, such as heaters, which is not optimal for low-power IoT monitoring systems. Memristor-based sensors (gasistors) have been investigated as innovative gas sensors owing to their advantages, including high response, low power consumption, and room-temperature (RT) operation. Based on IGZO, the proposed isopropanol alcohol (IPA) gas sensor demonstrates a detection speed of 105 s and a high response of 55.15 for 50 ppm of IPA gas at RT. Moreover, rapid recovery to the initial state was achievable in 50 µs using pulsed voltage and without gas purging. Finally, a low-power circuit module was integrated for wireless signal transmission and processing to ensure IoT compatibility. The stability of sensing results from gasistors based on IGZO has been demonstrated, even when integrated into IoT systems. This enables energy-efficient gas analysis and real-time monitoring at ~0.34 mW, supporting recovery via pulse bias. This research offers practical insights into IoT gas detection, presenting a wireless sensing system for sensitive, low-powered sensors.

**Keywords:** isopropanol alcohol gas; gas sensor; Internet of Things; monitoring; memristor

## 1. Introduction

While the world is presently experiencing a catastrophic energy crisis, the ongoing advancement of the Internet of Things (IoT) necessitates the integration of an immense quantity of sensors, which consumes an enormous amount of energy [1–5]. In considering the vast quantity of sensors that must be integrated into such a network, there is an immediate demand for sensors with the following attributes: micro- or nanoscale dimensions, continuously improving sensitivity and detectivity, significantly reduced response times, and power consumption that is orders of magnitude lower than that of existing commercial devices. Conventional gas sensors are still energized by a voluminous and inflexible external power source, which not only results in the expansion of the system's overall dimensions but also significantly compromises the device's portability and comfort. To optimize power efficiency and ensure extended device life in gas monitoring technology, low-powered functionalities are anticipated to be integrated into the sensor [6–8]. Chen H. et al., for instance, reported that the energy-storage capability of FMCPIB-based devices enables them to function as photo capacitors to detect $NO_2$ for an additional 1.7 h in the dark without requiring an external power supply [8]. According to Cho et al., the operational power of the gas sensor composed of ZnO nanowires was ~184 µW, indicating that it can be utilized in practical IoT devices [9].

Integrating IoT systems into industrial safety protocols to monitor volatile organic compounds (VOCs) to avert potential industrial catastrophes has received considerable attention. These cutting-edge devices can identify and measure the levels of diverse species, including ethanol, $NO_2$, and CO, in both biological fluids and the atmosphere [10–12]. They operate within field management systems that are low-power and high-density. Among various types of gas sensors, gas sensors based on metal oxide semiconductors have attracted considerable attention for detecting VOCs due to their rapid response time and broad sensitivity to various target gases [13,14]. The miniaturization and low cost of manufacturing metal oxide semiconductor gas sensors enable the implementation of high-density arrays in systems. Furthermore, it has been reported that applying nanomaterial-based metal oxides enhances the properties of sensors [15,16]. According to the results of Le et al., porous $CoFe_2O_4$ nanorods exhibited a large surface area for the reaction, resulting in an increased reactivity to acetone at 350 °C [17]. These oxides are distinguished by their comparatively high mobility of field effects, safety features, low-leakage current, and room-temperature (RT) process availability. As a result of their compatibility, ease of synthesis, and capacity for low-temperature processing, metal oxides have attracted significant interest for implementation in IoT systems [18]. The emphasis is especially placed on amorphous indium gallium zinc oxide (a-IGZO) due to its potential as an active channel material. This is primarily due to the favorable attributes of its amorphous phase, including exceptional uniformity, long-term stability, and flexibility. Cho and N. G. demonstrated that, at an operating temperature of 300 °C, a gas sensor comprising a semiconductor a-IGZO tube network showed n-type gas-sensing properties and a 3.7-fold increased gas response in comparison to a planar a-IGZO thin film ($R_{Gas}/R_{Air}$ = 29.4) [19]. However, notwithstanding their numerous advantages, the critical high-temperature operation of these gas sensors based on metal oxides has restricted their use in ubiquitous gas monitoring systems. Ensuring high gas sensitivity and fostering surface redox reactions are contingent upon this temperature range. Additional devices, such as ultraviolet or heaters, are necessary to apply this energy, increasing the volume and power consumption. Consequently, further improvements are required to implement IoT monitoring techniques for low-power consumption and operating at RT.

Gasistors, memristor-based gas sensors that combine a gas sensor and a memristor, have been recently introduced. Characterized by their distinctive detecting mechanism that diverges from traditional gas sensor approaches that rely on variations in the width of the depletion region on a surface, gasistors have garnered significant interest owing to their low power consumption, high sensitivity, and rapid recovery time [20–22]. They can be a feasible resolution as they boast a fast response time at RT and a high sensitivity towards the target gas. It is widely acknowledged that a variation in the resistance of a gasistor can be attributed to the creation or rupture of vacancy paths, which consist of conducting filaments (CFs) and oxygen and nitrogen vacancies in the thin film [23,24]. By employing nano-sized CFs for gas-sensing, gasistors can identify a target gas with a minimal sensing current at RT and produce a robust response even for trace gases, thereby conserving energy. Chue et al. reported a memristive gas sensor of $TiO_2$ nanorods at RT exhibiting rapid response and quick recovery time. [25]. Qiu et al. realized an extremely fast response and recovery speed of 1.2 s through an ultrasensitive gas sensor developed from a $SnS/TiO_2$-based memristor [26]. Furthermore, in one of our prior investigations [27], we documented a rapidly recovering and detecting IGZO-based gasistor at RT, in addition to a velocity of 1 s/90 ns. However, IGZO-based gasistors for VOC gas monitoring systems have not yet been investigated, and their research on low-power monitoring systems for IoT is limited.

In this study, we examined isopropanol alcohol (IPA) gas, one of the VOCs in IGZO-based gasistors. First, the unit device's sensitivity and response characteristics concerning IPA's concentration were initially assessed. Applying a pulse voltage allows the gasistor to be restored to its initial state even after the CFs rupture in response to the gas. During this investigation, the gasistor was returned to its original condition by applying a pulse with a width of 50 µs and an amplitude of 1 V. Furthermore, an IGZO-based gasistor was incorporated into the system based on the evaluated properties, and the results of the measurements were observed remotely using a mobile device. Consequently, the IGZO-based gasistor embedded in the IoT system achieved accurate measurement of the applied concentration and restored the initial state by applying voltage from a mobile device, as shown in Figure 1. The potential of implementing real-time monitoring using gasistor-based gas monitoring systems with operating voltages as low as 0.34 mW suggests the challenge posed by the high-power consumption of conventional semiconductor-based gas sensors could be overcome.

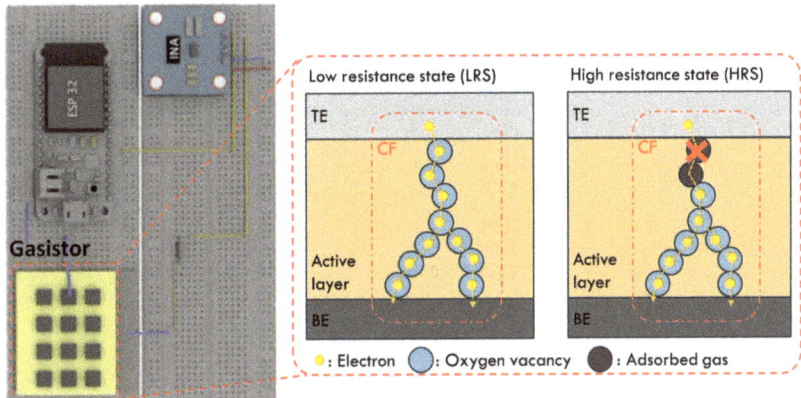

**Figure 1.** Conceptualization of an artificial IGZO-based gasistor embedded in an IoT gas monitoring system. The schematic illustration on the right shows the gasistor-based IPA gas monitoring system as well as the reaction and recovery mechanisms of the gasistor devices. The image on the left depicts a mobile device that employs an IoT to monitor IPA gas levels in the air in real time.

## 2. Materials and Methods

First, to prepare the IGZO-based gasistor, the Si substrate was cleaned using acetone, methanol, and deionized water for 10 min, respectively. The Pt bottom electrode (BE), which was 100 nm thick, was deposited on the cleaned $SiO_2$/Si substrate using an electron-beam evaporation system. A 100 nm IGZO was then deposited on the Pt/$SiO_2$/Si substrate using an RF sputtering system. The IGZO deposition was performed at a 100 W sputtering power in 20 sccm of Ar ambient at a base pressure of 20 mTorr and a working pressure of 5 mTorr. The compositional ratio of the sputtering target was $In_2O_3$:$Ga_2O_3$:ZnO = 1:1:1 (VTEX) with 99.99% purity. Subsequently, the 8 nm Ti top electrode (TE) was deposited on IGZO/Pt/$SiO_2$/Si substrates using an RF sputtering system. A Field Emission Scanning Electron Microscope (FE-SEM, SU8010, Hitachi, Tokyo, Japan) was used to obtain a cross-section image of the gasistor to validate that the device had been deposited as planned. In order to evaluate the material characteristics of the IGZO film, the deposited IGZO film was analyzed using an X-ray diffractometer (XRD, Empyrean, Panalytical, Malvern, UK). The IGZO thin film's chemical states were analyzed utilizing X-ray photoelectron spectroscopy (XPS, NEXSA G2, Thermo Fisher, Waltham, MA, USA).

To evaluate the resistive switching (RS) capacity, we measured the electrical characteristics of the IGZO-based gasistor using a Keithley 4200 Semiconductor Characterization System (SCS). To demonstrate the gas-sensing capability of the IGZO-based gasistor at RT, the current of the IGZO-based gasistor was monitored at a read voltage of 0.2 V while inducing IPA gas. The IGZO-based gasistor was placed on the ground plate inside the gas-sensing chamber and electrically connected to a Keithley 4200 SCS. In the gas-sensing test, a 30 $cm^3$ aluminum chamber was used, which was isolated from the outside environment.

The chamber was evacuated with air for 50 s before the injection of IPA gas. Next, air and IPA gas were introduced into the chamber to assess the response properties. The target and air gas circulated until there was no further current variation for 50 s, allowing the gasistor to react completely. The concentration of IPA gas was adjusted from 10 to 50 ppm. The object was returned to its original state by implementing a pulse bias. We optimized the pulse voltage and applied a pulse with a size of 1 V and a width of 50 µs. Each of these procedures was replicated for each gas concentration procedure. To examine the response characteristics of the concentration, we adjusted the air and target gas flow rates while maintaining a constant total flow rate of 500 sccm. The humidity was fixed and maintained at 36% relative humidity (RH).

The microcontroller (ESP 32) and current sensor (INA 219) were connected to a breadboard for monitoring and communication. IPA gas was detected within the chamber by the IGZO-based gasistor, which subsequently notified the mobile device of the current change. The mobile device determined the concentration of IPA gas in the chamber based on the real-time values received from the IGZO-based gasistor. The IoT module transmitted the current change value of the gasistor in the chamber to the mobile device every 2 s, allowing for real-time monitoring of IPA gas. The incoming data included current, voltage, power consumption, and concentration on the mobile device through the Blynk application.

## 3. Results and Discussion

Before describing the resistive switching (RS) and gas-sensing characteristics of the IGZO-based gasistor, the material properties of IGZO were investigated. A cross-sectional FE-SEM image, shown in Figure 2a, was measured to evaluate the fabricated device structures, confirming each layer. Furthermore, to analyze the structural characteristics of the IGZO film, the deposited IGZO film was analyzed using an X-ray diffractometer (PANalytical, empyrean). As shown in Figure 2b, the IGZO/Pt sample shows diffraction peaks with 2θ values at around 31.46°, 38.30°, and 44.85°. A diffraction peak located at 31.46° (JCPDS 01-070-3626; [28]) shows the (009) IGZO film, which is consistent with other previous studies [29]. At 38.30° and 44.85°, the crystalline phases were identified as the (111) and (200) planes, respectively, of face-centered cubic bulk metallic counterparts [30]. The mean crystallite size of the (009) IGZO film was 1.47, calculated using Bragg's law. To confirm the chemical composition of the IGZO, we investigated the XPS peaks to analyze changes in the binding energy of atomic orbitals, which can be related to changes in the chemical environment of the atoms. Figure 2c–f show that O, In, Ga, and Zn oxidation states are analyzed using their respective high-resolution spectra. As obtained through XPS analysis, the O 1s spectra of a-IGZO can be separated into three synthetic signals, as shown in Figure 2c. Our analysis showed that the binding energies for the three signals were 530.1, 531.7, and 532.3 eV. The binding energy peaks reveal fully oxidized oxygen (M-O) at 530.1 eV, oxygen vacancies at 531.7 eV, and oxygen in the hydroxide state (M-OH) at 532.3 eV, as indicated by the binding energy peaks [31,32]. Two peaks centered at binding energies of 444 eV and 451.5 eV, respectively, in the XPS data for the In 3d signal are shown in Figure 2d. These peaks are the In 3d5/2 and In 3d3/2 doublets, with an orbital splitting of 6.5 eV. The peaks of Ga 2p and Zn 2p was found to be around 26.9 and 23 eV, respectively, which is consistent with previous studies [33].

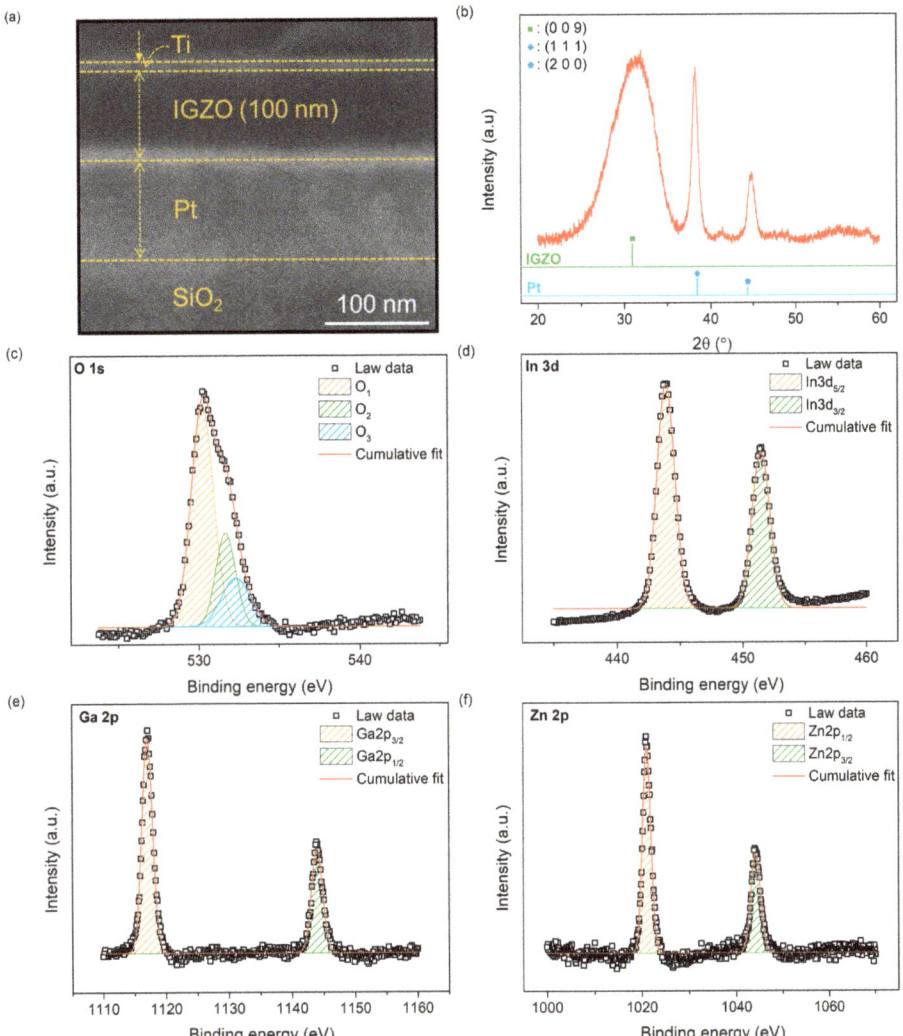

**Figure 2.** Material characteristics of IGZO-based gasistor. (**a**) Cross-section FE-SEM image with a scale bar of 100 nm. (**b**) XRD result of IGZO film. XPS spectra peaks of (**c**) O 1s, (**d**) In 3d, (**e**) Ga 2p, and (**f**) Zn 2p in IGZO film.

Subsequently, the RS of the IGZO-based gasistor was accessed by measuring the current–voltage (I-V) curve characteristics of a DC voltage sweep with a Keithley 4200 SCS. The morphology of the CF was also examined using a conductive atomic force microscope (C-AFM). C-AFMs enable a more dependable evaluation of the evolutionary behavior of a single filament due to the rarity of multi-filament events (which are frequent at the device level) in such a tiny probing area [34]. According to the CF model, the resistance of the IGZO can be changed in the opposite direction from HRS to LRS by implementing a setup or reset procedure. In this study, the IGZO film was used as an insulator. An initial DC bias sweep was performed on the IGZO-based gas resistance between BE and TE to determine the optimal voltage for CF formation, as shown in Figure 3a. During this process, oxygen ions ($O^{2-}$) in the IGZO film migrated towards TE while oxygen vacancies ($V_o^{2+}$) persisted. As the concentration of $V_o^{2+}$ exceeded a threshold, it underwent a self-reorganization process, resulting in the formation of CFs within the IGZO film, as shown in the magnified inset of Figure 3a. Next, we measured the I-V curve of the IGZO-based resistor under DC bias sweeps (0 V → −0.7 V → 0 V → +1.2 V), as shown in Figure 3b. A reset operation occurred when the resistance state of the resistor was changed from LRS to HRS due to a DC bias sweep being conducted from 0 V to −0.7 V. The sudden decrease in current during the reset procedure can be ascribed to the Joule heating-induced rupture of the CFs at −0.62 V. Conversely, following the re-sweeping of a DC bias from −0.7 V to 1.2 V, the resistance state of the gasistor was changed to LRS, an operation known as the set process. The abrupt surge in current magnitude at 0.89 V can be attributed to the reformation of the CFs occurring within the IGZO film. As depicted in Figure 3c, the current at LRS and HRS was monitored for $10^4$ s at a read voltage ($V_{read}$) of 0.2 V to determine whether the CFs were maintained in the ambient. Consequently, we noted that the CFs remained stable for $10^4$ s in the gasistor, suggesting that they remained unreactive with the surrounding gases before injecting the IPA gas. As shown in Figure 3d, the endurance test was performed at the $V_{read}$ for 200 DC cycles per resistance state to further assess the gasistor's dependability. The HRS exhibited variability following more than 100 cycles; this is an intrinsic characteristic of oxide-based memristors, which, according to the principles of atomic motion physics [35], are susceptible to state variability and inhomogeneity issues. Nonetheless, the large ratio between HRS and LRS prevents problems such as read errors stemming from this volatility [36]. As a result, a 100 nm thick IGZO-based gasistor shows stable operation without any failures during 200 cycles.

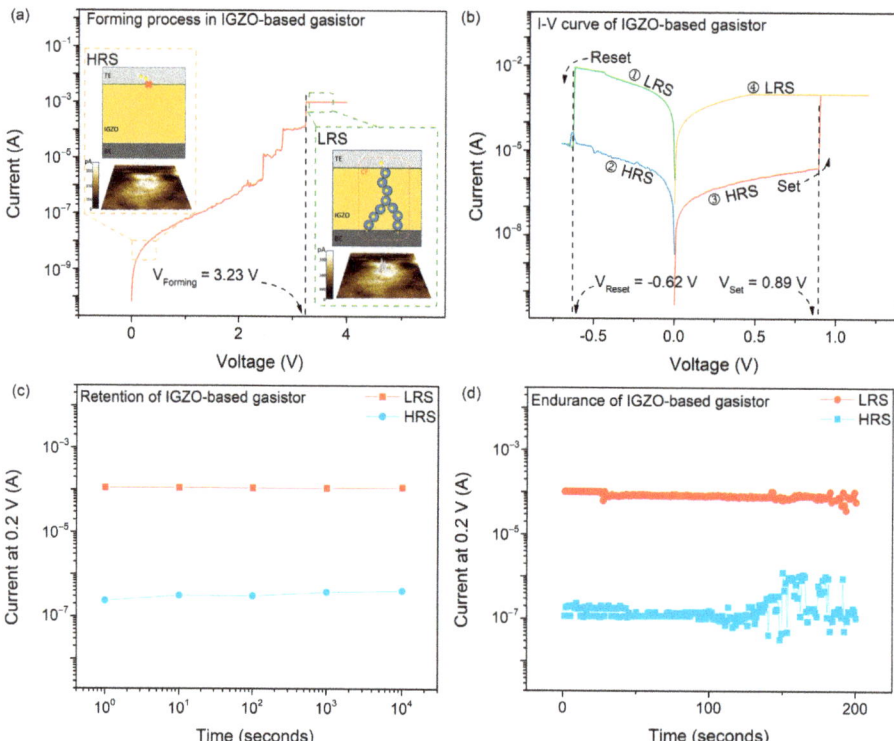

**Figure 3.** Schematic illustration and RS characteristics of an IGZO-based gasistor. (**a**) The forming process of the IGZO-based gasistor, with the enlarged figure showing the atomic structure changes in the IGZO-based gasistor before and after the CF forming process measured by C-AFM; (**b**) I–V curve; (**c**) retention; and (**d**) endurance characteristics of the IGZO-based gasistor.

Next, the proposed gasistor's transient response for IPA gas is monitored to study the gas-sensing characteristics, as shown in Figure 4a. Temperature and humidity substantially impact the gas-sensing abilities of semiconductor gas sensors, as is common knowledge [37,38]. On the other hand, prior research has demonstrated that gasistors exhibit relatively low sensitivity to changes in temperature and humidity [21]. Despite this immunity, humidity was maintained at 36% RH during gas sensing, and temperature was fixed to RT to exclude the effects of temperature and humidity. Then, to assess the stability, the sensing characteristics according to the number of RS operations were examined, as shown in Figure 4b. Thus, it was observed that the response time and current change in the device that initially formed CFs and reacted to 50 ppm gas and the device that broke and reformed CFs after conducting 200 DC cycles were nearly identical. We thus confirmed that reliability sensing can be achieved despite the carrying out of numerous operations. To validate the recovery time of the proposed gasistor, test sensing results are shown in Figure 4c. Following the injection of dry air gas for 50 s, IPA gas was injected with air gas at point A. Following a 500 s release of the target gas, IPA gas was purged at point B. The point required to recover 90% of the total resistance change was applied to determine the recovery times [39]. We observed that the IGZO-based gasistor, as proposed, returned to its original state approximately 543 s after the purging of IPA at point C. Nevertheless, previous work has demonstrated that CFs can be reformed via voltage application, thereby avoiding slow recovery times. Therefore, we applied a pulsed bias to restore it to its initial state to accelerate the process following the gas reaction. In addition, we examined whether a pulse voltage could restore the gasistor in the chamber to its initial state without purging. Figure 4d shows this with a continuous injection of 50 ppm of IPA gas after 50 s. Then, the initial state was reinstated 100 s after the reaction with the gas by applying a pulse voltage without gas purging. Consequently, it was validated that the initial condition could be reinstated by applying a pulse voltage, even in an environment containing gas. Furthermore, the subsequent reaction time decreased due to the absence of gas purging. This demonstrated that the formation of CFs in response to voltage application could restore the system to its initial state, even in an environment with continuous gas flow. Figure 4e shows the transient response to the IPA gas, where the gas concentration ranged from 10 to 50 ppm. The resistive state stability was observed for 50 s while air gas was introduced. Following the injection of IPA gas for 500 s, the IPA gas supply was discontinued, and a recovery pulse voltage was concurrently applied to reinstate the initial state. As a result, we observed trends where the current decreased from the initial state as the gas when the concentration of the IPA gas increased. When 10 ppm of IPA gas was introduced into the IGZO-based gasistor, IPA was adsorbed onto the surface of the IGZO film. The adsorbed IPA gas possessed a negative charge due to its property as a reducing gas [40–42]. Following adsorption with a negative charge, the IPA gas underwent a chemical reaction with $V_o^{2+}$, creating ruptured CFs. As a result, the reduction in current observed was ascribed to the existence of these fragmented CFs. The observed decrease in current with increasing IPA concentration can be attributed to the increase in the destructed CFs caused by the adsorbed IPA gas. To conduct a quantitative assessment of the response (S) characteristic, $S = R_{Gas}/R_{Air}$ was subsequently examined, where $R_{Gas}$ and $R_{Air}$ represent the resistances of the gasistor before and after IPA flow, respectively, during air injection, as shown in Figure 4f. The results demonstrated that the response increased with IPA concentration, reaching its maximum value of 55.15 at 50 ppm. The high response of pristine $In_2O_3$ to isopropanol could be attributed to the low C-C bonding energy of isopropanol (345 KJ·mol$^{-1}$) compared to the (O-H) bond in methanol (458.8 kJ·mol$^{-1}$) and the (C=O) bond in acetone (798.9 kJ·mol$^{-1}$) [43,44]. The response time graphed against IPA concentration is shown in Figure 4g. With increasing gas concentrations, the device's response time decreases. This could be attributed to the increased gas concentration, which causes a corresponding rise in the diffusion rate from the TE to the CFs. At a concentration of 10 ppm, this test required 384 s to detect the current change. At 50 ppm (the maximum concentration we assessed), the current change occurred in 105 s.

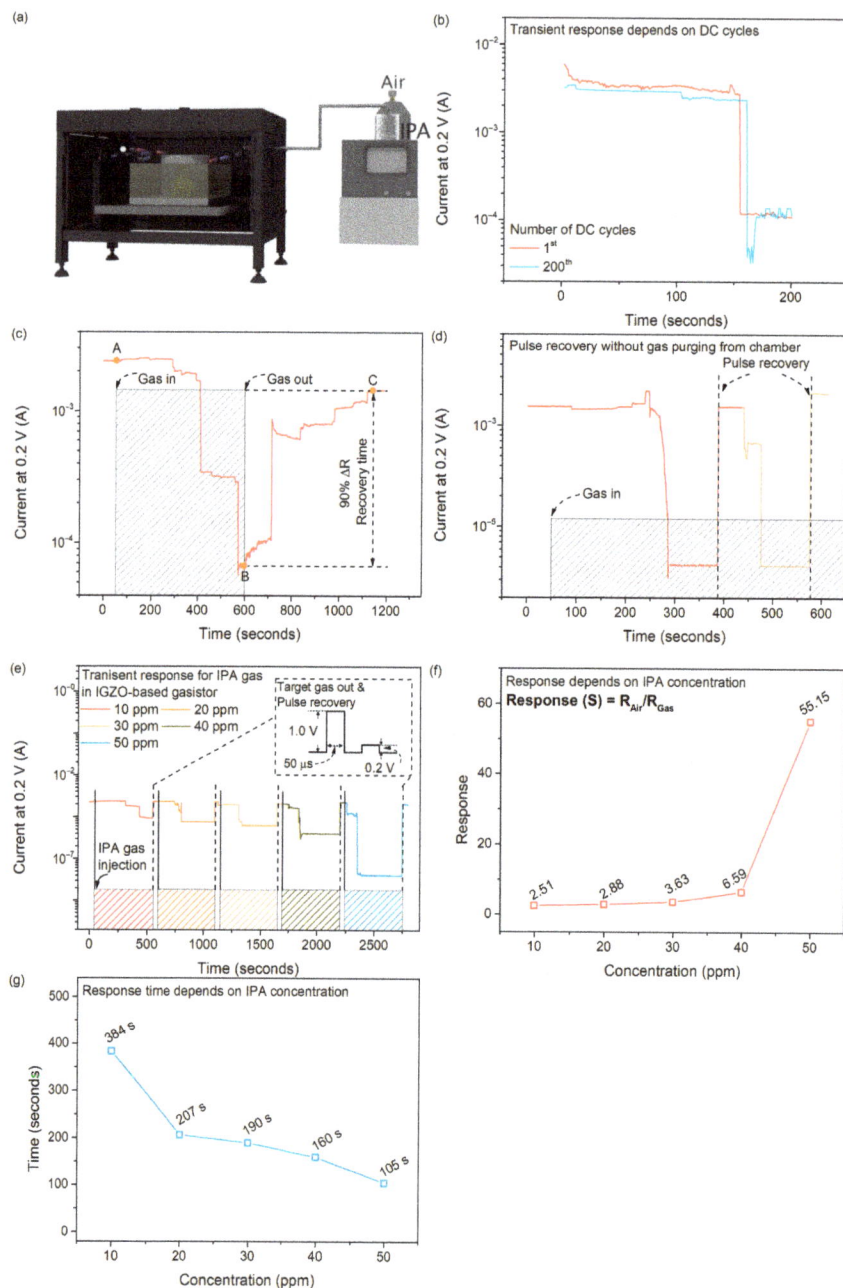

**Figure 4.** (**a**) Schematic structure of the gas-sensing environment. (**b**) A transient response based on the number of DC cycles to verify the stability of a gasistor. (**c**) Transient response testing to validate recovery time of IGZO-based gasistor. (**d**) Pulse recovery test without gas purging from chamber. (**e**) Transient response characteristics of IGZO-based gasistor for IPA gas depends on concentration. (**f**) Response and (**g**) response time depends on IPA gas concentrations in IGZO-based gasistor.

Finally, an IoT real-time monitoring system for IPA gas was integrated with an IGZO-based gasistor. This system incorporates an Arduino microprocessor, a current sensor, and an IGZO-based gasistor for IPA gas monitoring, as shown in Figure 5a. Figure 5b shows the effective monitoring of an unidentified reference gas through the utilization of the proposed system in conjunction with a mobile phone. This demonstrates that the IoT can measure IPA concentration in real time. In this context, we have effectively shown an IPA gas-sensing system for real-time monitoring and IoT systems. Users can obtain accurate concentration measurements via a smartphone application developed by the system. Table 1 shows a power consumption comparison of VOC gas sensors. Regarding power consumption, the evaluated monitoring system exhibits a negligible ~0.34 mW, which is notably low compared to established semiconductor gas sensors that rely on microheaters [45,46]. However, our proposed system's power consumption is more significant than that of semiconductor gas sensors that utilize self-heating nanowires [47,48]. Nanostructures with a high surface-to-volume ratio are crucial for fabricating gas-sensing devices with exceptional performance [49,50]. In addition, nanostructures facilitate the efficient and rapid diffusion of gases through their network, consequently enhancing the surface area accessible for gas sensing. Nevertheless, the fabrication process for these gas sensors based on nanostructures or nanomaterials is exceedingly complicated, whereas the proposed gasistor retains the benefit of being simple to manufacture. Although gasistor-based systems are still in their infancy, it is possible to improve their power consumption by optimizing word lines and bits to reduce series resistance. Furthermore, gasistor can benefit from applying nanostructures, membranes, and nanomaterials typically used in semiconductor gas sensors, allowing for additional power consumption enhancements.

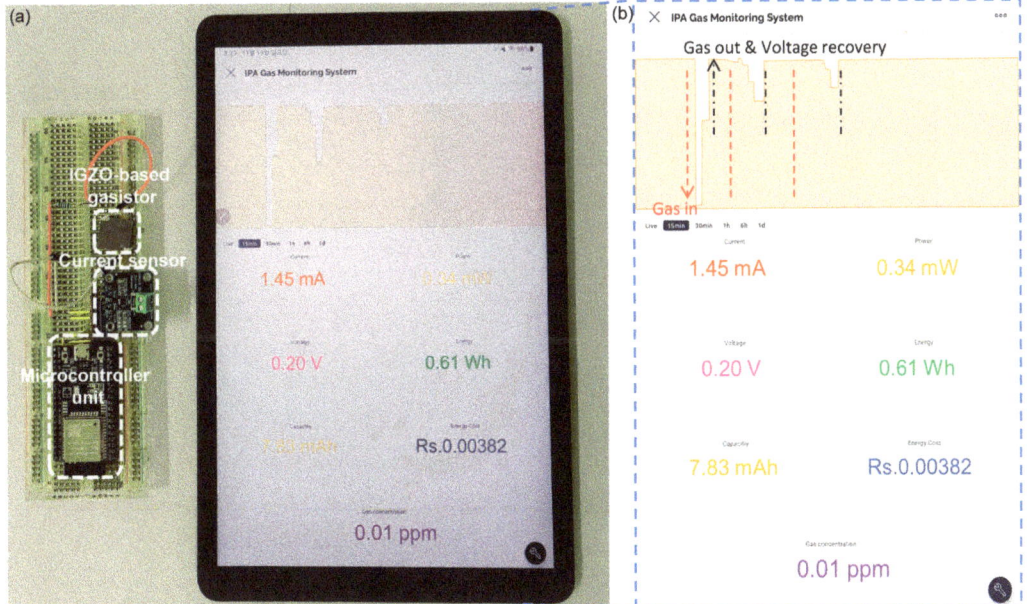

**Figure 5.** (**a**) Schematic representation of an IoT IPA gas monitoring system with an IGZO-based gasistor. (**b**) Real-time monitoring characteristics using an IoT system with an IGZO-based gasistor depend on IPA gas concentrations.

Table 1. Comparison of VOC gas sensors.

| Target Gas | Sensing Material | Driven Source | Concentration (ppm) | Response ($R_{gas}/R_{air}$) | Response Time | Recovery Time | Power Consumption | Ref. |
|---|---|---|---|---|---|---|---|---|
| Benzene | SnO$_2$@Au | 415 °C | 0.005 | ~4 | ~50 s | ~50 s | 8.9 mW | [46] |
| CH$_4$ | SiNWs/TiO$_2$ | RT | 120 | 1.5 | 75 s | 191 s | - | [49] |
| Methanol | sulfonated RGO hydrogel | RT | 0.005 | 1.40 | - | - | 20 nW | [50] |
| VOC | Fe$_3$O$_4$ | 300 °C | 0.6 | ~1.2 | 6.1 s | 10.7 s | 92 mW | [51] |
| IPA | a-IGZO | RT | 10 | 2.51 | 384 s | 50 μs | 0.34 mW | This work |

## 4. Conclusions

Gas sensors are indispensable constituents in many applications, encompassing environmental monitoring and portable IoT systems. However, achieving sustainable measurements and minimizing power consumption pose a significant challenge. Gasistors have been studied because of their beneficial attributes, including rapid response time and operation at RT. Through the integration of an IGZO-based gasistor into an IPA real-time gas monitoring system, we successfully demonstrated the sensor's low power consumption. The proposed IPA sensor based on IGZO exhibited a notable response of 55.15 and a rapid detection speed of 105 s at RT. Furthermore, recovery to the initial state was possible in 50 μs even when pulsed voltage was applied in the presence of gas. To ensure compatibility with the IoT system, the proposed device incorporated a low-power circuit module designed to wirelessly transmit, modulate, and process signals. Thus, the integration of an IGZO-based gasistor into the proposed IoT system enabled gas analysis and real-time monitoring with a power consumption of less than ~0.34 mW. The primary purpose of this module with an IGZO-based gasistor is to facilitate the incorporation of remote and early warning systems in the event of a gas release while monitoring gas concentrations in the IoT system. This research paper contributes insights into gas detection by demonstrating a practical approach to constructing a wireless sensing system with a sensitive gas sensor that is both low-powered and energy-efficient.

**Author Contributions:** Conceptualization, H.-D.K., M.C. and D.L.; methodology, M.C, D.L. and H.-D.K.; validation, D.L.; investigation, M.C.; resources, H.-D.K.; data curation, H.-D.K., M.C. and D.L.; writing—original draft preparation, M.C.; writing—review and editing, D.L. and H.-D.K.; visualization, M.C.; supervision, H.-D.K.; project administration, H.-D.K.; funding acquisition, H.-D.K. All authors have read and agreed to the published version of the manuscript.

**Funding:** This work was supported in part by the Basic Science Research Program via the National Research Foundation of Korea (NRF) funded by the Ministry of Education under grant NRF-2022R1F1A1060655 and the Nano-Material Technology Development Program (NRF, Ministry of Science, ICT, and Future Planning (2009-0082580)); and in part by the Korea Institute for Advancement of Technology (KIAT) grant funded by the Korea Government (MOTIE) via the Competency Development Program for Industry Specialists under grant P0020966.

**Data Availability Statement:** Data are contained within the article.

**Conflicts of Interest:** The authors declare no conflicts of interest.

## References

1. Kaspar, C.; Ravoo, B.; van der Wiel, W.G.; Wegner, S.; Pernice, W. The rise of intelligent matter. *Nature* **2021**, *594*, 345–355. [CrossRef]
2. Shi, Y.; Wang, F.; Tian, J.; Li, S.; Fu, E.; Nie, J.; Lei, R.; Ding, Y.; Chen, X.; Wang, Z.L. Self-powered electro-tactile system for virtual tactile experiences. *Sci. Adv.* **2021**, *7*, eabe2943. [CrossRef] [PubMed]
3. Waseem, A.; Abdullah, A.; Bagal, I.V.; Ha, J.-S.; Lee, J.K.; Ryu, S.-W. Self-powered and flexible piezo-sensors based on conductivity-controlled GaN nanowire-arrays for mimicking rapid-and slow-adapting mechanoreceptors. *Npj Flex. Electron.* **2022**, *6*, 58. [CrossRef]

4. Yang, Q.; Yang, S.; Qiu, P.; Peng, L.; Wei, T.-R.; Zhang, Z.; Shi, X.; Chen, L. Flexible thermoelectrics based on ductile semiconductors. *Science* **2022**, *377*, 854–858. [CrossRef] [PubMed]
5. Zhai, K.; Wang, H.; Ding, Q.; Wu, Z.; Ding, M.; Tao, K.; Yang, B.R.; Xie, X.; Li, C.; Wu, J. High-Performance Strain Sensors Based on Organohydrogel Microsphere Film for Wearable Human–Computer Interfacing. *Adv. Sci.* **2023**, *10*, 2205632. [CrossRef] [PubMed]
6. Liu, D.; Chen, Q.; Chen, A.; Wu, J. Self-powered gas sensor based on SiNWs/ITO photodiode. *RSC Adv.* **2019**, *9*, 23554–23559. [CrossRef] [PubMed]
7. Jia, Y.; Zhang, Z.; Xiao, L.; Lv, R. Carbon nanotube-silicon nanowire heterojunction solar cells with gas-dependent photovoltaic performances and their application in self-powered $NO_2$ detecting. *Nanoscale Res. Lett.* **2016**, *11*, 299. [CrossRef]
8. Chen, H.; Zhang, M.; Xing, B.; Fu, X.; Bo, R.; Mulmudi, H.K.; Huang, S.; Ho-Baillie, A.W.; Catchpole, K.R.; Tricoli, A. Superior self-charged and-powered chemical sensing with high performance for $NO_2$ detection at room temperature. *Adv. Opt. Mater.* **2020**, *8*, 1901863. [CrossRef]
9. Cho, I.; Sim, Y.C.; Cho, M.; Cho, Y.-H.; Park, I. Monolithic micro light-emitting diode/metal oxide nanowire gas sensor with microwatt-level power consumption. *ACS Sens.* **2020**, *5*, 563–570. [CrossRef]
10. Ahn, J.; Kim, H.; Kim, E.; Ko, J. VOCkit: A low-cost IoT sensing platform for volatile organic compound classification. *Ad Hoc Netw.* **2021**, *113*, 102360. [CrossRef]
11. Agrawal, A.V.; Kumar, N.; Kumar, M. Strategy and future prospects to develop room-temperature-recoverable $NO_2$ gas sensor based on two-dimensional molybdenum disulfide. *Nano-Micro Lett.* **2021**, *13*, 38. [CrossRef] [PubMed]
12. Parmar, G.; Lakhani, S.; Chattopadhyay, M.K. An IoT based low cost air pollution monitoring system. In Proceedings of the 2017 International Conference on Recent Innovations in Signal Processing and Embedded Systems (RISE), Bhopal, India, 27–29 October 2017; IEEE: New York, NY, USA, 2017; pp. 524–528.
13. Mirzaei, A.; Leonardi, S.; Neri, G. Detection of hazardous volatile organic compounds (VOCs) by metal oxide nanostructures-based gas sensors: A review. *Ceram. Int.* **2016**, *42*, 15119–15141. [CrossRef]
14. Acharyya, S.; Nag, S.; Kimbahune, S.; Ghose, A.; Pal, A.; Guha, P.K. Selective discrimination of VOCs applying gas sensing kinetic analysis over a metal oxide-based chemiresistive gas sensor. *ACS Sens.* **2021**, *6*, 2218–2224. [CrossRef] [PubMed]
15. Gao, Y.; Kong, Q.; Zhang, J.; Xi, G. General fabrication and enhanced VOC gas-sensing properties of hierarchically porous metal oxides. *RSC Adv.* **2017**, *7*, 35897–35904. [CrossRef]
16. Marikutsa, A.; Novikova, A.; Rumyantseva, M.; Khmelevsky, N.; Gaskov, A. Comparison of Au-functionalized semiconductor metal oxides in sensitivity to VOC. *Sens. Actuators B Chem.* **2021**, *326*, 128980. [CrossRef]
17. Le, D.T.T.; Long, N.D.H.; Xuan, C.T.; Van Toan, N.; Hung, C.M.; Van Duy, N.; Theu, L.T.; Dinh, V.A.; Hoa, N.D. Porous $CoFe_2O_4$ nanorods: VOC gas-sensing characteristics and DFT calculation. *Sens. Actuators B Chem.* **2023**, *379*, 133286.
18. Yabuta, H.; Sano, M.; Abe, K.; Aiba, T.; Den, T.; Kumomi, H.; Nomura, K.; Kamiya, T.; Hosono, H. High-mobility thin-film transistor with amorphous $InGaZnO_4$ channel fabricated by room temperature rf-magnetron sputtering. *Appl. Phys. Lett.* **2006**, *89*, 112123. [CrossRef]
19. Cho, N.G.; Kim, I.-D. $NO_2$ gas sensing properties of amorphous $InGaZnO_4$ submicron-tubes prepared by polymeric fiber templating route. *Sens. Actuators B Chem.* **2011**, *160*, 499–504. [CrossRef]
20. Lee, D.; Jung, J.; Kim, S.; Kim, H.-D. Gas detection and recovery characteristics at room temperature observed in a $Zr_3N_4$-based memristor sensor array. *Sens. Actuators B Chem.* **2023**, *376*, 132993. [CrossRef]
21. Chae, M.; Lee, D.; Kim, S.; Kim, H.-d. NO Sensing Properties of BN-based Memristor Sensor Array for Real-time NO Monitoring-Systems. *Sens. Actuators B Chem.* **2023**, *394*, 134373. [CrossRef]
22. Chae, M.; Lee, D.; Jung, J.; Kim, H.-D. Enhanced memristor-based gas sensor for fast detection using a porous carbon nanotube top electrode with membrane. *Cell Rep. Phys. Sci.* **2023**, *4*, 101659. [CrossRef]
23. Lee, D.; Bae, D.; Chae, M.; Kim, H.-D. High sensitivity of isopropyl alcohol gas sensor based on memristor device operated at room temperature. *J. Korean Phys. Soc.* **2022**, *80*, 1065–1070. [CrossRef]
24. Lee, D.; Yun, M.J.; Kim, K.H.; Kim, S.; Kim, H.-D. Advanced Recovery and High-Sensitive Properties of Memristor-Based Gas Sensor Devices Operated at Room Temperature. *ACS Sens.* **2021**, *6*, 4217–4224. [CrossRef] [PubMed]
25. Chun, S.Y.; Song, Y.G.; Kim, J.E.; Kwon, J.U.; Soh, K.; Kwon, J.Y.; Kang, C.Y.; Yoon, J.H. Artificial Olfactory System Based on a Chemi-memristive Device. *Adv. Mater.* **2023**, *35*, 2302219. [CrossRef] [PubMed]
26. Qiu, P.; Qin, Y.; Xia, Q. Ultrasensitive gas sensor developed from $SnS/TiO_2$-based memristor for dilute methanol detection at room temperature. *Sens. Actuators B Chem.* **2023**, *392*, 134038. [CrossRef]
27. Lee, D.; Jung, J.; Kim, K.H.; Bae, D.; Chae, M.; Kim, S.; Kim, H.-d. Highly Sensitive Oxygen Sensing Characteristics Observed in IGZO Based Gasistor in a Mixed Gas Ambient at Room Temperature. *ACS Sens.* **2022**, *7*, 2567–2576. [CrossRef]
28. Suko, A.; Jia, J.; Nakamura, S.-I.; Kawashima, E.; Utsuno, F.; Yano, K.; Shigesato, Y. Crystallization behavior of amorphous indium–gallium–zinc-oxide films and its effects on thin-film transistor performance. *Jpn. J. Appl. Phys.* **2016**, *55*, 035504. [CrossRef]
29. Jeong, H.-J.; Kim, Y.-S.; Jeong, S.-G.; Park, J.-S. Impact of Annealing Temperature on Atomic Layer Deposited In–Ga–Zn–O Thin-Film Transistors. *ACS Appl. Electron. Mater.* **2022**, *4*, 1343–1350. [CrossRef]
30. Popov, A.A.; Varygin, A.D.; Plyusnin, P.E.; Sharafutdinov, M.R.; Korenev, S.V.; Serkova, A.N.; Shubin, Y.V. X-ray diffraction reinvestigation of the Ni-Pt phase diagram. *J. Alloys Compd.* **2022**, *891*, 161974. [CrossRef]

31. Pujar, P.; Gandla, S.; Singh, M.; Gupta, B.; Tarafder, K.; Gupta, D.; Noh, Y.-Y.; Mandal, S. Development of low temperature stoichiometric solution combustion derived transparent conductive ternary zinc tin co-doped indium oxide electrodes. *RSC Adv.* **2017**, *7*, 48253–48262. [CrossRef]
32. Kim, D.-G.; Lee, T.-K.; Park, K.-S.; Chang, Y.-G.; Han, K.-J.; Choi, D.-K. Hydrogen behavior under X-ray irradiation for a-IGZO thin film transistors. *Appl. Phys. Lett.* **2020**, *116*, 013502. [CrossRef]
33. Sen, A.; Park, H.; Pujar, P.; Bala, A.; Cho, H.; Liu, N.; Gandla, S.; Kim, S. Probing the Efficacy of Large-Scale Nonporous IGZO for Visible-to-NIR Detection Capability: An Approach toward High-Performance Image Sensor Circuitry. *ACS Nano* **2022**, *16*, 9267–9277. [CrossRef] [PubMed]
34. Iglesias, V.; Jing, X.; Lanza, M. Combination of Semiconductor Parameter Analyzer and Conductive Atomic Force Microscope for Advanced Nanoelectronic Characterization. In *Conductive Atomic Force Microscopy: Applications in Nanomaterials*; Wiley: Hoboken, NJ, USA, 2017; pp. 225–241. [CrossRef]
35. Kim, K.M.; Yang, J.J.; Merced, E.; Graves, C.; Lam, S.; Davila, N.; Hu, M.; Ge, N.; Li, Z.; Williams, R.S. Low variability resistor–memristor circuit masking the actual memristor states. *Adv. Electron. Mater.* **2015**, *1*, 1500095. [CrossRef]
36. Bricalli, A.; Ambrosi, E.; Laudato, M.; Maestro, M.; Rodriguez, R.; Ielmini, D. Resistive switching device technology based on silicon oxide for improved ON–OFF ratio—Part I: Memory devices. *IEEE Trans. Electron Devices* **2017**, *65*, 115–121. [CrossRef]
37. Ling, Z.; Leach, C. The effect of relative humidity on the $NO_2$ sensitivity of a $SnO_2/WO_3$ heterojunction gas sensor. *Sens. Actuators B Chem.* **2004**, *102*, 102–106. [CrossRef]
38. Mo, Y.; Okawa, Y.; Tajima, M.; Nakai, T.; Yoshiike, N.; Natukawa, K. Micro-machined gas sensor array based on metal film micro-heater. *Sens. Actuators B Chem.* **2001**, *79*, 175–181. [CrossRef]
39. Ha, N.H.; Thinh, D.D.; Huong, N.T.; Phuong, N.H.; Thach, P.D.; Hong, H.S. Fast response of carbon monoxide gas sensors using a highly porous network of ZnO nanoparticles decorated on 3D reduced graphene oxide. *Appl. Surf. Sci.* **2018**, *434*, 1048–1054. [CrossRef]
40. Wang, H.; Li, Y.; Yang, M. Fast response thin film $SnO_2$ gas sensors operating at room temperature. *Sens. Actuators B Chem.* **2006**, *119*, 380–383. [CrossRef]
41. Akamatsu, T.; Itoh, T.; Izu, N.; Shin, W. NO and $NO_2$ sensing properties of $WO_3$ and $Co_3O_4$ based gas sensors. *Sensors* **2013**, *13*, 12467–12481. [CrossRef]
42. Li, S.-H.; Chu, Z.; Meng, F.-F.; Luo, T.; Hu, X.-Y.; Huang, S.-Z.; Jin, Z. Highly sensitive gas sensor based on $SnO_2$ nanorings for detection of isopropanol. *J. Alloys Compd.* **2016**, *688*, 712–717. [CrossRef]
43. Wang, S.-C.; Wang, X.-H.; Qiao, G.-Q.; Chen, X.-Y.; Wang, X.-Z.; Wu, N.-N.; Tian, J.; Cui, H.-Z. NiO nanoparticles-decorated ZnO hierarchical structures for isopropanol gas sensing. *Rare Met.* **2022**, *41*, 960–971. [CrossRef]
44. Bai, Y.; Fu, H.; Yang, X.; Xiong, S.; Li, S.; An, X. Conductometric isopropanol gas sensor: Ce-doped $In_2O_3$ nanosheet-assembled hierarchical microstructure. *Sens. Actuators B Chem.* **2023**, *377*, 133007. [CrossRef]
45. Cho, I.; Kang, K.; Yang, D.; Yun, J.; Park, I. Localized liquid-phase synthesis of porous $SnO_2$ nanotubes on MEMS platform for low-power, high performance gas sensors. *ACS Appl. Mater. Interfaces* **2017**, *9*, 27111–27119. [CrossRef] [PubMed]
46. Elmi, I.; Zampolli, S.; Cozzani, E.; Mancarella, F.; Cardinali, G. Development of ultra-low-power consumption MOX sensors with ppb-level VOC detection capabilities for emerging applications. *Sens. Actuators B Chem.* **2008**, *135*, 342–351. [CrossRef]
47. Ngoc, T.M.; Van Duy, N.; Hung, C.M.; Hoa, N.D.; Trung, N.N.; Nguyen, H.; Van Hieu, N. Ultralow power consumption gas sensor based on a self-heated nanojunction of $SnO_2$ nanowires. *RSC Adv.* **2018**, *8*, 36323–36330. [CrossRef] [PubMed]
48. Chikkadi, K.; Muoth, M.; Maiwald, V.; Roman, C.; Hierold, C. Ultra-low power operation of self-heated, suspended carbon nanotube gas sensors. *Appl. Phys. Lett.* **2013**, *103*, 223109. [CrossRef]
49. Liu, D.; Lin, L.; Chen, Q.; Zhou, H.; Wu, J. Low power consumption gas sensor created from silicon nanowires/$TiO_2$ core–shell heterojunctions. *ACS Sens.* **2017**, *2*, 1491–1497. [CrossRef]
50. Wu, J.; Tao, K.; Guo, Y.; Li, Z.; Wang, X.; Luo, Z.; Feng, S.; Du, C.; Chen, D.; Miao, J. A 3D chemically modified graphene hydrogel for fast, highly sensitive, and selective gas sensor. *Adv. Sci.* **2017**, *4*, 1600319. [CrossRef]
51. Hsiao, Y.-J.; Nagarjuna, Y.; Tsai, C.-A.; Wang, S.-C. High selectivity $Fe_3O_4$ nanoparticle to volatile organic compound (VOC) for MEMS gas sensors. *Mater. Res. Express* **2020**, *7*, 065013. [CrossRef]

**Disclaimer/Publisher's Note:** The statements, opinions and data contained in all publications are solely those of the individual author(s) and contributor(s) and not of MDPI and/or the editor(s). MDPI and/or the editor(s) disclaim responsibility for any injury to people or property resulting from any ideas, methods, instructions or products referred to in the content.

*Article*

# Biosensor-Based Multimodal Deep Human Locomotion Decoding via Internet of Healthcare Things

Madiha Javeed [1,*], Maha Abdelhaq [2,*], Asaad Algarni [3] and Ahmad Jalal [1]

1. Department of Computer Science, Air University, Islamabad 44000, Pakistan; ahmadjalal@mail.au.edu.pk
2. Department of Information Technology, College of Computer and Information Sciences, Princess Nourah bint Abdulrahman University, P.O. Box 84428, Riyadh 11671, Saudi Arabia
3. Department of Computer Sciences, Faculty of Computing and Information Technology, Northern Border University, Rafha 91911, Saudi Arabia; asaad.algarni@nbu.edu.sa
* Correspondence: 191880@students.au.edu.pk or mjaved84@gmail.com (M.J.); msabdelhaq@pnu.edu.sa (M.A.)

**Abstract:** Multiple Internet of Healthcare Things (IoHT)-based devices have been utilized as sensing methodologies for human locomotion decoding to aid in applications related to e-healthcare. Different measurement conditions affect the daily routine monitoring, including the sensor type, wearing style, data retrieval method, and processing model. Currently, several models are present in this domain that include a variety of techniques for pre-processing, descriptor extraction, and reduction, along with the classification of data captured from multiple sensors. However, such models consisting of multiple subject-based data using different techniques may degrade the accuracy rate of locomotion decoding. Therefore, this study proposes a deep neural network model that not only applies the state-of-the-art Quaternion-based filtration technique for motion and ambient data along with background subtraction and skeleton modeling for video-based data, but also learns important descriptors from novel graph-based representations and Gaussian Markov random-field mechanisms. Due to the non-linear nature of data, these descriptors are further utilized to extract the codebook via the Gaussian mixture regression model. Furthermore, the codebook is provided to the recurrent neural network to classify the activities for the locomotion-decoding system. We show the validity of the proposed model across two publicly available data sampling strategies, namely, the HWU-USP and LARa datasets. The proposed model is significantly improved over previous systems, as it achieved 82.22% and 82.50% for the HWU-USP and LARa datasets, respectively. The proposed IoHT-based locomotion-decoding model is useful for unobtrusive human activity recognition over extended periods in e-healthcare facilities.

**Keywords:** human activity recognition; Internet of Healthcare Things; locomotion prediction; multimodal systems; recurrent neural network; RGB; wearable sensors

**Citation:** Javeed, M.; Abdelhaq, M.; Algarni, A.; Jalal, A. Biosensor-Based Multimodal Deep Human Locomotion Decoding via Internet of Healthcare Things. *Micromachines* 2023, 14, 2204. https://doi.org/10.3390/mi14122204

Academic Editors: Lei Jing, Yoshinori Matsumoto and Zhan Zhang

Received: 30 October 2023
Revised: 28 November 2023
Accepted: 30 November 2023
Published: 3 December 2023

**Copyright:** © 2023 by the authors. Licensee MDPI, Basel, Switzerland. This article is an open access article distributed under the terms and conditions of the Creative Commons Attribution (CC BY) license (https://creativecommons.org/licenses/by/4.0/).

## 1. Introduction

Recent trends in the Internet of Healthcare Things (IoHT) have boosted wearable and visual-technology-based human locomotion decoding. This boost converts the healthcare industry from cure to prevention [1–4]. Various IoHT devices are available for healthcare and research, including smart devices, inertial units, and cameras. Data from such IoHT devices have been extracted, processed, and analyzed for human locomotion decoding. For ambient assisted living, sensor-based data have been used to support and supervise people, also known as human activity recognition (HAR) [5–7]. Applications of such HAR systems include injury recognition, medical analysis, long-term or short-term care, health monitoring, and independent quality of life [8–12].

These HAR systems can use machine learning or deep learning techniques to decode the activities of daily living by extracting data from motion, ambient, or vision-based sensors [13–16]. Modern smart devices manipulate the data and thus cannot be utilized for locomotion decoding [17–20]. Some HAR systems have less efficiency due to errors induced by the data acquisition that must be resolved using a robust filter [21–23]. Exiting feature extraction methods cannot perform well for HAR systems and provide less efficient results [24–28]. Therefore, a multimodal sensor-based human-locomotion-decoding (HLD) system consisting of motion, ambient, and vision sensors is proposed in this paper. The key contributions of this research are as follows:

- An innovative multimodal system for locomotion decoding via multiple sensors fused to enhance the HAR performance [29–31];
- The effective and novel filtration of the inertial sensor data [32–34] by using a proposed state-of-the-art Quaternion-based filter;
- A novel approach to filtering the ambient-based data that includes infrared cameras and switches attached to the environment;
- Hand-crafted contemporary descriptor extraction methods [35–38] are proposed and applied to acquire related descriptors [39–42] using novel techniques;
- Efficient ambient sensor descriptor extraction based on a unique and novel graph representation;
- The proficient recognition of activities [43–46] for locomotion decoding via detection through a recurrent neural network (RNN).

Section 2 explains the sensor-based activity recognition systems presented in the literature. Next, Section 3 details the proposed locomotion-decoding system for the IoHT industry [47–51]. The experiments performed over the selected datasets using the proposed method and their results, along with a comparison of the baseline system and previous state-of-the-art models, are discussed in Section 4. The conclusion of the whole paper is presented in Section 5.

## 2. Related Work

Locomotion decoding with a combination of IoHT-based sensors can be utilized for different applications [52–55], including the execution and tagging of data, which associates the meanings of sensor data interpretations by using symbols. A single sensor is not enough to provide the semantic meaning of a situation. Therefore, multimodal sensor-based systems serve this purpose. For this resolution, multiple systems have been proposed in history to evaluate the effectiveness, completeness, and reliability of such sensor-based decoding systems.

### 2.1. Sensor-Based Locomotion Decoding

In [56], Franco et al. propose a multimodal system for locomotor activity recognition. They used RGB video and other sensors for data acquisition. Histograms of oriented gradient (HOG) descriptors and skeleton-based information were extracted from the RGB data frames to capture the most prominent body postures. For the activity classification, a voting system was defined to obtain votes from support vector machine (SVM) and random forest classifiers. However, the proposed system could not achieve higher results due to the absence of a filtration technique for the data. Another system is proposed in [57] that collects motion sensor data. Next, data are processed using a linear interpolation filter and segmentation. Features are extracted using four different extraction techniques and normalized using the z-score function. Then, features are selected via correlation and evolutionary search algorithms. Further, the class imbalance is removed using the synthetic minority over-sampling method. Features are fused, and multi-view stacking is utilized to classify humans.

### 2.2. Multimodal Locomotion Decoding

In [58], the authors propose a robust human activity recognition method. They used multimodal data based on wearable inertial and RGB-D sensors. The inertial data were pre-processed using magnitude computation and noise removal techniques, and dense HOGs were extracted from video data. Time domain features are extracted from inertial signals, and bag-of-words encoding is utilized for video frame sequences. Furthermore, the features are fused, and K-nearest-neighbor and support vector machines are used for the human activity classification.

A long short-term memory (LSTM) network-based system is proposed in [59]. To recognize activities of daily living, the authors used a deep learning model via data acquired from real-world and synthetic environments. The sensors were attached to the wrists, ankles, and waist to detect activities, including eating and driving. Each sensor's accuracy was observed to elaborate the custom weights for each sensor fusion. This study recommended using one sensor on the upper body parts and one sensor on the lower body parts to obtain reasonable results. However, due to the restricted data used and limited weight learning in the system, the method cannot adapt to changes over time.

In [60], a system of Marfusion based on a convolutional neural network (CNN) and attention mechanism is proposed. Features are extracted from multimodal sensors and a set of CNNs is utilized for each sensor. Next, a dot-product, scaled, self-attention process is applied to give weight to each sensor. Then, CNN and attention-based modules are utilized for feature fusion with different parameters. Further, fully connected batch normalization, dropout, ReLU, and softmax layers are used for the classification via the obtainment of the probabilities for different activities. The proposed system gave an acceptable performance but experimented with limited human locomotion. Therefore, the results are not robust for real-time environments.

## 3. Materials and Methods

The proposed locomotion-decoding architecture is described in Figure 1. The input data for the proposed IoHT-based system were taken from two publicly available datasets named Logistic Activity Recognition Challenge (LARa) [61] and Heriot-Watt University/University of Sao Paulo (HWU-USP) [62], which are present in the form of time series in a time segment of size W from S sensors. Sensors of three types were used: physical signals $\{pi\}$, ambient signals $\{pa\}$, and visual frame sequences $\{pv\}$. Algorithm 1 demonstrates the complete IoHT-based HLD system. The input $\{pi, pa, pv\}$ from the S sensors was pre-processed for each time segment of a W size. Next, the descriptors were extracted and optimized $\{V_i^*, K_i^*, S_i^*, A_i^*\}$ for each W segment. Further, the descriptors were trained by using an RNN and tested the remaining descriptors to recognize activities $\{A^*\}$ to decode human locomotion. All these phases of the IoHT-based HLD system are further explained in the next subsections.

---

**Algorithm 1 HLD Algorithm**

---

**Input:** physical IMU signals $\{p_i\}$, ambient signals $\{p_a\}$, visual frame sequences $\{p_v\}$;
**Output:** recognized activities $\{A^*\}$;

1. Pre-process $\{p_i, p_a, p_v\}$ for each segment W in *Module I*;
2. Extract descriptors $\{V_i^*, K_i^*, S_i^*, A_i^*\}$ for W in *Module II*;
3. Optimize descriptors for W in *Module III*;
4. Train descriptors over classifier to obtain $f(X,\theta)$;
5. Test remaining descriprtors to obtain $\{\theta,\theta^*\}$;
6. Recognize activities $\{A^*\}$;

---

*3.1. Pre-Processing Motion and Ambient Data*

A novel quaternion-based filter is proposed in this study to pre-process the physical-motion [63] and ambient data from the sensor inertial measurement units (IMUs). The signals are clarified via low- and high-pass Butterworth filters [64,65] for further processing. Next, the signals are normalized using the Euclidean distance [66,67]:

$$Norm = \sqrt{LPF_1 + LPF_2 + LPF_3} + \sqrt{HPF_1 + HPF_2 + HPF_3} \qquad (1)$$

where $LPF_1$, $LPF_2$, and $LPF_3$ denote the filtered values for the x-, y-, and z-axes via the Butterworth filter, respectively. $HPF_1$, $HPF_2$, and $HPF_3$ represent the filtered values of the x-, y-, and z-axes through the Butterworth filter, respectively.

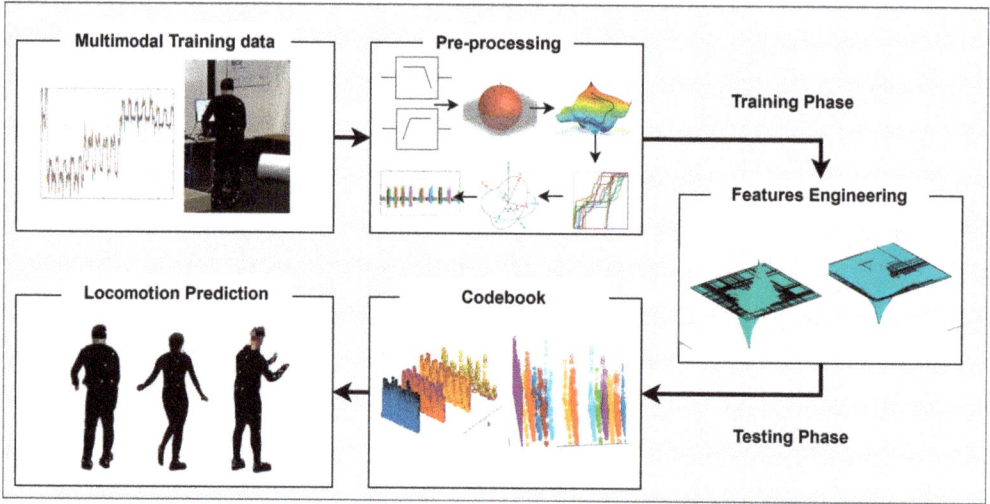

**Figure 1.** Architecture diagram of proposed IoHT-based human-locomotion-decoding system.

Then, for the accelerometer signals, gravity from a stationary activity, such as lying down, is extracted as the minimum gravity ($g_m$) and average gravity ($g_a$). Then, the gravitational error ($g_e$) [68,69] is removed from the accelerometer signals, giving more accurate and error-free signals for further processing. Similarly, the earth's magnetic field is used to remove the magnetic errors from magnetometer signals [70,71].

After normalization, discrete wavelet transform [72] is applied to the gyroscope signals to transform them into quaternions in order to avoid the gimbal lock problem. Later, the derivative of the quaternions is considered, and gradient descent is applied to attain the minimum rate of change. Further, a local minimum [73] is selected, and the gyroscope signals are normalized using the Euler angles:

$$Axz = atan2(z, x), \quad (2)$$

$$Ayz = atan2(z, y), \quad (3)$$

$$Axy = atan2(y, x), \quad (4)$$

where $Axz$, $Ayz$, and $Axy$ are the Euler angles. Lastly, all three pre-processed signals are normalized together. Figure 2 explains the pre-processing step for the physical-motion module in detail.

*3.2. Pre-Processing Visual Data*

For the pre-processing, videos from both datasets were converted into frame sequences. A delta of 50 was chosen to restrict the number of pre-processing sequences to avoid redundant data processing. Next, we retrieved a background image from both data sequences. Then, the background was removed by subtracting the background image from the original frame sequences [74,75]. The background subtraction from the original image sequence is displayed in Figure 3. Discrete wavelet transform was used over the frame sequences to reduce the noise present.

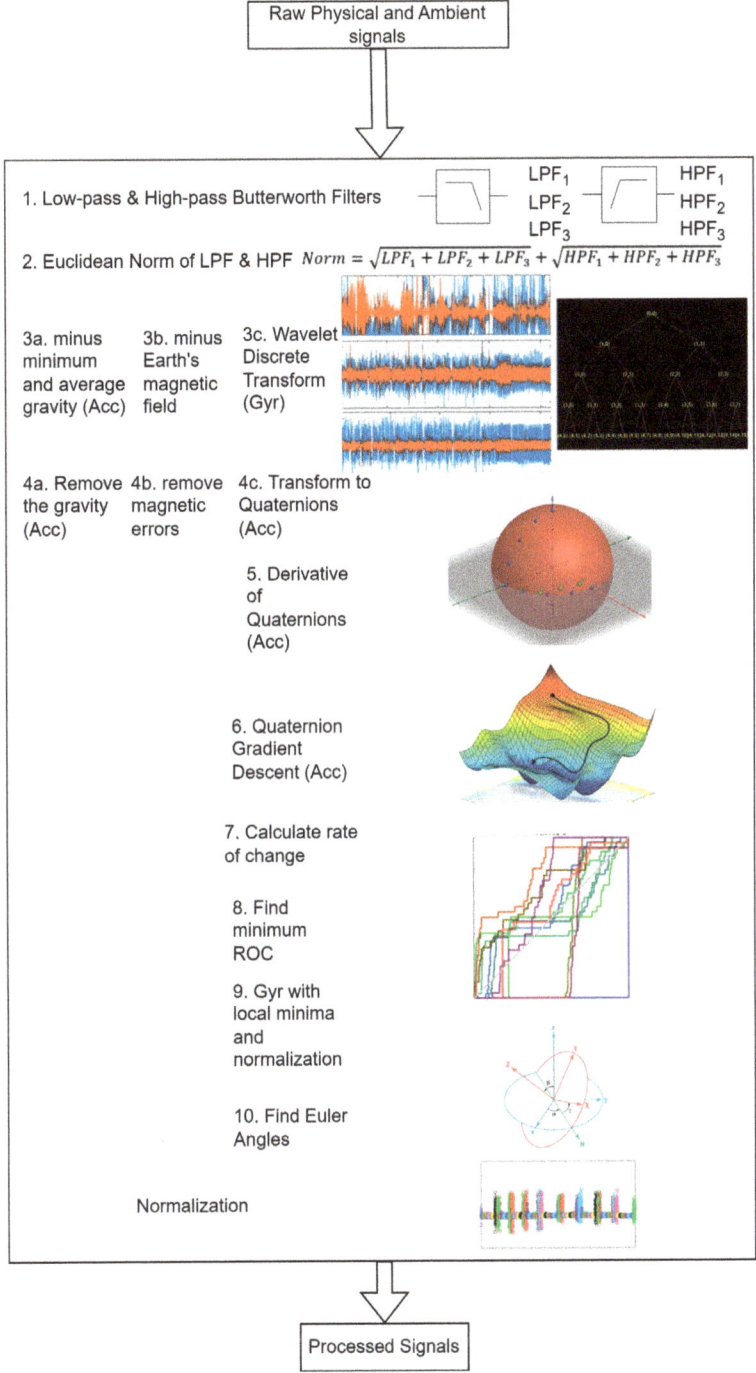

**Figure 2.** Pre-processing module proposed for physical-motion and ambient data.

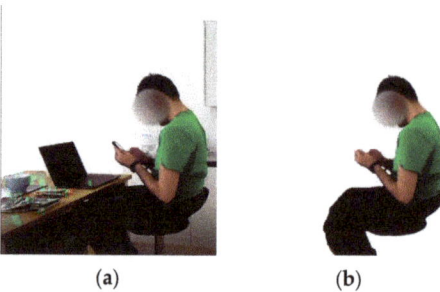

**Figure 3.** (a) Before background deduction and (b) after background deduction of a frame sequence from HWU-USP dataset.

Skeleton modeling was performed through blob and centroid techniques for human detection in the frame sequences. First, the blobs were defined from the human movable parts, which was followed by taking the centroids and deciding on five types of skeleton body points—head, shoulders, elbows, wrists, torso, knees, and ankles [76]. Figure 4 shows the skeleton points extracted for drinking tea and reading a newspaper.

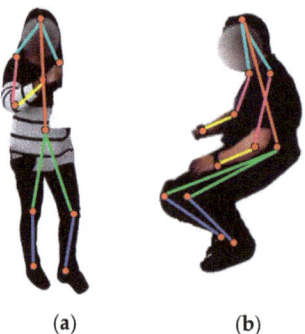

**Figure 4.** Skeleton point decoding from frame sequences of (a) drinking tea and (b) reading a newspaper.

### 3.3. Data Segmentation

Next, to deal with the dimensions of the datasets, this study segmented the motion and ambient pre-processed data into overlapped [77] and time-based [78] segments, whereas the vision-based data were segmented through event-based segments. For all three types of data $\{p_i^*, p_a^*, p_v^*\}$, Figure 5 shows the segmentation process by using manifold locomotion activities.

### 3.4. Motion Descriptor Extraction

The pre-processed and segmented motion-based data were further provided to two different techniques for the descriptor extraction, including Gaussian Markov random field (GMRF) and a novel contribution in the form of a multisynchrosqueezing transform (MSST)-based spatial–temporal graph.

GMRF can take multidimensional data, and a stochastic process becomes Gaussian when all its distributions are Gaussian-normalized [79]. Equations (5) and (6) show the expectation function ($\widetilde{\mu}_t$) and covariance function ($\widetilde{\Sigma}s, t$) using $s$ samples and $t$ times. Figure 6 presents the results for the GMRF for a window of kinematic physical data on HWU-USP.

$$\widetilde{\mu}_t = E\widetilde{X}_t, \tag{5}$$

$$\widetilde{\sum} s,t = \text{cov}\left(\widetilde{X}_s, \widetilde{X}_t\right). \tag{6}$$

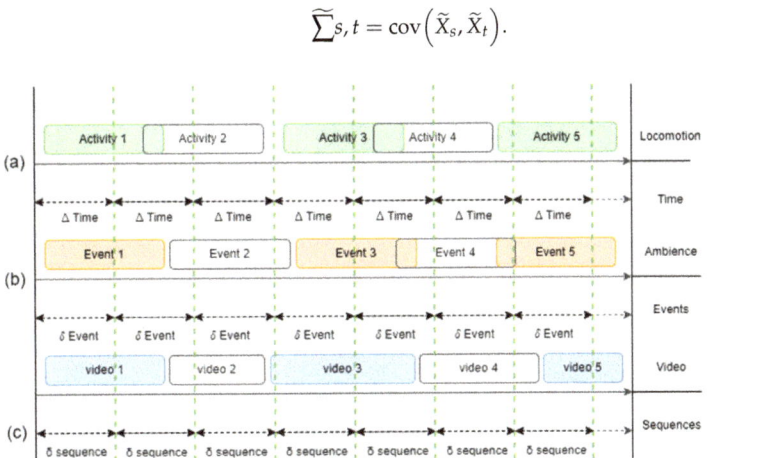

**Figure 5.** Segmentation for (**a**) motion, (**b**) ambient, and (**c**) visual data.

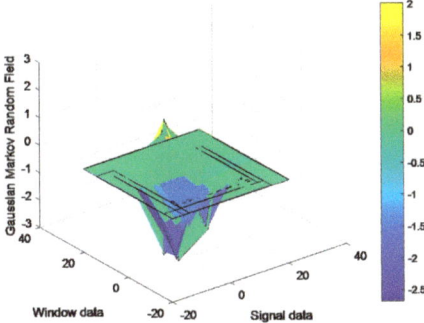

**Figure 6.** Results of GMRF application on HWU-USP dataset.

MSST represents multiple synchrosqueezing transforms iteratively [80] and is calculated as follows:

$$Ts^{[M]}(t,\gamma) = \int_{-\infty}^{+\infty} Ts^{M-1}(t,\gamma)\delta(\gamma - \hat{\omega}(t,\omega))d\omega, \tag{7}$$

where $M$ gives the iteration number $\leq 2$ and $Ts^{[M]}(t,\gamma)$ is the spread time–frequency coefficient. The short-time periodogram is further calculated as follows:

$$p(s,f) = \frac{1}{T}\left|Y(s,f)\right|^2 \tag{8}$$

where $p(s,f)$ is the result of frequency ($f$) and time ($s$). $T$ shows the window length. Further, the spatial–temporal graph was constructed using six nodes or frequencies. Figure 7 shows the novel spatial–temporal graph extracted from MSST for a random static pattern.

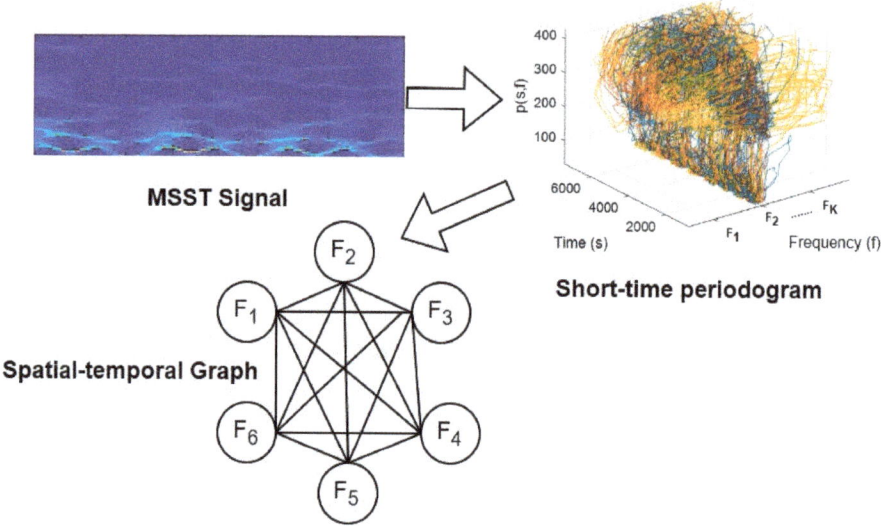

**Figure 7.** Process of constructing novel spatial–temporal graph from MSST.

*3.5. Ambient Descriptor Extraction*

A graph-based representation has been proposed as a novel descriptor extraction for ambient sensor pre-processing [81]. For each sensor attached to the ambient, a graph ($R$) is produced using a descriptor matrix ($M$) and adjacency matrix ($K$) given by the following:

$$R = (M, K) \qquad (9)$$

where $M$ is the descriptor matrix consisting of the sensor type, number of neighbors, and sensor orientation. $K$ contains the number of adjacent sensors for each node and the names of neighboring sensors. Figure 8 presents the details of the proposed graph-based ambient descriptors.

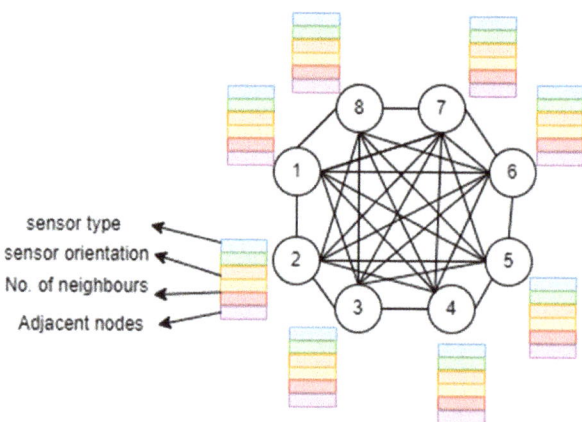

**Figure 8.** Proposed novel graph-based ambient feature extraction.

## 3.6. Vision Descriptor Extraction

In thermal descriptors, the movement from one frame to another is captured in the form of thermal maps. More movement is described using higher heat values in yellow, and less movement is shown using red or black [82]. In Equation (10), $x$ represents a one-dimensional vector comprising the extracted values, $i$ represents the index value, and $R$ denotes the RGB value. Figure 9 presents the heat map for the full-body frame sequence.

$$TM(x) = \sum_{i=0}^{k} \ln R(i). \tag{10}$$

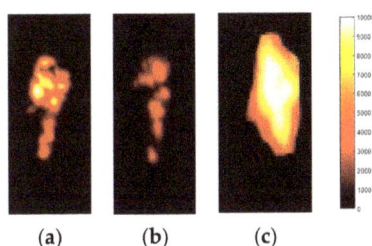

(a)     (b)     (c)

**Figure 9.** Thermal heat map extracted for activities including (**a**) drinking tea, (**b**) opening a drawer, and (**c**) reading a newspaper.

The full-body descriptor extraction method for visual data utilized is called the saliency map (SM) approach. It is computationally expensive to process an entire frame simultaneously; therefore, the SM approach suggests sequentially looking at or fixating on the salient locations of a frame. The fixated region is analyzed, and then attention is redirected to other salient regions using saccade movements requiring more focus [83]. The SM approach is a successful and biologically plausible technique for modeling visual attention. The generalized Gaussian distribution shown in Equation (11) is used to model each of these:

$$P(f_i) = \frac{\theta_i}{2\sigma_i \gamma\left(\theta_i^{-1}\right)} \exp\left(-\left|\frac{f_i}{\sigma_i}\right|^{\theta_i}\right), \tag{11}$$

where $\theta_i > 0$ is the shape parameter, $\sigma_i > 0$ provides the scale parameter, and $\gamma$ gives the gamma function. Figure 10 presents the results of SMs applied over a full-body frame sequence.

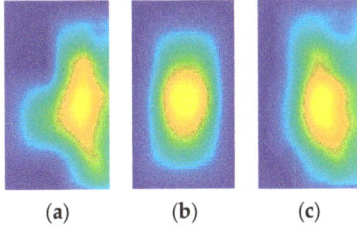

(a)     (b)     (c)

**Figure 10.** Results of saliency maps applied over full-body frame sequences for (**a**) drinking tea, (**b**) opening a drawer, and (**c**) reading a newspaper.

The orientation descriptor technique is the first descriptor extraction technique for the skeleton body points. Five skeleton body points are used to make triangles and obtain angles from them. The tangent angle in Equation (12) is measured between the three sides of each triangle [84]:

$$\tan \theta = \frac{u \cdot v}{|u||v|}, \tag{12}$$

where $u \cdot v$ is the dot product of vectors $u$ and $v$ that are any two sides of a triangle. Figure 11 demonstrates the examples of triangles formed by combining two human skeleton body points in some activities, such as drinking tea and reading a newspaper.

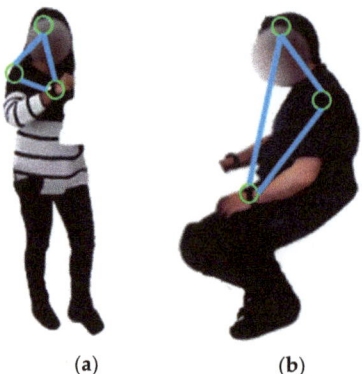

**Figure 11.** The triangular shape is formed by combining human skeleton body points for (**a**) drinking tea and (**b**) reading a newspaper.

The second descriptor extraction technique used for the skeleton body points is the spider local image feature (SLIF) technique. A spiderweb representation emulates the skeleton body point nodes as web intersection points in a frame sequence [85,86]. The position of each node $(n, m)$ is denoted by a set of two-dimensional coordinates, as follows:

$$x_{n,m} = \left( \frac{m \cdot cos\left(\frac{2\pi n}{N}\right)}{M}, \frac{m \cdot sin\left(\frac{2\pi n}{N}\right)}{M} \right), \qquad (13)$$

where the first and second terms represent the horizontal and vertical coordinates, respectively. For a set of previously defined skeleton body points, the SLIFs are extracted by selectively extracting pixel information from around the neighborhood of each point and applying a spiderweb over the point. Figure 12 shows a spiderweb applied over two sample frame sequences.

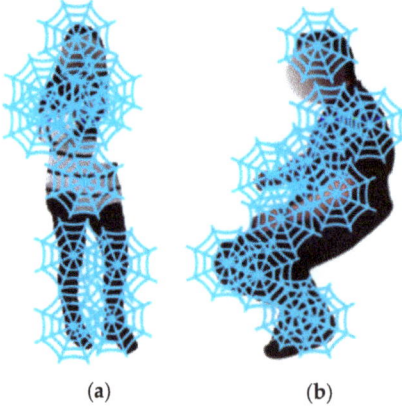

**Figure 12.** Spiderweb applied for (**a**) drinking tea and (**b**) reading a newspaper over HWU-USP dataset.

## 3.7. Codebook Generation

A Gaussian mixture model (GMM) codebook is used to encode the descriptors extracted from previous subsections. An expectation maximization (EM) algorithm is used in the GMM to present complex descriptors. This algorithm approximates the parameter set ($\Theta$) and aids in calculating the maximum likelihood through an initial parameter set ($\Theta 1$), and then continuously applies the $E$ and $M$ steps. Then, it produces $\{\Theta 1, \Theta 2, \ldots, \Theta m, \ldots\}$ and both $E$ and $M$ steps as follows:

$$\gamma^m\left(z_k^j \mid x_j, \Theta^m\right) = \frac{\omega_k^m f(x_j|\mu_k^m, \Sigma_k^m)}{\sum_{i=1}^{K} \omega^m f(x_j| \mu_i^m, \Sigma_i^m)}, \quad (14)$$

$$\Sigma_k^{m-1} = \frac{\sum_{j=1}^{N} \gamma^m(z_k^j|x_j, \Theta^m)\left(x_j - \mu_k^{m+1}\right)\left(x_j - \mu_k^{m+1}\right)^T}{\sum_{j=1}^{N} \gamma^m(z_k^j|x_j, \Theta^m)}. \quad (15)$$

where $\gamma^m\left(z_k^j \mid x_j, \Theta^m\right)$ gives the probability of the *j*th example and the *k*th Gaussian at the *m*th iteration with weights ($\omega_k^m$), means ($\mu_k^m$), and covariance ($\Sigma_k^m$) values. Similarly, a single generalized signal is extracted from the set of descriptors given using Gaussian mixture regression (GMR). Henceforth, a smooth signal via regression can be taken out by coding the temporal signal features [87] through a mixture of Gaussians. Each vector of the signals' GMM is taken as the input (xI) and output (xO) using GMR via this method.

## 3.8. Locomotion Decoding

A simple feedforward neural network poorly handles the sequence of data. It never forms a cycle between two hidden layers, and information always flows in one direction, never going back. It comprises an input layer, a hidden layer, and an output layer. An RNN [88] also contains these three layers, but it focuses on considering the current state along with the previous state in the form of output from the previously hidden layer via memory. Thus, the current state and previous state are used to produce output for the next time step, as shown in Figure 13. An activation function is also used to calculate the current state; we used tan $h$ as the activation function. Due to the input pattern change, the RNN performs better by incorporating backpropagation.

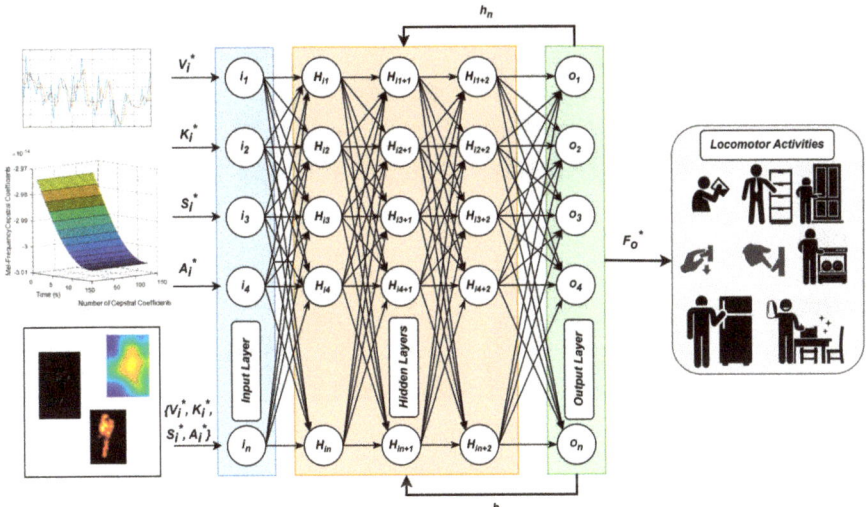

**Figure 13.** RNN incorporated into the proposed IoHT-based HLD system.

## 4. Performance Evaluation

In order to evaluate the IoHT-based HLD system, the following datasets and evaluation criteria were used.

### 4.1. Dataset Descriptions

Several publicly available datasets are present for human locomotion decoding via activity recognition. However, they can be different in terms of the number of subjects, number of activities performed, environmental setup, number of sensors, type of sensors, and sampling rates. In the proposed IoHT-based HLD system, we used two publicly available datasets, HWU-USP [62] and LARa [61], captured in diverse environmental setups and three different sensor modalities to make the system more robust. A 10-fold cross-validation technique was utilized to evaluate the proposed system. The following sections give details on the datasets mentioned above:

HWU-USP: A dataset recorded in a "living-lab" was selected for this study. It contains recordings from binary switches, PIR sensors, RGBD cameras installed over a robot, and IMU devices. The camera color is VGA 640 × 480 at 25 fps. A total of 16 participants performed nine activities with 144 instances with an average length of 48 s [62]. The participants were voluntary and healthy with neither functional nor visual impairments. The dataset contains activities of daily living with either periodical patterns or long-term dependencies and, hence, it is different from other multimodal environments. A variety of activities have been performed, such as *making a cup of tea*, *making a sandwich*, *making a bowl of cereal*, *setting the table*, *using a laptop*, *using a phone*, *reading a newspaper*, *cleaning the dishes*, and *tidying the kitchen*. Figure 14 represents the sample of activities performed by one of the participants in the HWU-USP dataset;

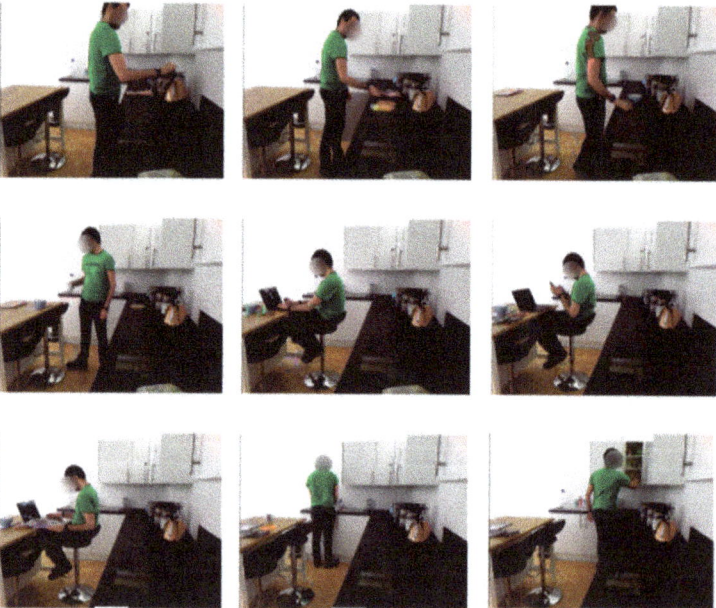

**Figure 14.** Activities performed by subject in HWU-USP dataset [61].

LARa: This dataset consists of an OmoCap system, a VICON system of 38 infrared cameras, three sets of IMU devices, and 30 recordings of 2 min for each of the 14 subjects. A wide range of participants were selected, including both male and female, ranging in age from 22 to 59 years, weighing from 48 to 100 lbs, left- and right-handed, and with heights from 163 to 185 cm. The dataset was recorded in a total of seven sessions of 758 min of recording. Acceleration sensors recorded the locomotion at a rate of 100 Hz [61]. The dataset is unbalanced regarding the annotations due to the complex process. The dataset is based on the activities performed in a logistics-based context. An expert trained the subjects in advance to recordings. A total of eight activities were recorded for each subject, including standing, walking, carting, handling (upwards), handling (centered), handling (downwards), synchronization, and none. Figure 15 gives a few sample frame sequences from the dataset.

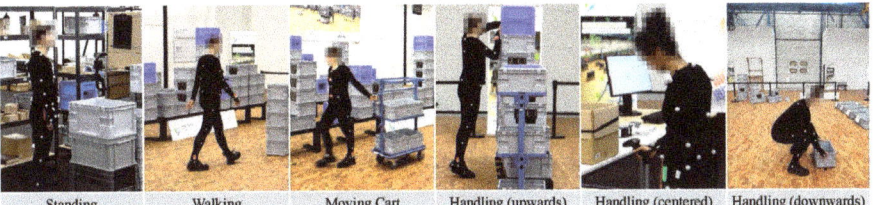

**Figure 15.** Sample frame sequences from the LARa dataset [62].

### 4.2. Experiment 1: Evaluation Protocol

Evaluation metrics can be used to evaluate the performance of the chosen deep learning classifier, including the accuracy, precision, and F1-score [89]. Table 1 shows the evaluation metrics derived from the experimental results. In our study, these metrics were chosen where the accuracy was the ratio between the decoded samples and the total number of samples. The three metrics can be defined as follows:

$$Acc = \frac{TP + TN}{TP + FN + FP + TN}, \tag{16}$$

$$rec = \frac{TP}{TP + FN}, \tag{17}$$

$$F1 - score = \frac{2 \times (rec \cdot pre)}{(rec + pre)}, \tag{18}$$

where $TP, TN$ are the true-positive and true-negative values, $FP, FN$ give the false-positive and false-negative values, and $pre$ is the precision, which can be calculated as follows:

$$pre = \frac{TP}{TP + FP} \tag{19}$$

**Table 1.** Comparative analysis of proposed IoHT-based HLD system with other deep learning approaches using accuracy, recall, precision, and F1-score for the two benchmarked datasets.

| Performance | Proposed System with First Novelty | Proposed System with Second Novelty | Proposed System with Both Novelties | CNN | LSTM |
|---|---|---|---|---|---|
| HWU-USP | | | | | |
| Accuracy | 78.89% | 80.00% | **82.22%** | 72.22% | 70.00% |
| Recall | 0.79 | 0.80 | **0.82** | 0.72 | 0.70 |
| Precision | 0.79 | 0.81 | **0.83** | 0.73 | 0.71 |
| F1-Score | 0.79 | 0.81 | **0.82** | 0.72 | 0.70 |
| LARa | | | | | |
| Accuracy | 80.00% | 77.50% | **82.50%** | 78.75% | 76.25% |
| Recall | 0.80 | 0.77 | **0.82** | 0.78 | 0.76 |
| Precision | 0.80 | 0.78 | **0.83** | 0.79 | 0.76 |
| F1-Score | 0.80 | 0.77 | **0.82** | 0.79 | 0.76 |

## 4.3. Experiment 2: Comparison with Baseline HLD Systems

In the first experiment, we tested to highlight the importance of novel techniques introduced in this system. The first novelty is the motion and ambient data filtration technique that can handle sensor signal-based errors, biasness, and drift. The second novelty is ambient and motion descriptor extraction through a graph-based approach that helps extract robust descriptors related to the data type. The comparative results for the proposed IoHT-based HLD system with the first novelty, second novelty, and both together are given in Table 1, along with a comparison of the same system classification through the CNN [90] and LSTM [91].

We used the scikit-learn library to train all three classifiers. We set the learning rate for the CNN to 0.001, and the maximum epoch number was 200. The input layer contained the descriptors extracted. Then, we proposed three convolution layers with the ReLU activation function. Next, the pooling layer was utilized after each convolution layer. A flattened layer was also used to flatten the shape of the layers. Further, a fully connected layer with two hidden layers and a softmax layer were also used to test the trained data through output. For the LSTM, we used the architecture proposed in [92], where an input layer, a few LSTM-based temporal models, a flattened layer, and a fully connected network were used to recognize the ADL. Table 2 shows the confidence levels of extracted skeleton body-points compared to the ground truth values over HWU-USP and LARa datasets.

**Table 2.** Confidence levels for skeleton body points over HWU-USP and LARa datasets.

| Skeleton Body Points | Confidence Levels for HWU-USP | Confidence Levels for LARa |
|---|---|---|
| Head | 0.95 | 0.94 |
| Shoulders | 0.92 | 0.90 |
| Elbows | 0.88 | 0.89 |
| Wrists | 0.91 | 0.90 |
| Torso | 0.85 | 0.88 |
| Knees | 0.89 | 0.92 |
| Ankles | 0.95 | 0.94 |
| **Mean Confidence** | **0.90** | **0.91** |

## 4.4. Experiment 3: Comparison with Other Works Utilizing Filtration and Descriptors

This section will focus on comparing the two novelties with the existing techniques by comparing them with the proposed HLD system. Figure 16 compares the accuracies of the proposed HLD mechanism and other existing techniques [93–95] that also used data filtration along with feature extraction. In [93], the authors utilized a combination of IMU, mechanomyography, and electromyography sensors and filtered them using median, band-pass, and moving-average filters to remove noise. Next, they made windows of 5 s each from the data and applied different techniques for the feature extraction, including peak-to-peak, abrupt changes, skewness, and mean frequency. Further, to reduce the features' vector dimension, they propose a multi-layer sequential forward selection method followed by classification via the random forest.

Haresamudram et al. present a self-supervised technique called masked reconstruction for HAR in [94]. They used small-labeled datasets and filtered data using transformer encoders. Then, they trained the network using different features and transfer learning mechanisms. In [95], a similar method to filter the data from motion, ambient, and vision-based sensors is proposed. The authors extracted features such as dynamic time warping, hidden Markov random fields, Mel-frequency cepstral coefficients, a gray-level co-variance matrix, and geodesic distance. Further, these features were optimized using a genetic algorithm and the system-recognized activities via a hidden Markov model-based classifier. As can be observed in Figure 16, the proposed HLD system with two novelties outperformed the existing works in terms of accuracy, sensitivity, and specificity.

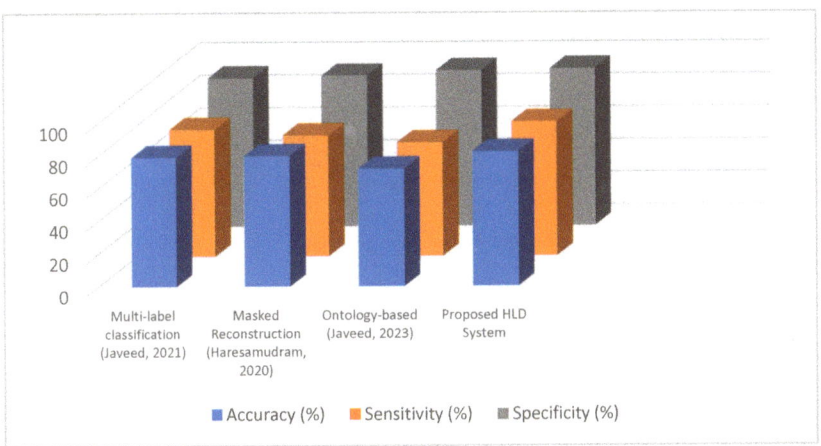

**Figure 16.** Comparison of previous works [93–95] with proposed HLD systems over the two novelties proposed.

*4.5. Experiment 4: Comparisons with Existing Works*

This section gives a comparison of our proposed IoHT-based HLD method with other previous state-of-the-art systems. We compared the proposed HLD system with methodologies that have hand-crafted descriptor extraction techniques, multiple datasets, machine learning, and applied deep learning techniques. Table 3 summarizes the comparison of the proposed system with other systems based on the classifiers, descriptor domain, modality, and accuracy achieved.

**Table 3.** Comparative analysis of proposed IoHT-based HLD system in terms of accuracy with existing work in the literature.

| Ref. | Classifier | Descriptor Domain | Modality | Accuracy |
|---|---|---|---|---|
| [96] | Random Forest | Time-based | Multiple | 81.00 |
| [97] | CNN-LSTM | Deep-learning-based | Multiple | 75.00 |
| [98] | HMM | Machine learning | Single | 78.33 |
| [99] | Multi-Layer Perceptron | Frequency and time | Single | 74.20 |
| [100] | Multi-Layer Perceptron | Entropy | Multiple | 75.50 |
| [101] | Markov Chain | Multi-features | Multiple | 74.94 |
| [102] | Recurrent Neural Network | Convolutional | Multiple | 82.00 |
| [103] | Recurrent Neural Network | Raw | Single | 80.43 |
| **Proposed** | **Recurrent Neural Network** | **Energy, Graph, Frequency, and Time** | **Multiple** | **82.36** |

The comparison between multiple human activity recognition models is explained in the table. It focuses on the classifiers used to recognize these activities. The descriptors extracted for classification are also presented. Different models acquired either single- or multiple-sensor-based raw data. Single-sensor-based means that the data were acquired from one sensor type. In contrast, multimodal-sensor-based means that the data were gathered from multiple sensor types. The accuracies of each system compared are given in the table.

## 5. Discussion

Although human locomotion decoding was achieved successfully using the proposed IoHT-based HLD system, this study also has a few limitations. The skeleton body points extracted can be obstructed in different human postures, which can cause limitations for accurate locomotion decoding. A couple of examples are highlighted in Figure 17 using red ellipses. The proposed filtration technique and descriptor extraction methodologies have to be assessed using some systems and datasets to verify the results. There is still a need to test this novel HLD system over different settings and datasets to validate the outcomes.

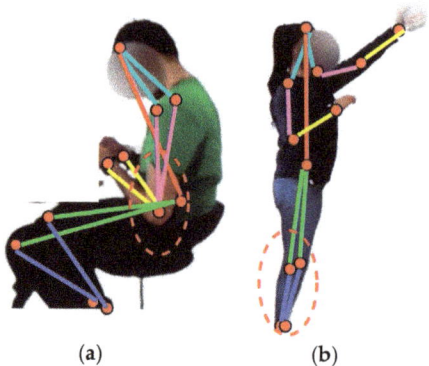

**Figure 17.** Samples of obstruction caused by human postures in activities over HWU-USP dataset: (**a**) using a phone and (**b**) taking out a bowl.

## 6. Conclusions

This article proposes a deep-learning-based human-locomotion-decoding system via novel filtration techniques and two innovative descriptor extraction mechanisms. The study compared two novelties of the proposed system using an RNN, a CNN, and LSTM. The RNN outperformed the other two deep learners concerning the accuracy of the IoHT-based HLD system. We have also shown that all the compared classifiers performed acceptably over the HWU-USP and LARa datasets. By comparing the three classifiers and other previous state-of-the-art methodologies, we conclude that the proposed IoHT-based HLD architecture enhances the accuracy rates for human locomotion decoding. Therefore, the proposed system has many applications in human activity decoding and can be scaled for more practical solutions in smart homes, ambient assisted living, and care-based facilities. In the future, we can compare and improve the results of the current study using different settings, datasets, and deep learning techniques.

**Author Contributions:** Conceptualization: M.J., M.A. and A.A.; methodology: M.A. and M.J.; software: M.A. and A.J.; validation: M.J. and M.A.; formal analysis: A.J. and A.A.; resources: M.J., A.A., M.A. and A.J.; writing—review and editing: M.J., M.A. and A.J.; funding acquisition: M.J., M.A., A.A. and A.J. All authors have read and agreed to the published version of the manuscript.

**Funding:** This research was supported by Princess Nourah bint Abdulrahman University Researchers Supporting Project Number (PNURSP2023R97), Princess Nourah bint Abdulrahman University, Riyadh, Saudi Arabia.

**Data Availability Statement:** Data are contained within the article.

**Acknowledgments:** Princess Nourah bint Abdulrahman University Researchers Supporting Project Number (PNURSP2023R97), Princess Nourah bint Abdulrahman University, Riyadh, Saudi Arabia.

**Conflicts of Interest:** The authors declare no conflict of interest.

## References

1. Ramanujam, E.; Perumal, T.; Padmvavathi, S. Human Activity Recognition with Smartphone and Wearable Sensors Using Deep Learning Techniques: A Review. *IEEE Sens. J.* **2021**, *21*, 13029–13040. [CrossRef]
2. Ouyed, O.; Allili, M.S. Group-of-features relevance in multinomial kernel logistic regression and application to human interaction recognition. *Expert Syst. Appl.* **2020**, *148*, 113247. [CrossRef]
3. Abid Hasan, S.M.; Ko, K. Depth edge detection by image-based smoothing and morphological operations. *J. Comput. Des. Eng.* **2016**, *3*, 191–197. [CrossRef]

4. Batool, M.; Jalal, A.; Kim, K. Telemonitoring of daily activity using Accelerometer and Gyroscope in smart home environments. *J. Electr. Eng. Technol.* **2020**, *15*, 2801–2809. [CrossRef]
5. Javeed, M.; Mudawi, N.A.; Alabduallah, B.I.; Jalal, A.; Kim, W. A Multimodal IoT-Based Locomotion Classification System Using Features Engineering and Recursive Neural Network. *Sensors* **2023**, *23*, 4716. [CrossRef] [PubMed]
6. Shen, X.; Du, S.-C.; Sun, Y.-N.; Sun, P.Z.H.; Law, R.; Wu, E.Q. Advance Scheduling for Chronic Care Under Online or Offline Revisit Uncertainty. *IEEE Trans. Autom. Sci. Eng.* **2023**, 1–14. [CrossRef]
7. Wang, N.; Chen, J.; Chen, W.; Shi, Z.; Yang, H.; Liu, P.; Wei, X.; Dong, X.; Wang, C.; Mao, L.; et al. The effectiveness of case management for cancer patients: An umbrella review. *BMC Health Serv. Res.* **2022**, *22*, 1247. [CrossRef]
8. Hu, S.; Chen, W.; Hu, H.; Huang, W.; Chen, J.; Hu, J. Coaching to develop leadership for healthcare managers: A mixed-method systematic review protocol. *Syst. Rev.* **2022**, *11*, 67. [CrossRef]
9. Azmat, U.; Ahmad, J. Smartphone inertial sensors for human locomotion activity recognition based on template matching and codebook generation. In Proceedings of the IEEE International Conference on Communication Technologies, Rawalpindi, Pakistan, 21–22 September 2021.
10. Lv, Z.; Chen, D.; Feng, H.; Zhu, H.; Lv, H. Digital Twins in Unmanned Aerial Vehicles for Rapid Medical Resource Delivery in Epidemics. *IEEE Trans. Intell. Transp. Syst.* **2022**, *23*, 25106–25114. [CrossRef] [PubMed]
11. İnce, Ö.F.; Ince, I.F.; Yıldırım, M.E.; Park, J.S.; Song, J.K.; Yoon, B.W. Human activity recognition with analysis of angles between skeletal joints using a RGB-depth sensor. *ETRI J.* **2020**, *42*, 78–89. [CrossRef]
12. Cheng, B.; Zhu, D.; Zhao, S.; Chen, J. Situation-Aware IoT Service Coordination Using the Event-Driven SOA Paradigm. *IEEE Trans. Netw. Serv. Manag.* **2016**, *13*, 349–361. [CrossRef]
13. Sun, Y.; Xu, C.; Li, G.; Xu, W.; Kong, J.; Jiang, D.; Tao, B.; Chen, D. Intelligent human computer interaction based on non-redundant EMG signal. *Alex. Eng. J.* **2020**, *59*, 1149–1157. [CrossRef]
14. Muneeb, M.; Rustam, H.; Ahmad, J. Automate Appliances via Gestures Recognition for Elderly Living Assistance. In Proceedings of the 2023 4th International Conference on Advancements in Computational Sciences (ICACS), Lahore, Pakistan, 20–22 February 2023; pp. 1–6. [CrossRef]
15. Nguyen, N.; Bui, D.; Tran, X. A novel hardware architecture for human detection using HOG-SVM co-optimization. In Proceedings of the APCCAS, Bangkok, Thailand, 11–14 November 2019. [CrossRef]
16. Nadeem, A.; Jalal, A.; Kim, K. Automatic human posture estimation for sport activity recognition with robust body parts detection and entropy markov model. *Multimed. Tools Appl.* **2021**, *80*, 21465–21498. [CrossRef]
17. Zank, M.; Nescher, T.; Kunz, A. Tracking human locomotion by relative positional feet tracking. In Proceedings of the IEEE Virtual Reality (VR), Arles, France, 23–27 March 2015. [CrossRef]
18. Jalal, A.; Mahmood, M. Students' behavior mining in e-learning environment using cognitive processes with information technologies. *Educ. Inf. Technol.* **2019**, *24*, 2797–2821. [CrossRef]
19. Batool, M.; Jalal, A.; Kim, K. Sensors Technologies for Human Activity Analysis Based on SVM Optimized by PSO Algorithm. In Proceedings of the 2019 International Conference on Applied and Engineering Mathematics (ICAEM), Taxila, Pakistan, 27–29 August 2019; pp. 145–150. [CrossRef]
20. Prati, A.; Shan, C.; Wang, K.I.-K. Sensors, vision and networks: From video surveillance to activity recognition and health monitoring. *J. Ambient Intell. Smart Environ.* **2019**, *11*, 5–22. [CrossRef]
21. Wang, Y.; Xu, N.; Liu, A.-A.; Li, W.; Zhang, Y. High-Order Interaction Learning for Image Captioning. *IEEE Trans. Circuits Syst. Video Technol.* **2022**, *32*, 4417–4430. [CrossRef]
22. Zhang, C.; Xiao, P.; Zhao, Z.-T.; Liu, Z.; Yu, H.; Hu, X.-Y.; Chu, H.-B.; Xu, J.-J.; Liu, M.-Y.; Zou, Q.; et al. A Wearable Localized Surface Plasmons Antenna Sensor for Communication and Sweat Sensing. *IEEE Sens. J.* **2023**, *23*, 11591–11599. [CrossRef]
23. Lin, Q.; Xiongbo, G.; Zhang, W.; Cai, L.; Yang, R.; Chen, H.; Cai, K. A Novel Approach of Surface Texture Mapping for Cone-beam Computed Tomography in Image-guided Surgical Navigation. *IEEE J. Biomed. Health Inform.* **2023**, 1–10. [CrossRef]
24. Hu, Z.; Ren, L.; Wei, G.; Qian, Z.; Liang, W.; Chen, W.; Lu, X.; Ren, L.; Wang, K. Energy Flow and Functional Behavior of Individual Muscles at Different Speeds During Human Walking. *IEEE Trans. Neural Syst. Rehabil. Eng.* **2023**, *31*, 294–303. [CrossRef]
25. Zhang, R.; Li, L.; Zhang, Q.; Zhang, J.; Xu, L.; Zhang, B.; Wang, B. Differential Feature Awareness Network within Antagonistic Learning for Infrared-Visible Object Detection. *IEEE Trans. Circuits Syst. Video Technol.* **2023**. [CrossRef]
26. Mahmood, M.; Ahmad, J.; Kim, K. WHITE STAG model: Wise human interaction tracking and estimation (WHITE) using spatio-temporal and angular-geometric (STAG) descriptors. *Multimed. Tools Appl.* **2020**, *79*, 6919–6950. [CrossRef]
27. Zheng, M.; Zhi, K.; Zeng, J.; Tian, C.; You, L. A hybrid CNN for image denoising. *J. Artif. Intell. Technol.* **2022**, *2*, 93–99. [CrossRef]
28. Gao, Z.; Pan, X.; Shao, J.; Jiang, X.; Su, Z.; Jin, K.; Ye, J. Automatic interpretation and clinical evaluation for fundus fluorescein angiography images of diabetic retinopathy patients by deep learning. *Br. J. Ophthalmol.* **2022**, *107*, 1852–1858. [CrossRef]
29. Wang, Y.; Qi, F.; Wipf, D.P.; Cai, C.; Yu, T.; Li, Y.; Zhang, Y.; Yu, Z.; Wu, W. Sparse Bayesian Learning for End-to-End EEG Decoding. *IEEE Trans. Pattern Anal. Mach. Intell.* **2023**, *45*, 15632–15649. [CrossRef] [PubMed]
30. Lu, S.; Liu, S.; Hou, P.; Yang, B.; Liu, M.; Yin, L.; Zheng, W. Soft Tissue Feature Tracking Based on Deep Matching Network. *Comput. Model. Eng. Sci.* **2023**, *136*, 363–379. [CrossRef]
31. Sreenu, G.; Saleem Durai, M.A. Intelligent video surveillance: A review through deep learning techniques for crowd analysis. *J. Big Data* **2019**, *6*, 48. [CrossRef]

32. Xu, H.; Pan, Y.; Li, J.; Nie, L.; Xu, X. Activity recognition method for home-based elderly care service based on random forest and activity similarity. *IEEE Access* **2019**, *7*, 16217–16225. [CrossRef]
33. Beddiar, D.R.; Nini, B.; Sabokrou, M.; Hadid, A. Vision-based human activity recognition: A survey. *Multimed. Tools Appl.* **2020**, *79*, 30509–30555. [CrossRef]
34. Hu, X.; Kuang, Q.; Cai, Q.; Xue, Y.; Zhou, W.; Li, Y. A Coherent Pattern Mining Algorithm Based on All Contiguous Column Bicluster. *J. Artif. Intell. Technol.* **2022**, *2*, 80–92. [CrossRef]
35. Quaid, M.A.K.; Ahmad, J. Wearable sensors based human behavioral pattern recognition using statistical features and reweighted genetic algorithm. *Multimed. Tools Appl.* **2020**, *79*, 6061–6083. [CrossRef]
36. Ahmad, F. Deep image retrieval using artificial neural network interpolation and indexing based on similarity measurement. *CAAI Trans. Intell. Technol.* **2022**, *7*, 200–218. [CrossRef]
37. Zhang, J.; Ye, G.; Tu, Z.; Qin, Y.; Qin, Q.; Zhang, J.; Liu, J. A spatial attentive and temporal dilated (SATD) GCN for skeleton-based action recognition. *CAAI Trans. Intell. Technol.* **2022**, *7*, 46–55. [CrossRef]
38. Lu, S.; Yang, J.; Yang, B.; Yin, Z.; Liu, M.; Yin, L.; Zheng, W. Analysis and Design of Surgical Instrument Localization Algorithm. *Comput. Model. Eng. Sci.* **2023**, *137*, 669–685. [CrossRef]
39. Zhang, J.; Zhu, C.; Zheng, L.; Xu, K. ROSEFusion: Random optimization for online dense reconstruction under fast camera motion. *ACM Trans. Graph.* **2021**, *40*, 1–17. [CrossRef]
40. Meng, J.; Li, Y.; Liang, H.; Ma, Y. Single-image Dehazing based on two-stream convolutional neural network. *J. Artif. Intell. Technol.* **2022**, *2*, 100–110. [CrossRef]
41. Ma, K.; Li, Z.; Liu, P.; Yang, J.; Geng, Y.; Yang, B.; Guan, X. Reliability-Constrained Throughput Optimization of Industrial Wireless Sensor Networks With Energy Harvesting Relay. *IEEE Internet Things J.* **2021**, *8*, 13343–13354. [CrossRef]
42. Zhuang, Y.; Jiang, N.; Xu, Y.; Xiangjie, K.; Kong, X. Progressive Distributed and Parallel Similarity Retrieval of Large CT Image Sequences in Mobile Telemedicine Networks. *Wirel. Commun. Mob. Comput.* **2022**, *2022*, 6458350. [CrossRef]
43. Miao, Y.; Wang, X.; Wang, S.; Li, R. Adaptive Switching Control Based on Dynamic Zero-Moment Point for Versatile Hip Exoskeleton Under Hybrid Locomotion. *IEEE Trans. Ind. Electron.* **2023**, *70*, 11443–11452. [CrossRef]
44. He, B.; Lu, Q.; Lang, J.; Yu, H.; Peng, C.; Bing, P.; Li, S.; Zhou, Q.; Liang, Y.; Tian, G. A New Method for CTC Images Recognition Based on Machine Learning. *Front. Bioeng. Biotechnol.* **2020**, *8*, 897. [CrossRef]
45. Li, Z.; Kong, Y.; Jiang, C. A Transfer Double Deep Q Network Based DDoS Detection Method for Internet of Vehicles. *IEEE Trans. Veh. Technol.* **2023**, *72*, 5317–5331. [CrossRef]
46. Hassan, F.S.; Gutub, A. Improving data hiding within colour images using hue component of HSV colour space. *CAAI Trans. Intell. Technol.* **2022**, *7*, 56–68. [CrossRef]
47. Zheng, W.; Xun, Y.; Wu, X.; Deng, Z.; Chen, X.; Sui, Y. A Comparative Study of Class Rebalancing Methods for Security Bug Report Classification. *IEEE Trans. Reliab.* **2021**, *70*, 1658–1670. [CrossRef]
48. Zheng, C.; An, Y.; Wang, Z.; Wu, H.; Qin, X.; Eynard, B.; Zhang, Y. Hybrid offline programming method for robotic welding systems. *Robot. Comput.-Integr. Manuf.* **2022**, *73*, 102238. [CrossRef]
49. Zhang, X.; Wang, Y.; Yang, M.; Geng, G. Toward Concurrent Video Multicast Orchestration for Caching-Assisted Mobile Networks. *IEEE Trans. Veh. Technol.* **2021**, *70*, 13205–13220. [CrossRef]
50. Qi, M.; Cui, S.; Chang, X.; Xu, Y.; Meng, H.; Wang, Y.; Yin, T. Multi-region Nonuniform Brightness Correction Algorithm Based on L-Channel Gamma Transform. *Secur. Commun. Netw.* **2022**, *2022*, 2675950. [CrossRef]
51. Zhao, W.; Lun, R.; Espy, D.D.; Reinthal, M.A. Rule based real time motion assessment for rehabilitation exercises. In Proceedings of the IEEE Symposium Computational Intelligence in Healthcare and E-Health, Orlando, FL, USA, 9–12 December 2014. [CrossRef]
52. Hao, S.; Jiali, P.; Xiaomin, Z.; Xiaoqin, W.; Lina, L.; Xin, Q.; Qin, L. Group identity modulates bidding behavior in repeated lottery contest: Neural signatures from event-related potentials and electroencephalography oscillations. *Front. Neurosci.* **2023**, *17*, 1184601. [CrossRef]
53. Barnachon, M.; Bouakaz, S.; Boufama, B.; Guillou, E. Ongoing human action recognition with motion capture. *Pattern Recognit.* **2014**, *47*, 238–247. [CrossRef]
54. Lu, S.; Yang, B.; Xiao, Y.; Liu, S.; Liu, M.; Yin, L.; Zheng, W. Iterative reconstruction of low-dose CT based on differential sparse. *Biomed. Signal Process. Control* **2023**, *79*, 104204. [CrossRef]
55. Ordóñez, F.; Roggen, D. Deep Convolutional and LSTM Recurrent Neural Networks for Multimodal Wearable Activity Recognition. *Sensors* **2016**, *16*, 115. [CrossRef]
56. Franco, A.; Magnani, A.; Maio, D. A multimodal approach for human activity recognition based on skeleton and RGB data. *Pattern Recognit. Lett.* **2020**, *131*, 293–299. [CrossRef]
57. Nweke, H.F.; Teh, Y.W.; Mujtaba, G.; Alo, U.R.; Al-Garadi, M.A. Multi-sensor fusion based on multiple classifier systems for human activity identification. *Hum. Cent. Comput. Inf. Sci.* **2019**, *9*, 34. [CrossRef]
58. Ehatisham-Ul-Haq, M.; Javed, A.; Azam, M.A.; Malik, H.M.A.; Irtaza, A.; Lee, I.H.; Mahmood, M.T. Robust Human Activity Recognition Using Multimodal Feature-Level Fusion. *IEEE Access* **2019**, *7*, 60736–60751. [CrossRef]
59. Chung, S.; Lim, J.; Noh, K.J.; Kim, G.; Jeong, H. Sensor Data Acquisition and Multimodal Sensor Fusion for Human Activity Recognition Using Deep Learning. *Sensors* **2019**, *19*, 1716. [CrossRef] [PubMed]
60. Zhao, Y.; Guo, S.; Chen, Z.; Shen, Q.; Meng, Z.; Xu, H. Marfusion: An Attention-Based Multimodal Fusion Model for Human Activity Recognition in Real-World Scenarios. *Appl. Sci.* **2022**, *12*, 5408. [CrossRef]

61. Niemann, F.; Reining, C.; Rueda, F.M.; Nair, N.R.; Steffens, J.A.; Fink, G.A.; Hompel, M.T. LARa: Creating a Dataset for Human Activity Recognition in Logistics Using Semantic Attributes. *Sensors* **2020**, *20*, 4083. [CrossRef] [PubMed]
62. Ranieri, C.M.; MacLeod, S.; Dragone, M.; Vargas, P.A.; Romero, R.A.F. Activity Recognition for Ambient Assisted Living with Videos, Inertial Units and Ambient Sensors. *Sensors* **2021**, *21*, 768. [CrossRef]
63. Bersch, S.D.; Azzi, D.; Khusainov, R.; Achumba, I.E.; Ries, J. Sensor data acquisition and processing parameters for human activity classification. *Sensors* **2014**, *14*, 4239–4270. [CrossRef]
64. Huang, H.; Liu, L.; Wang, J.; Zhou, Y.; Hu, H.; Ye, X.; Liu, G.; Xu, Z.; Xu, H.; Yang, W.; et al. Aggregation caused quenching to aggregation induced emission transformation: A precise tuning based on BN-doped polycyclic aromatic hydrocarbons toward subcellular organelle specific imaging. *Chem. Sci.* **2022**, *13*, 3129–3139. [CrossRef]
65. Schrader, L.; Vargas Toro, A.; Konietzny, S.; Rüping, S.; Schäpers, B.; Steinböck, M.; Krewer, C.; Müller, F.; Güttler, J.; Bock, T. Advanced sensing and human activity recognition in early intervention and rehabilitation of elderly people. *Popul. Ageing* **2020**, *13*, 139–165. [CrossRef]
66. Lee, M.; Kim, S.B. Sensor-Based Open-Set Human Activity Recognition Using Representation Learning with Mixup Triplets. *IEEE Access* **2022**, *10*, 119333–119344. [CrossRef]
67. Patro, S.G.K.; Mishra, B.K.; Panda, S.K.; Kumar, R.; Long, H.V.; Taniar, D.; Priyadarshini, I. A Hybrid Action-Related K-Nearest Neighbour (HAR-KNN) Approach for Recommendation Systems. *IEEE Access* **2020**, *8*, 90978–90991. [CrossRef]
68. Li, J.; Tian, L.; Wang, H.; An, Y.; Wang, K.; Yu, L. Segmentation and recognition of basic and transitional activities for continuous physical human activity. *IEEE Access* **2019**, *7*, 42565–42576. [CrossRef]
69. Chen, D.; Wang, Q.; Li, Y.; Li, Y.; Zhou, H.; Fan, Y. A general linear free energy relationship for predicting partition coefficients of neutral organic compounds. *Chemosphere* **2020**, *247*, 125869. [CrossRef] [PubMed]
70. Hou, X.; Zhang, L.; Su, Y.; Gao, G.; Liu, Y.; Na, Z.; Xu, Q.; Ding, T.; Xiao, L.; Li, L.; et al. A space crawling robotic bio-paw (SCRBP) enabled by triboelectric sensors for surface identification. *Nano Energy* **2023**, *105*, 108013. [CrossRef]
71. Hou, X.; Xin, L.; Fu, Y.; Na, Z.; Gao, G.; Liu, Y.; Xu, Q.; Zhao, P.; Yan, G.; Su, Y.; et al. A self-powered biomimetic mouse whisker sensor (BMWS) aiming at terrestrial and space objects perception. *Nano Energy* **2023**, *118*, 109034. [CrossRef]
72. Mi, W.; Xia, Y.; Bian, Y. Meta-analysis of the association between aldose reductase gene (CA)n microsatellite variants and risk of diabetic retinopathy. *Exp. Ther. Med.* **2019**, *18*, 4499–4509. [CrossRef] [PubMed]
73. Ye, X.; Wang, J.; Qiu, W.; Chen, Y.; Shen, L. Excessive gliosis after vitrectomy for the highly myopic macular hole: A Spectral Domain Optical Coherence Tomography Study. *Retina* **2023**, *43*, 200–208. [CrossRef] [PubMed]
74. Chen, C.; Liu, S. Detection and Segmentation of Occluded Vehicles Based on Skeleton Features. In Proceedings of the 2012 Second International Conference on Instrumentation, Measurement, Computer, Communication and Control, Harbin, China, 8–10 December 2012; pp. 1055–1059. [CrossRef]
75. Chen, C.; Jafari, R.; Kehtarnavaz, N. A survey of depth and inertial sensor fusion for human action recognition. *Multimed. Tools Appl.* **2017**, *76*, 4405–4425. [CrossRef]
76. Amir, N.; Ahmad, J.; Kibum, K. Human Actions Tracking and Recognition Based on Body Parts Detection via Artificial Neural Network. In Proceedings of the 2020 3rd International Conference on Advancements in Computational Sciences (ICACS), Lahore, Pakistan, 17–19 February 2020.
77. Zhou, B.; Wang, C.; Huan, Z.; Li, Z.; Chen, Y.; Gao, G.; Li, H.; Dong, C.; Liang, J. A Novel Segmentation Scheme with Multi-Probability Threshold for Human Activity Recognition Using Wearable Sensors. *Sensors* **2022**, *22*, 7446. [CrossRef]
78. Yao, Q.-Y.; Fu, M.-L.; Zhao, Q.; Zheng, X.-M.; Tang, K.; Cao, L.-M. Image-based visualization of stents in mechanical thrombectomy for acute ischemic stroke: Preliminary findings from a series of cases. *World J. Clin. Cases* **2023**, *11*, 5047–5055. [CrossRef]
79. Su, W.; Ni, J.; Hu, X.; Fridrich, J. Image Steganography With Symmetric Embedding Using Gaussian Markov Random Field Model. *IEEE Trans. Circuits Syst. Video Technol.* **2021**, *31*, 1001–1015. [CrossRef]
80. Li, X.; Zhao, H.; Yu, L.; Chen, H.; Deng, W.; Deng, W. Feature Extraction Using Parameterized Multisynchrosqueezing Transform. *IEEE Sens. J.* **2022**, *22*, 14263–14272. [CrossRef]
81. Jin, K.; Gao, Z.; Jiang, X.; Wang, Y.; Ma, X.; Li, Y.; Ye, J. MSHF: A Multi-Source Heterogeneous Fundus (MSHF) Dataset for Image Quality Assessment. *Sci. Data* **2023**, *10*, 286. [CrossRef] [PubMed]
82. Amir, N.; Ahmad, J.; Kim, K. Accurate Physical Activity Recognition using Multidimensional Features and Markov Model for Smart Health Fitness. *Symmetry* **2020**, *12*, 1766.
83. Kanan, C.; Cottrell, G. Robust Classification of Objects, Faces, and Flowers Using Natural Image Statistics. In Proceedings of the 2010 IEEE Computer Society Conference on Computer Vision and Pattern Recognition, San Francisco, CA, USA, 13–18 June 2010; pp. 2472–2479. [CrossRef]
84. Arbain, N.A.; Azmi, M.S.; Muda, A.K.D.; Radzid, A.R.; Tahir, A. A Review of Triangle Geometry Features in Object Recognition. In Proceedings of the 2019 9th Symposium on Computer Applications & Industrial Electronics (ISCAIE), Kota Kinabalu, Malaysia, 27–28 April 2019; pp. 254–258. [CrossRef]
85. Fausto, F.; Cuevas, E.; Gonzales, A. A New Descriptor for Image Matching Based on Bionic Principles. *Pattern Anal. Appl.* **2017**, *20*, 1245–1259. [CrossRef]
86. Yu, Y.; Yang, J.P.; Shiu, C.-S.; Simoni, J.M.; Xiao, S.; Chen, W.-T.; Rao, D.; Wang, M. Psychometric testing of the Chinese version of the Medical Outcomes Study Social Support Survey among people living with HIV/AIDS in China. *Appl. Nurs. Res.* **2015**, *28*, 328–333. [CrossRef] [PubMed]

87. Ali, H.H.; Moftah, H.M.; Youssif, A.A.A. Depth-based human activity recognition: A comparative perspective study on feature extraction. *Future Comput. Inform. J.* **2018**, *3*, 51–67. [CrossRef]
88. Nguyen, H.-C.; Nguyen, T.-H.; Scherer, R.; Le, V.-H. Deep Learning for Human Activity Recognition on 3D Human Skeleton: Survey and Comparative Study. *Sensors* **2023**, *23*, 5121. [CrossRef]
89. Singh, S.P.; Sharma, M.K.; Lay-Ekuakille, A.; Gangwar, D.; Gupta, S. Deep ConvLSTM With Self-Attention for Human Activity Decoding Using Wearable Sensors. *IEEE Sens. J.* **2021**, *21*, 8575–8582. [CrossRef]
90. Farag, M.M. Matched Filter Interpretation of CNN Classifiers with Application to HAR. *Sensors* **2022**, *22*, 8060. [CrossRef] [PubMed]
91. Husni, N.L.; Sari, P.A.R.; Handayani, A.S.; Dewi, T.; Seno, S.A.H.; Caesarendra, W.; Glowacz, A.; Oprzędkiewicz, K.; Sułowicz, M. Real-Time Littering Activity Monitoring Based on Image Classification Method. *Smart Cities* **2021**, *4*, 1496–1518. [CrossRef]
92. Khatun, M.A.; Abu Yousuf, M.; Ahmed, S.; Uddin, Z.; Alyami, S.A.; Al-Ashhab, S.; Akhdar, H.F.; Khan, A.; Azad, A.; Moni, M.A. Deep CNN-LSTM With Self-Attention Model for Human Activity Recognition Using Wearable Sensor. *IEEE J. Transl. Eng. Health Med.* **2022**, *10*, 2700316. [CrossRef]
93. Javeed, M.; Jalal, A.; Kim, K. Wearable Sensors based Exertion Recognition using Statistical Features and Random Forest for Physical Healthcare Monitoring. In Proceedings of the 2021 International Bhurban Conference on Applied Sciences and Technologies (IBCAST), Islamabad, Pakistan, 12–16 January 2021; pp. 512–517. [CrossRef]
94. Haresamudram, H.; Beedu, A.; Agrawal, V.; Grady, P.L.; Essa, I. Masked Reconstruction Based Self-Supervision for Human Activity Recognition. In Proceedings of the 24th annual International Symposium on Wearable Computers, Cancun, Mexico, 12–16 September 2020.
95. Javeed, M.; Mudawi, N.A.; Alazeb, A.; Alotaibi, S.S.; Almujally, N.A.; Jalal, A. Deep Ontology-Based Human Locomotor Activity Recognition System via Multisensory Devices. *IEEE Access* **2023**, *11*, 105466–105478. [CrossRef]
96. Cosoli, G.; Antognoli, L.; Scalise, L. Wearable Electrocardiography for Physical Activity Monitoring: Definition of Validation Protocol and Automatic Classification. *Biosensors* **2023**, *13*, 154. [CrossRef]
97. Ehatisham-ul-Haq, M.; Murtaza, F.; Azam, M.A.; Amin, Y. Daily Living Activity Recognition In-The-Wild: Modeling and Inferring Activity-Aware Human Contexts. *Electronics* **2022**, *11*, 226. [CrossRef]
98. Jalal, A.; Kamal, S.; Kim, D. A depth video sensor-based life-logging human activity recognition system for elderly care in smart indoor environments. *Sensors* **2014**, *14*, 11735–11759.
99. Aşuroğlu, T. Complex Human Activity Recognition Using a Local Weighted Approach. *IEEE Access* **2022**, *10*, 101207–101219. [CrossRef]
100. Azmat, U.; Ahmad, J.; Madiha, J. Multi-sensors Fused IoT-based Home Surveillance via Bag of Visual and Motion Features. In Proceedings of the 2023 International Conference on Communication, Computing and Digital Systems (C-CODE), Islamabad, Pakistan, 17–18 May 2023; pp. 1–6. [CrossRef]
101. Ahmad, J.; Kim, Y.-H.; Kim, Y.-J.; Kamal, S.; Kim, D. Robust human activity recognition from depth video using spatiotemporal multi-fused features. *Pattern Recognit.* **2017**, *61*, 295–308.
102. Boukhechba, M.; Cai, L.; Wu, C.; Barnes, L.E. ActiPPG: Using deep neural networks for activity recognition from wrist-worn photoplethysmography (PPG) sensors. *Smart Health* **2019**, *14*, 100082. [CrossRef]
103. Sánchez-Caballero, A.; Fuentes-Jiménez, D.; Losada-Gutiérrez, C. Real-time human action recognition using raw depth video-based recurrent neural networks. *Multimed. Tools Appl.* **2023**, *82*, 16213–16235. [CrossRef]

**Disclaimer/Publisher's Note:** The statements, opinions and data contained in all publications are solely those of the individual author(s) and contributor(s) and not of MDPI and/or the editor(s). MDPI and/or the editor(s) disclaim responsibility for any injury to people or property resulting from any ideas, methods, instructions or products referred to in the content.

*Article*

# HRBUST-LLPED: A Benchmark Dataset for Wearable Low-Light Pedestrian Detection

Tianlin Li, Guanglu Sun *, Linsen Yu and Kai Zhou

School of Computer Science and Technology, Harbin University of Science and Technology, No. 52 Xuefu Road, Nangang District, Harbin 150080, China; 2010400006@stu.hrbust.edu.cn (T.L.); yulinsen@hrbust.edu.cn (L.Y.); zhoukaiace@hotmail.com (K.Z.)
* Correspondence: sunguanglu@hrbust.edu.cn; Tel.: +86-451-8639-0660

**Abstract:** Detecting pedestrians in low-light conditions is challenging, especially in the context of wearable platforms. Infrared cameras have been employed to enhance detection capabilities, whereas low-light cameras capture the more intricate features of pedestrians. With this in mind, we introduce a low-light pedestrian detection (called HRBUST-LLPED) dataset by capturing pedestrian data on campus using wearable low-light cameras. Most of the data were gathered under starlight-level illumination. Our dataset annotates 32,148 pedestrian instances in 4269 keyframes. The pedestrian density reaches high values with more than seven people per image. We provide four lightweight, low-light pedestrian detection models based on advanced YOLOv5 and YOLOv8. By training the models on public datasets and fine-tuning them on the HRBUST-LLPED dataset, our model obtained 69.90% in terms of AP@0.5:0.95 and 1.6 ms for the inference time. The experiments demonstrate that our research can assist in advancing pedestrian detection research by using low-light cameras in wearable devices.

**Keywords:** low-light pedestrian detection dataset; lightweight model; wearable devices

## 1. Introduction

Over the past two decades, there has been a significant advancement in IoT and artificial intelligence technologies. As a result, researchers have turned their attention to developing intelligent wearable assistive systems that are made up of wearable cameras, sensors, computing components, and machine learning models [1]. This has led to an increase in studies aimed at assisting visually impaired individuals in various areas, such as travel [2], food [3], and screen detection [4]. Other areas of research include human-pet [5] or human-machine [6,7] interaction. Wearable devices combined with computer vision models are being used to help users observe things that are typically difficult to see. Despite the numerous studies on object detection using wearable devices, research on detecting humans in a scene is still limited, making it challenging to apply in areas such as nighttime surveillance, fire rescue, and forest inspections.

Since the maturity of convolutional neural networks in 2012, object detection algorithms have experienced vigorous development [8]. Single-stage object detection models represented by SSD [9] and YOLO [10], as well as two-stage object detection models depicted by Faster R-CNN [11] and FPN [12], have been proposed, achieving excellent results in terms of speed and accuracy. The maturity of object detection algorithms has also ushered in pedestrian detection algorithms into the era of deep learning. In order to fulfill the training data needs of machine learning and deep learning models, some usual pedestrian detection datasets have been proposed, like Caltech [13] and KITTI [14]. In recent years, datasets such as CityPersons [15], CrowdHuman [16], WIDER Pedestrian, WiderPerson [17], EuroCity [18], and TJU-Pedestrian [19] have been collected from cities, the countryside, and broader environments using vehicle-mounted cameras or surveillance cameras. These datasets enable the trained models to adapt to a broader range of scenarios. The EuroCity

**Citation:** Li, T.; Sun, G.; Yu, L.; Zhou, K. HRBUST-LLPED: A Benchmark Dataset for Wearable Low-Light Pedestrian Detection. *Micromachines* **2023**, *14*, 2164. https://doi.org/10.3390/mi14122164

Academic Editor: Viviana Mulloni

Received: 15 September 2023
Revised: 26 October 2023
Accepted: 7 November 2023
Published: 28 November 2023

**Copyright:** © 2023 by the authors. Licensee MDPI, Basel, Switzerland. This article is an open access article distributed under the terms and conditions of the Creative Commons Attribution (CC BY) license (https://creativecommons.org/licenses/by/4.0/).

and TJU-Pedestrian datasets also include pedestrian data with low illumination conditions, aiming to achieve good recognition performance in terms of pedestrian detection models in nighttime scenarios. However, conventional cameras struggle to capture clear images under low-light conditions, significantly impacting data annotation and model recognition performance, as shown in Figure 1a.

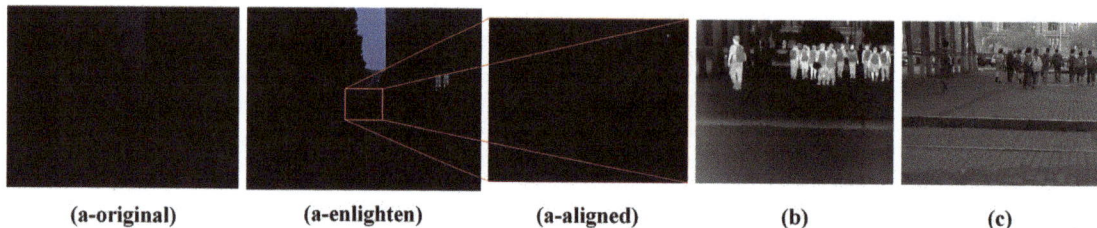

**Figure 1.** In starlight-level illumination environments, the imaging effects of visible light cameras, infrared cameras, and low-light cameras are as follows: (**a-original**) represents an image captured directly with a mobile phone; (**a-enlighten**) represents the image enhanced using the Zero DCE++ model; (**a-aligned**) represents the image in the enhanced version with the corresponding resolution. (**b**) represents an image captured with an infrared camera, and (**c**) illustrates an image captured with a low-light camera.

Humans emit heat, which can be captured using infrared cameras in colder environments to distinguish pedestrians from the background, as shown in Figure 1b. As a result, OSU [20] proposed an infrared dataset collected during the daytime, and TNO [21] also provided a dataset that combines infrared and visible light captured at night. Later, with the development of research on autonomous driving, datasets such as CVC-14 [22], KAIST [23], and FLIR were introduced, which consist of pedestrian data captured using vehicle-mounted visible light-infrared cameras for modal alignment. Subsequently, the LLVIP dataset [24] was introduced to advance research on multi-spectral fusion and pedestrian detection under low-light conditions. Although infrared images can separate individuals from the background, they have a limited imaging distance and contain less detailed textures, making it difficult to distinguish pedestrians with high overlap.

Low-light cameras with CMOS chips specially designed to capture long-wavelength light waves can achieve precise imaging under starlight-level illumination conditions, as shown in Figure 1c. By considering the helpfulness of low-light images for pedestrian detection in low-light environments, we constructed the Low-Light Pedestrian Detection (HRBUST-LLPED, collected by Harbin University of Science and Technology) dataset in this study. The dataset consists of 150 videos captured under low-light conditions, from which 4269 keyframes were extracted and annotated with 32,148 pedestrians. In order to meet the requirements of wearable devices, we developed wearable low-light pedestrian detection models based on small and nano versions of YOLOv5 and YOLOv8. When considering the fact that the information captured by low-light cameras is relatively limited compared to visible-light cameras, we first trained the models on the KITTI, KAIST, LLVIP, and TJU-Pedestrian datasets separately and then fine-tuned them using our dataset. As a result, our trained models achieved satisfactory results in speed and accuracy.

Contributions. Our contributions cover several aspects.

(1) We have expanded the focus of pedestrian detection to low-light images and have constructed a low-light pedestrian detection dataset using a low-light camera. The dataset contains denser pedestrian instances compared to existing pedestrian detection datasets.

(2) We have provided lightweight, wearable, low-light pedestrian detection models based on the YOLOv5 and YOLOv8 frameworks, considering the lower computational

power of wearable platforms when compared to GPUs. We have improved the model's performance by modifying the activation layer and loss functions.

(3) We first pretrained our models on four visible light pedestrian detection datasets and then fine-tuned them on our constructed HRBUST-LLPED dataset. We achieved a performance of 69.90% in terms of AP@0.5:0.95 and an inference time of 1.6 ms per image.

In Section 2, we discuss the research progress on pedestrian detection datasets, object detection, and object detection using wearable devices. In Section 3, we present the details of our dataset. In Section 4, we describe the training methods and underlying architecture of our wearable low-light pedestrian detection model. In Section 5, we provide the experimental setup and evaluate the performance of our model. Finally, in Section 6, we conclude our work.

## 2. Related Work

This section will review some commonly used pedestrian detection datasets and then discuss some popular object detection models. Finally, we will introduce some research on applying object detection models to wearable devices.

*2.1. Pedestrian Detection Datasets*

The first dataset for pedestrian detection is MIT [25], which was proposed in 2000 and has images at a resolution of only $128 \times 64$. In the following years, the INRIA [26], Daimler [27], and TUD-Brussels [28] datasets were successively introduced, with increased image resolutions of $640 \times 480$ and varying numbers of images, ranging from hundreds to tens of thousands. Manually designed features characterized these earlier datasets.

However, with the development of neural networks and deep learning, early pedestrian detection datasets were too small to provide enough data for model fitting. In 2010, the Caltech [13] dataset was proposed, which consisted of 11 videos. The training set included six videos, with every third frame selected, while the test set comprised five videos, with every 30th frame chosen. Subsequently, in 2012, the KITTI [14] dataset was introduced, collected from onboard vehicle dashcams. The training set consisted of 7481 images, and the test set contained 7518 images, with a resolution of $1240 \times 376$. CityPersons [15], derived from the Cityscapes [29] dataset from 2017, comprised 2975 images in the training set, 500 images in the validation set, and 1575 images in the test set, with a resolution of $2048 \times 1024$. Crowd-Human [16], introduced in 2018, annotated complete pedestrians in the images and labeled the visible parts and heads of each pedestrian. WIDER Pedestrian, a 2019 dataset, focused on vehicle and pedestrian detection and drew images primarily from roadside surveillance and onboard cameras. WiderPerson [17], proposed in 2019, concentrated on pedestrian detection in outdoor scenes rather than traffic scenarios, with the data mainly collected from web sources. The EuroCity [18] dataset, introduced in 2019, captured pedestrian data under various lighting conditions, including daytime and nighttime, from multiple cities in Europe, making it the first all-weather pedestrian detection dataset. TJU-Pedestrian [19], a pedestrian detection dataset proposed by Tianjin University in 2020, covers two main scenarios: road traffic and the campus and consists of pedestrian data from both day and night. It is currently one of the most comprehensive pedestrian detection datasets available.

All the pedestrian detection datasets mentioned above were captured using visible light cameras. However, capturing pedestrians clearly in low-light scenarios is challenging using conventional visible light cameras. In 2015, the KAIST dataset [23] introduced the first pedestrian detection dataset that includes both infrared and visible light modalities, taking into account the illumination variations between day and night that can affect the performance of automated systems. The KAIST dataset primarily provides 12 pairs of fully aligned visible light and infrared videos, with six pairs used for training and six pairs for testing. The CVC-14 dataset [22], proposed in 2016, is a dataset that simultaneously collects visible light and infrared images, primarily targeting autonomous driving scenarios. It consists of 7085 training images and 1433 test images with a resolution of $640 \times 512$. One of the drawbacks of this dataset is the lack of complete alignment between the visible light and infrared modalities. The FLIR Thermal dataset, introduced in 2018, mainly contains 9711 infrared images and 9233 RGB images for model training and validation. It aims to promote the use of FLIR's infrared cameras in autonomous driving systems. The LLVIP dataset [24], proposed in 2022, constructs a dataset that corresponds to visible light and infrared images. It is primarily used for image fusion, pedestrian detection, and modality transfer tasks.

*2.2. Object Detection*

The development of object detection models is based on the success of AlexNet and VGGNet. Early object detection models can be divided into two categories: two-stage models and one-stage models.

The pioneering work of the two-stage models is the R-CNN [30] model in 2014, which is also the first deep learning-based object detection model. The main idea of the two-stage object detection model is to select rough regions of interest that may contain objects after extracting image features through convolution and then classify specific objects from the aligned regions of interest. Subsequently, Fast R-CNN [31] and Faster R-CNN [11] improve the object selection process, reducing computational costs and enabling end-to-end training of the entire network. In order to better detect multi-scale objects, MS-CNN [32] and FPN [12] were introduced into two-stage object detection algorithms, improving their accuracy and adaptability to complex scenes.

One-stage object detection models directly divide extracted feature maps into $N \times N$ regions to predict objects for each region, eliminating the need to predict regions of interest and improving the speed of object detection models. Popular models in this category are the YOLO [10] and SSD [9], and YOLO has evolved from YOLOv1 to YOLOv8. Although YOLOv1 [10] reduces computational complexity by directly regressing the position of the bounding box, it is not effective in detecting dense and small objects. YOLOv2 [33] and YOLOv3 [34] improved on this basis and combined the PassThrough method, anchor point, and FPN to predict multi-scale targets. YOLOv4 [35] and YOLOv5 [36] further improve performance by integrating the CSP and SPP structures, adaptive anchor calculation, and Focus operations. YOLOv6 [37] and YOLOv7 [38] introduced RepVGG and ELAN modules to improve performance, respectively, whereas YOLOv8 improved the CSP structure, prediction head, and loss function to further improve calculation speed and accuracy.

*2.3. Object Detection on Wearable Devices*

Research on object detection models for wearable devices can be roughly divided into two categories based on the data source: intelligent assistive devices based on everyday life images and assistive devices based on AR platforms.

In AR-based research, Eckert et al. [39] combined HoloLens with YOLOv2 to design an assistive system that provides audio navigation for the visually impaired. Bahri et al. [40] deployed and evaluated YOLOv3 and Faster R-CNN on the HoloLens platform. Park et al. [41] used the Mask R-CNN model to achieve object segmentation in AR devices and evaluated its application in matching, inspecting, and maintaining items in intelligent manufacturing. Park et al. [6] employed RetinaNet for object detection and used AR technology as a platform to provide intelligent assistance for collaborative robots. Dimitropoulos et al. [7] utilized AR and deep learning for wearable recognition, enabling interaction between human operators and robots through voice or gestures.

In the research based on everyday life images, Han et al. [3] used Faster R-CNN for food object detection, implemented it in glasses as a wearable device, and automatically captured images of dining scenes while estimating the number of chewing motions. Kim et al. [5] utilized wearable devices to recognize and predict dog behavior and evaluated the performance of Faster R-CNN and YOLOv3/v4. Li et al. [4] employed wearable sensors and the YOLOv5 model to detect electronic screens, evaluating the duration of electronic screen usage. Arifando et al. [42] proposed a lightweight and high-precision bus detection model based on YOLOv5 to assist visually impaired individuals in boarding the correct bus and getting off at the right bus stop. Maya-Martínez et al. [43] introduced a pedestrian detection architecture based on Tiny YOLOv3, implemented in low-cost wearable devices as an alternative solution for assisting visually impaired individuals. Although these studies provide a good basis for pedestrian detection using wearable devices, they do not consider pedestrian detection at night in low light.

## 3. The HRBUST-LLEPD Dataset

This section will first provide a detailed introduction to the proposed Low-Light Pedestrian Detection (HRBUST-LLPED) dataset, including how we capture the visual data, preprocess and select the keyframes, and annotate the pedestrians. Then, we will briefly compare our constructed dataset with other commonly used datasets and analyze the advantages and disadvantages of the HRBUST-LLPED dataset.

### 3.1. Dataset Build

Data Capture: The low-light camera we used is the Iraytek PF6L, with an output resolution of $720 \times 576/8$ μm, a focal length of F25mm/F1.4, and a theoretical illuminance resolution of 0.002Lux. The camera is attached to a helmet. We wear the helmet to capture data to simulate the real perspective of humans. We mainly shoot campus scenes from winter to summer. The collection time in winter was 18:00–22:00, and in summer, it was between 20:00 and 22:00. In total, we collected 150 videos with a frame rate of 60 Hz. The length of the videos ranged from 33 s to 7 min and 45 s, with an average length of 95 s and a total of 856,183 frames.

Data Process: First of all, when considering the thermal stability and high sensitivity of CCD (CMOS) in the video acquisition process of low-light cameras, noise will inevitably be introduced into the video. Additionally, since the frame rate of the video is 60 Hz and the difference in pedestrian poses between adjacent frames is minimal, we first use a smoothing denoising technique with the neighboring two frames to enhance the current frame. Next, when considering that the pedestrian gaits in the video are usually slow, there is significant redundancy in the pedestrian poses. Therefore, we select one frame every 180 frames (i.e., 3 s per frame) as a keyframe. We then remove frames that do not contain pedestrian targets and search for clear frames within a range of ±7 frames of the blurred frames to update the keyframes. Ultimately, we obtained 4269 frames as the image data for constructing our low-light pedestrian detection dataset.

Data Annotation: We used the labelImg tool to annotate the processed image data manually. For each person present in the image with less than 90% occlusion (i.e., except for cases where only a tiny portion of the lower leg or arm is visible), we labeled them as "Pedestrian". We cross-referenced the uncertain pedestrian annotations with the original videos to avoid missing pedestrians due to visual reasons or mistakenly labeling trees as pedestrians. As a result, we obtained a total of 32,148 pedestrian labels.

### 3.2. Dataset Analysis

In this section, we will analyze the performance of our dataset and compare it to other datasets. We will evaluate the strengths and limitations of our dataset.

Analyze HRBUST-LLPED itself

Our dataset consists of 4269 images annotated with 32,148 pedestrians. We evenly divided the dataset into training and testing sets with a 5:1 image ratio. Specifically, we took five consecutive images for training and one for testing. As a result, the constructed training set contains 3558 images with annotations for 26,774 pedestrians, and the testing set contains 711 images with annotations for 5374 pedestrians. Since both the training and testing sets were uniformly selected, we primarily focus on analyzing the training set, as shown in Figure 2. Figure 2a displays the distribution of pedestrian instances in the dataset, whereas Figure 2b illustrates the sizes and quantities of the bounding boxes. Figure 2c demonstrates the relationships between the center point's x-co-ordinate, y-coordinate, width, and the height of the bounding boxes. From the figure, we observe that the pedestrian boxes are distributed throughout the entire image. The distribution of the x-co-ordinate of the center point is relatively uniform, while the y-co-ordinate is more concentrated in the middle and lower parts of the image. The width mainly ranges from 5% to 20% of the image, and the height mainly ranges from 10% to 40% of the image. These characteristics align with the nature of pedestrian data, indicating that the dataset covers a wide range of pedestrian positions.

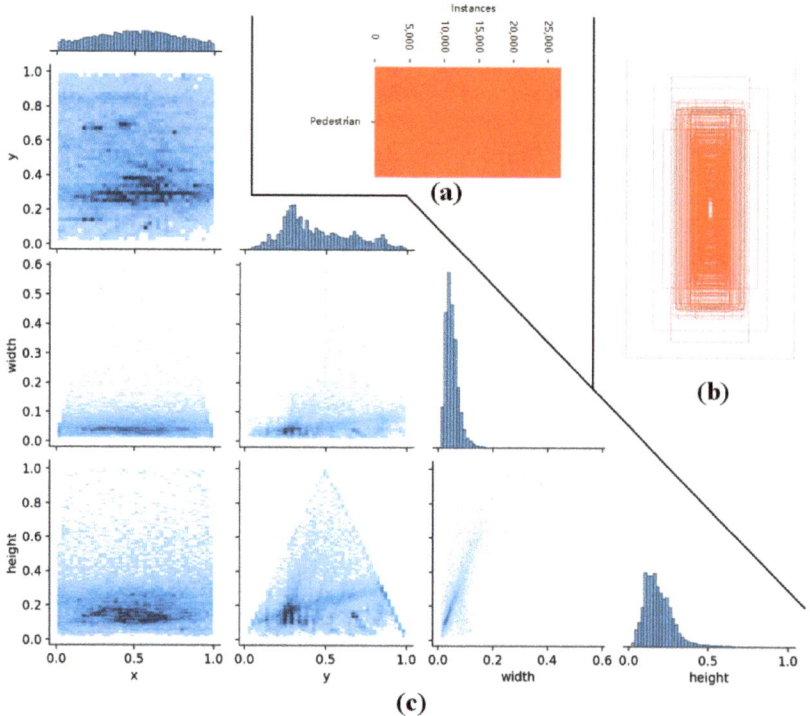

**Figure 2.** The distribution of the HRBUST-LLEPD training set. (**a**) represents the distribution of pedestrian instances in the dataset. (**b**) illustrates the sizes and quantities of the bounding boxes. (**c**) shows the distribution of the bounding box in the training set.

Compare with other datasets

Compared with other pedestrian detection datasets. We used the processed image dataset instead of the raw video as the training data, and when training the model from scratch using only our dataset, this results in low accuracy and poor robustness. Therefore, we extracted pedestrian labels and images containing only pedestrian labels from commonly used KITTI, TJU-PED, LLVIP, and KAIST datasets to build dedicated pretrained pedestrian detection datasets. The details of each processed dataset are presented in Table 1. For the specific handling of publicly available datasets, we applied the following procedures:

KITTI: This dataset contains nine categories, including Pedestrian, Truck, Car, Cyclist, DontCare, Misc, Van, Tram, and Person_sitting. We retained only the Pedestrian labels and extracted the corresponding training set images based on those labels.

TJU-PED: The original dataset of TJU-PED is TJU-DHD, which includes two scenes, "Traffic" and "Campus". The Traffic scene has five labels: Pedestrian, Rider, Car, Van, and Truck. The Campus scene consists of two labels: Pedestrian and Rider. As our research focuses on pedestrians, and there are differences between riders and pedestrians in terms of appearance, we retained only the Pedestrian labels and their corresponding images from both scenes, resulting in the TJU-PED-Traffic and TJU-PED-Campus datasets. Finally, we merged the datasets to obtain the TJU-PED dataset.

LLVIP: This dataset includes infrared and visible light modalities. Due to the significant differences between the infrared and visible light modalities, we only used the visible light modality for pretraining, resulting in the LLVIP-PED dataset.

KAIST: This dataset provides 12 pairs of visible-infrared videos. For the visible light videos, we selected every third frame as a key frame from the training set, retaining the images with pedestrian labels. We selected one image from the testing set every 30 frames, retaining the images with pedestrian labels. This process yielded the KAIST-PED dataset.

Table 1. Comparison of pedestrian data between HRBUST-LLPED and the other datasets.

| | Num. of Train Images | Num. of Test Images | Num. of Train Instances | Num. of Test Instances | Resolution | Image Type | Day/ Night | Pedestrians per Image |
|---|---|---|---|---|---|---|---|---|
| KITTI-PED | 1796 | - | 4708 | - | 1238 × 374 | Visible | Day | 2.62 |
| KAIST-PED | 7595 | 1383 | 24,304 | 4163 | 640 × 512 | Visible | Day | 3.17 |
| LLVIP-PED | 12,025 | 3463 | 34,135 | 8302 | 1280 × 1024 | Visible | Night | 2.74 |
| TJU-PED -Traffic | 13,858 | 2136 | 27,650 | 5244 | 1624 × 1200 | Visible | Day, Night | 2.06 |
| TJU-PED -Campus | 39,727 | 5204 | 234,455 | 36,161 | 640 × 480 ∼ 5248 × 3936 | Visible | Day, Night | 6.02 |
| TJU-PED | 53,585 | 7340 | 262,105 | 41,405 | 640 × 480 ∼ 5248 × 3936 | Visible | Day, Night | 4.98 |
| HRBUST- LLPED (ours) | 3558 | 711 | 26,774 | 5374 | 720 × 576 | **Low-light** | Night | 7.53 |

Object Size: We compared the distribution of objects of different sizes in HRBUST-LLPED and several other datasets. We defined the object sizes following the same criteria as the MS COCO dataset. Objects with an area smaller than $32 \times 32$ square pixels were classified as small objects, objects with an area ranging from $32 \times 32$ to $96 \times 96$ square pixels were classified as medium-sized objects, and objects with an area larger than $96 \times 96$ square pixels were classified as large objects. The specific distribution of object sizes in the dataset is shown in Figure 3. It can be observed that the HRBUST-LLPED dataset covers large, medium, and small objects, with a focus on medium-sized objects. Notably, in the HRBUST-LLPED dataset, the distribution of objects of different sizes is consistent between the training and testing sets. We consider this consistency to be beneficial for training the models.

Pedestrian Density: Pedestrian density is a crucial factor influencing the performance of pedestrian detection models. High pedestrian density implies many pedestrians overlapping with each other, and excessive occlusion can impact the model's ability to recognize pedestrians. Therefore, we evaluated the density of pedestrians in different datasets. Specifically, we counted the number of pedestrians in each image based on the dataset's annotations. We categorized the images into three groups: images with 0 to 5 pedestrians, images with 6 to 10 pedestrians, and images with 11 or more pedestrians. The results are presented in Figure 4. From the figure, we can observe that in the HRBUST-LLPED dataset, images with 6 to 10 pedestrians have the highest proportion, and there is also a significant proportion of images with 11 or more pedestrians. When combining Figure 4 with Table 1, it can be noticed that the HRBUST-LLPED dataset has a relatively high number of pedestrians, with the highest mean pedestrians per image. The distribution of pedestrian numbers is also relatively uniform.

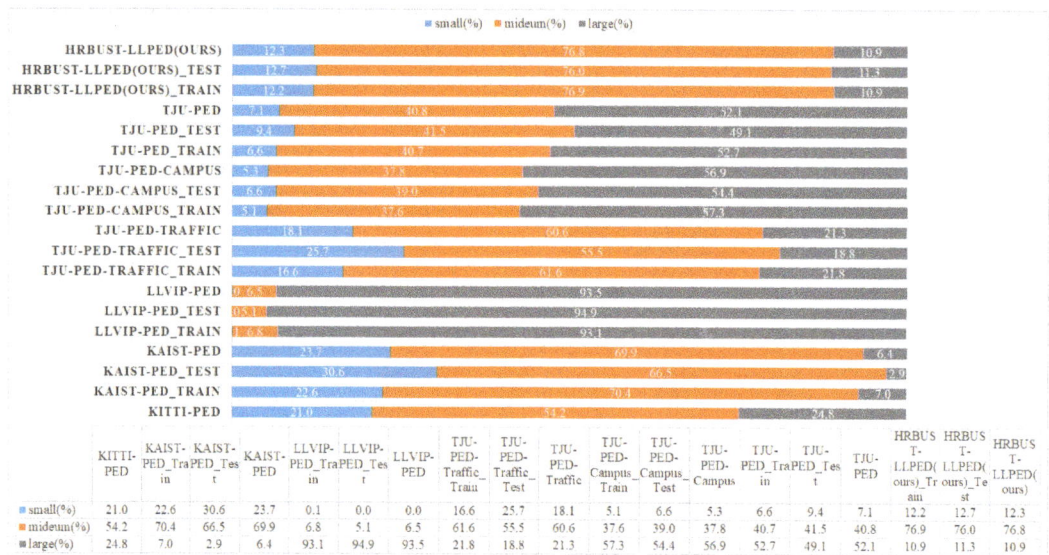

Figure 3. The distribution of pedestrian size in different datasets.

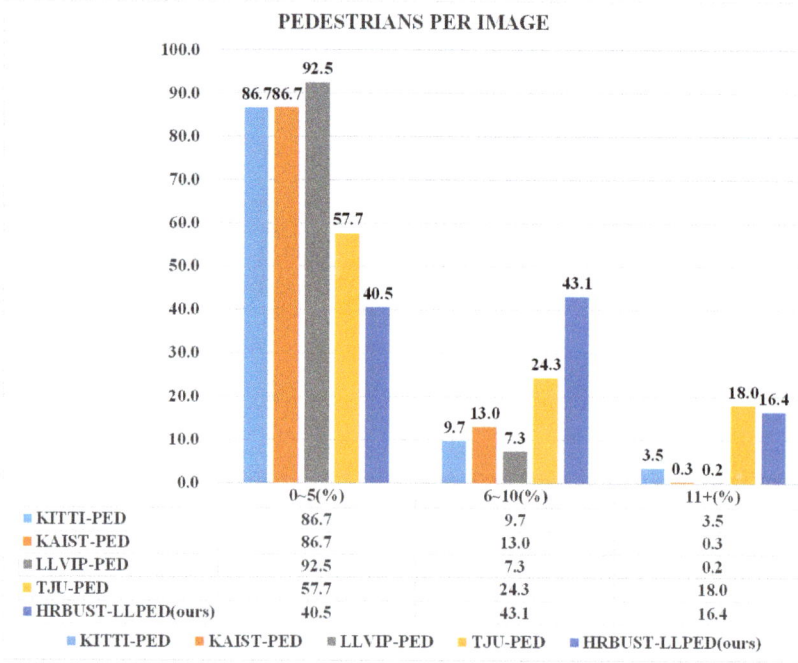

Figure 4. The distribution of pedestrian numbers per image in different datasets.

Therefore, when compared to the other datasets, the HRBUST-LLEPD dataset has the following advantages:

- HRBUST-LLPED is a pedestrian detection dataset for low-light conditions (starlight-level illumination). It captures clear images using low-light cameras, making it suitable for developing pedestrian detection algorithms in low-light environments.

- The dataset contains abundant pedestrian annotations, covering pedestrians of various sizes. Each image includes a substantial number of pedestrians, with significant occlusion between the pedestrians and between the pedestrians and the background. This enables comprehensive training and evaluation of the model's pedestrian recognition capability.
- The dataset captures scenes from different seasons, ranging from winter to summer, and includes weather conditions such as snow, sunny, and cloudy. This diversity in weather conditions ensures the model's robustness across different weather scenarios.

Nevertheless, our constructed dataset has a limitation due to the constraints of the data collection equipment, resulting in a relatively low resolution. Therefore, this dataset may not be suitable for training models with very deep architectures from scratch, as those models might struggle to fit the data fully.

## 4. Wearable Low-Light Pedestrian Detection

The training framework of this paper is illustrated in Figure 5. We first utilized the processed visible light pedestrian detection datasets from Section 3, namely, KITTI-PED, KAIST-PED, LLVIP-PED, TJU-PED, and their subsets, to train and test the pedestrian detection models. Subsequently, the trained wearable pedestrian detection models were transferred to our HRBUST-LLPED dataset for further training, resulting in wearable low-light pedestrian detection models with varying performance. Finally, we compared and analyzed the performances of the models to select high-accuracy and fast-speed models.

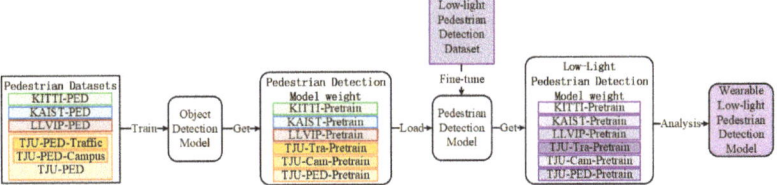

**Figure 5.** The training framework of the proposed model.

YOLOv5 and YOLOv8 are single-stage object detection models developed by the Ultralytics team. They are well-known for their high integration, fast deployment, and applicability to various downstream tasks. One significant characteristic of these two models is that they consist of five different model versions, ranging from small to extra large, denoted as nano (n), small (s), middle (m), large (l), and extra large (x), based on the depth and width of the models. These five model versions allow researchers to choose the depth and width of the model according to the specific requirements of their application scenarios. In this study, our application scenario focuses on wearable pedestrian detection in low-light conditions at night. Therefore, we select the nano and small versions as baselines for modification and experimentation. The overall structures of these two models are similar, as shown in Figure 6a, consisting of three components: backbone, neck structure, and detect head.

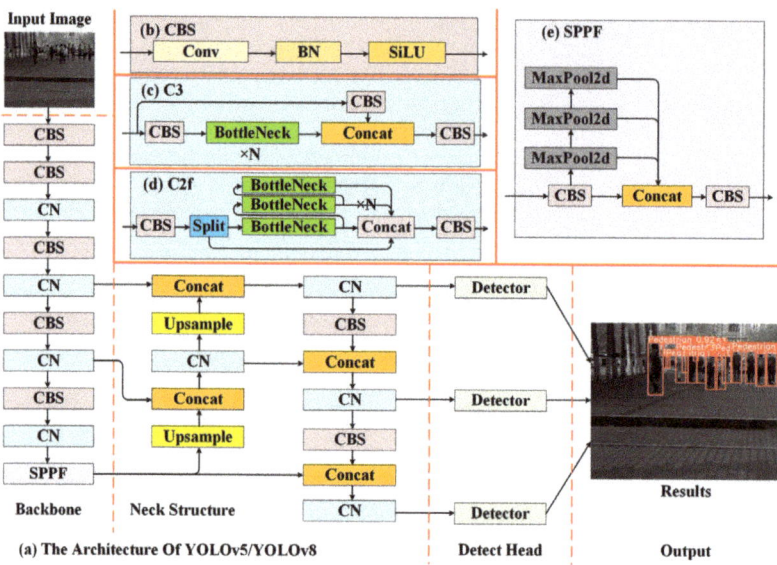

**Figure 6.** The architecture of YOLOv5/YOLOv8.

Backbone: Its primary function is to extract features from input images, primarily including CBS blocks, CN blocks, and SPPF blocks. The CBS block consists of a convolutional block, a BN layer, and a SiLU activation layer, as shown in Figure 6b. For the input feature $x$, the computation output of the SiLU activation function is given by Equation (1).

$$\text{SiLU}(x) = x * \text{sigmoid}(x) = \frac{x}{1+e^{-x}} \tag{1}$$

In order to implement the CN block, YOLOv5 utilizes the C3 block, which consists of three CBS blocks and n concatenated BottleNeck blocks, as shown in Figure 6c. YOLOv8, on the other hand, employs the C2f block, composed of three CBS blocks and n concatenated BottleNeck blocks, as illustrated in Figure 6d. The structure of the SPPF block is depicted in Figure 6e and includes two CSB blocks, three max-pooling layers, and a concatenation operation. In this study, to accelerate the detection speed of the model, the SiLU activation function is replaced with a more straightforward ReLU activation function.

Neck Structure: Its main function is to fuse the multi-scale features extracted from the backbone network, thereby interacting with semantic information of various granularities, drawing inspiration from the ideas of feature pyramids and pixel aggregation networks. In this section, besides implementing the CN block using their respective methods, YOLOv5 adds a $1 \times 1$ CBS block before each upsampling layer.

Detect Head: Its main function is to generate the detection results of the model based on the interacted feature information. In this aspect, YOLOv5 utilizes the Coupled head, which directly performs classification using a convolutional layer, as shown in Figure 7a. On the other hand, YOLOv8 employs the Decoupled head, as illustrated in Figure 7b, which separates the processes of classification and regression to improve the convergence speed and accuracy of the model.

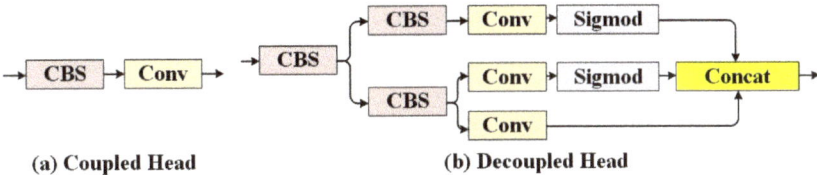

(a) Coupled Head    (b) Decoupled Head

**Figure 7.** The architecture of different Detect Heads.

Training: During the training process of the object detection model, the main focus is on training the model's ability to correctly classify the target's category and accurately regress its position. In the task of low-light pedestrian detection, when training the model's classification ability, the objective is to correctly classify pedestrians from the background, thereby utilizing pedestrian samples more effectively for model training supervision. Therefore, we employ Varifocal Loss to train the model's classification ability, which is calculated as follows:

$$\text{VFL}(p,q) = \begin{cases} -q(q\log(p) + (1-q)\log(1-p)) & q > 0 \\ -\alpha p^{\gamma} \log(1-p) & q = 0 \end{cases} \quad (2)$$

where p is the probability that the model predicts that the target is a pedestrian, and $q$ is the intersection over union (IoU) of the predicted bounding box and ground truth.

During the training process of regressing the target bounding boxes, we aim to make the predicted boxes as close as possible to the ground truth boxes. In other words, we aim to maximize the IoU between the predicted and ground truth boxes, minimize the distance between their center points, and ensure consistency in the aspect ratio. In order to achieve this, we utilize CIoU Loss to train the model's regression capability, which is calculated as follows:

$$\mathcal{L}_{CIoU} = 1 - IoU + \frac{\rho^2(\mathbf{b}, \mathbf{b}^{gt})}{c^2} + \alpha v \quad (3)$$

where $\rho$ represents the Euclidean distance, $\mathbf{b}$ and $\mathbf{b}^{gt}$ denote the predicted and ground truth center points, $c$ is the diagonal distance of the minimum bounding box, $\alpha$ is a balancing parameter calculated as in Equation (4), and $v$ represents the aspect ratio similarity, computed according to Equation (5).

$$\alpha = \frac{v}{(1 - IoU) + v} \quad (4)$$

$$v = \frac{4}{\pi^2}\left(\arctan\frac{w^{gt}}{h^{gt}} - \arctan\frac{w}{h}\right)^2 \quad (5)$$

Therefore, the formula of the Loss function $L$ of our model training is as follows:

$$\mathcal{L} = \mathcal{L}_{VFL} + \mathcal{L}_{CIoU} \quad (6)$$

## 5. Experiments

In this section, we utilize the low-light pedestrian detection dataset we constructed and publicly available data to train the pedestrian detection models described in Section 4. We then evaluate the accuracy and speed of the models. We aim to train a baseline model as a foundation for future research.

### 5.1. Evaluate Metric

Similar to the well-known object detection dataset MS COCO, we evaluate the performance of our models on pedestrian detection using precision (P), recall (R), AP@0.5, and AP@0.5:0.95. The calculation methods for the evaluation metrics in this paper are as follows:

True Positive (TP): The prediction is positive, and it is correct.
False Positive (FP): The prediction is positive, but it is wrong.
False Negative (FN): The prediction is negative, but it is wrong.
Precious (P): Describe the proportion of TP in the detection results, calculated as follows: $Precious = \frac{TP}{TP+FP}$.
Recall (R): Describe the detection rate of labeled pedestrians, calculated as follows: $Recall = \frac{TP}{TP+FN} = \frac{TP}{GT}$.
AP@0.5 (AP50): When the IoU threshold is set to 0.5, the area enclosed by the PR curve and the co-ordinate axis is calculated, i.e., $AP = \int_0^1 P(r)dr$.
AP@0.5:0.95 (AP): Similar to the MS COCO dataset, using 0.05 as the step for changing the IoU threshold, multiple APs are obtained by changing the IoU from 0.5 to 0.95, and then the average value is taken.
Missing Rate (MR): $= \frac{FN}{GT}$.
False Detection Rate (FDR): $= \frac{FP}{TP+FP}$.
Incorrect Localization Rate (ILR): The proportion of prediction boxes with an IoU threshold between 0.2 and 0.5 with ground truth boxes.

*5.2. Implementation Details*

The wearable low-light pedestrian detection models in this paper were implemented based on YOLOv5 and YOLOv8, with activation and loss function modifications to adjust training performance. We used four datasets: KITTI, TJU-PED, KAIST, and LLVIP. We loaded the pretrained weights of the general object detection dataset MS COCO to train our models. The training was conducted for 50 epochs with an initial learning rate of 0.001, which is reduced to 0.0005 after 20 epochs. For the KAIST dataset, we trained the model with an input resolution of 640 × 640 and set the batch size to 64. For the KITTI, TJU-PED, and LLVIP datasets, we trained the model with input resolutions of 640 × 640 and 1280 × 1280 and set the batch size to 64 and 16, respectively. Subsequently, when transferring the pretrained model to our dataset, we trained it for 50 epochs, with an initial learning rate of 0.001, which was reduced to 0.0005 after 20 epochs. The input resolution was 640 × 640, and the batch size was 64. We used two devices for training and testing the models. Both devices have an 11th Gen Intel Core i7-11800H @ 2.30 GHz CPU and 32 GB of memory. One device has an NVIDIA GeForce RTX 3080 Laptop GPU at 16 GB, whereas the other has two NVIDIA GeForce RTX 2080Ti GPUs at 11 GB.

*5.3. Experimental Results*

We first trained wearable pedestrian detection models based on YOLOv5s, YOLOv5n, YOLOv8s, and YOLOv8n directly on the HRBUST-LLPED dataset we constructed. In the subsequent text and tables, these four basic pedestrian detection models are abbreviated as YOLOv5s, YOLOv5n, YOLOv8s, and YOLOv8n, and the original YOLO model is marked with a "-o" suffix at the end (i.e., YOLOv5s-o, YOLOv5n-o, YOLOv8s-o, and YOLOv8n-o). The experimental setup was consistent with the experiment using the KAIST dataset for pretraining. The training results are shown in Table 2. The table shows that the YOLOv5s-o model achieves the best performance in terms of precious. The YOLOv5s model achieves the best performance in terms of recall rate and AP@0.5. On the other hand, the YOLOv8s model performs best in terms of AP@0.5:0.95. In summary, the YOLOv5s model can quickly fit the approximate distribution of our dataset, whereas the YOLOv8s model can more accurately identify the specific locations of pedestrians in the scene. Furthermore, the YOLOv8n model has the fastest inference speed, reaching 1.6 ms. From Tables 2 and 3, we can observe that our improved pedestrian detection models have a 1–2% improvement in terms of AP and 0.3–0.9 ms improvement in terms of inference time.

Table 2. Experimental results of wearable low-light pedestrian detection models and original YOLO models (marked with "-o") for training and testing using the HRBUST-LLPED Dataset.

| Method | Input Resolution | Train to LLPED, Test on LLPED | | | | Infer Time (ms) |
|---|---|---|---|---|---|---|
| | | P (%) | R (%) | AP50 (%) | AP (%) | |
| YOLOv5s-o | 640 × 640 | **93.05** | 84.17 | 92.91 | 59.04 | 3.5 |
| YOLOv8s-o | 640 × 640 | 92.41 | 83.12 | 92.35 | 61.82 | 4.3 |
| YOLOv5n-o | 640 × 640 | 90.98 | 80.67 | 90.57 | 55.88 | 2.9 |
| YOLOv8n-o | 640 × 640 | 91.82 | 80.01 | 89.89 | 59.20 | 2.4 |
| YOLOv5s | 640 × 640 | 92.72 | **85.83** | 93.93 | 61.49 | 3.1 |
| YOLOv8s | 640 × 640 | 92.20 | 85.59 | 93.44 | **63.64** | 3.4 |
| YOLOv5n | 640 × 640 | 92.19 | 82.58 | 91.86 | 57.91 | 2.6 |
| YOLOv8n | 640 × 640 | 92.01 | 82.32 | 91.50 | 60.99 | **1.6** |

Subsequently, we trained the models on four visible light pedestrian detection datasets (KITTI-PED, KAIST-PED, LLVIP-PED, and TJU-PED) and evaluated their performance on our constructed HRBUST-LLPED dataset. The experimental results are presented in Tables 3–8. Each table includes the model's name (column 1), the input resolution of the model during training (column 2), the results of the model trained and tested on the visible light pedestrian detection dataset (columns 3 to 6), the test results after transferring the model to the HRBUST-LLPED dataset (column 7 to 10), and the model's inference time (column 11). The inference time was measured on an NVIDIA GeForce RTX 3080 Laptop GPU with 16 GB of memory. The bold numbers indicate the optimal values for each evaluation metric. From Tables 3–8, it can be observed that our dataset is more difficult than the KITTI dataset. For the KAIST and LLVIP datasets, the pedestrians captured in the nighttime images are blurred and submerged in the background, making it more difficult to detect. When compared to the TJU series of datasets, our dataset still has a gap in terms of resolution diversity and data quantity, which is one of the issues we will address in our future work.

Table 3. Experimental results of wearable low-light pedestrian detection models trained on the LLVIP-PED Dataset and transfered to the HRBUST-LLPED Dataset.

| Method | Input Resolution | Trained and Tested on LLVIP-PED Dataset | | | | Transfered to and Tested on HRBUST-LLPED Dataset | | | | Infer Time (ms) |
|---|---|---|---|---|---|---|---|---|---|---|
| | | P (%) | R (%) | AP50 (%) | AP (%) | P (%) | R (%) | AP50 (%) | AP (%) | |
| YOLOv5s-o | 640 × 640 | 90.35 | 80.60 | 88.59 | 48.93 | 93.76 | 87.35 | 95.61 | 65.22 | 3.5 |
|  | 1280 × 1280 | 88.63 | 81.84 | 89.09 | 50.27 | 92.79 | 88.33 | 95.51 | 65.19 | 3.6 |
| YOLOv8s-o | 640 × 640 | 85.78 | 82.07 | 87.36 | 50.12 | 92.52 | 90.74 | 95.76 | 68.44 | 4.3 |
|  | 1280 × 1280 | 91.63 | 80.08 | 88.22 | 50.44 | 93.04 | 90.29 | 95.83 | 68.59 | 15.4 |
| YOLOv5n-o | 640 × 640 | 92.36 | 79.23 | 88.04 | 47.89 | 92.82 | 86.42 | 94.41 | 61.63 | 2.9 |
|  | 1280 × 1280 | 85.12 | 83.36 | 87.90 | 48.63 | 93.28 | 87.40 | 95.04 | 63.58 | 2.9 |
| YOLOv8n-o | 640 × 640 | 90.74 | 79.33 | 88.15 | 50.17 | 92.57 | 89.77 | 95.57 | 67.85 | 2.1 |
|  | 1280 × 1280 | 90.95 | 80.39 | 88.66 | 51.57 | 92.20 | 89.69 | 95.42 | 67.18 | 7.3 |
| YOLOv5s | 640 × 640 | **93.30** | 84.05 | **91.14** | 51.05 | 93.62 | 89.83 | 95.74 | 67.27 | 3.2 |
|  | 1280 × 1280 | 91.47 | 83.91 | 90.66 | **52.31** | **94.19** | 89.00 | 95.68 | 66.98 | 3.2 |
| YOLOv8s | 640 × 640 | 89.62 | 79.24 | 87.37 | 49.96 | 92.67 | 90.62 | 95.96 | 69.29 | 3.5 |
|  | 1280 × 1280 | 91.58 | 82.55 | 89.54 | 52.29 | 92.43 | **91.50** | **96.01** | **69.50** | 12.6 |
| YOLOv5n | 640 × 640 | 93.27 | 77.58 | 87.84 | 49.55 | 93.75 | 86.02 | 94.56 | 64.34 | 2.6 |
|  | 1280 × 1280 | 89.54 | 82.16 | 89.10 | 50.64 | 93.38 | 87.12 | 94.70 | 64.27 | 2.7 |
| YOLOv8n | 640 × 640 | 89.72 | 82.09 | 89.16 | 51.68 | 92.61 | 89.43 | 95.56 | 67.69 | **1.7** |
|  | 1280 × 1280 | 90.97 | 82.60 | 89.68 | 52.04 | 92.30 | 89.32 | 95.44 | 67.51 | 5.6 |

Table 4. Experimental results of wearable low-light pedestrian detection models trained on the KITTK-PED Dataset and transferred to the HRBUST-LLPED Dataset.

| Method | Input Resolution | Trained and Tested on KITTI-PED Dataset | | | | Transfered to and Tested on HRBUST-LLPED Dataset | | | | Infer Time (ms) |
|---|---|---|---|---|---|---|---|---|---|---|
| | | P (%) | R (%) | AP50 (%) | AP (%) | P (%) | R (%) | AP50 (%) | AP (%) | |
| YOLOv5s | 640 × 640 | 93.91 | 83.72 | 92.85 | 62.56 | **93.80** | 89.17 | 95.74 | 67.51 | 3.2 |
| | 1280 × 1280 | **99.04** | 91.16 | 97.29 | 75.69 | 93.16 | 89.41 | 95.77 | 67.58 | 3.2 |
| YOLOv8s | 640 × 640 | 94.61 | 88.43 | 95.12 | 74.85 | 93.30 | 91.05 | **96.30** | **69.73** | 3.4 |
| | 1280 × 1280 | 98.07 | **93.34** | **97.76** | **82.85** | 92.73 | **91.38** | 96.15 | 69.57 | 12.6 |
| YOLOv5n | 640 × 640 | 79.22 | 68.46 | 78.57 | 39.98 | 93.14 | 86.90 | 94.76 | 64.12 | 2.6 |
| | 1280 × 1280 | 96.59 | 85.82 | 94.87 | 66.18 | 92.53 | 87.37 | 94.71 | 64.17 | 2.6 |
| YOLOv8n | 640 × 640 | 92.30 | 79.69 | 90.93 | 66.31 | 92.49 | 89.36 | 95.44 | 67.90 | **1.7** |
| | 1280 × 1280 | 92.59 | 90.30 | 96.34 | 77.03 | 92.45 | 89.21 | 95.49 | 68.21 | 5.6 |

Table 5. Experimental results of wearable low-light pedestrian detection models trained on the KAIST-PED Dataset and transfered to the HRBUST-LLPED Dataset.

| Method | Input Resolution | Trained and Tested on KAIST-PED Dataset | | | | Transfered to and Tested on HRBUST-LLPED Dataset | | | | Infer Time (ms) |
|---|---|---|---|---|---|---|---|---|---|---|
| | | P (%) | R (%) | AP50 (%) | AP (%) | P (%) | R (%) | AP50 (%) | AP (%) | |
| YOLOv5s | 640 × 640 | 36.09 | 28.22 | **32.41** | 13.09 | **94.31** | 88.74 | 95.74 | 67.42 | 3.2 |
| YOLOv8s | 640 × 640 | **37.40** | 26.50 | 30.41 | **13.84** | 93.36 | **90.53** | **95.98** | **69.30** | 3.4 |
| YOLOv5n | 640 × 640 | 37.23 | **28.52** | 30.06 | 12.58 | 93.21 | 86.92 | 94.50 | 64.04 | 2.6 |
| YOLOv8n | 640 × 640 | 36.86 | 26.98 | 29.63 | 12.73 | 92.99 | 88.67 | 95.59 | 68.01 | **1.6** |

Table 6. Experimental results of wearable low-light pedestrian detection models trained on the TJU-PED-Traffic Dataset and transfer to the HRBUST-LLPED Dataset.

| Method | Input Resolution | Trained and Tested on TJU-PED-Traffic Dataset | | | | Transfered to and Tested on HRBUST-LLPED Dataset | | | | Infer Time (ms) |
|---|---|---|---|---|---|---|---|---|---|---|
| | | P (%) | R (%) | AP50 (%) | AP (%) | P (%) | R (%) | AP50 (%) | AP (%) | |
| YOLOv5s | 640 × 640 | 83.68 | 71.93 | 80.95 | 45.73 | 93.84 | 89.02 | 95.62 | 67.11 | 3.2 |
| | 1280 × 1280 | 88.52 | 77.84 | 87.89 | 53.62 | **94.69** | 88.97 | 95.88 | 67.75 | 3.2 |
| YOLOv8s | 640 × 640 | 84.76 | 73.20 | 82.87 | 48.43 | 92.65 | 90.97 | 96.08 | 69.56 | 3.4 |
| | 1280 × 1280 | **86.84** | **81.00** | 89.30 | **56.44** | 92.55 | **91.11** | **96.28** | **69.30** | 12.7 |
| YOLOv5n | 640 × 640 | 83.53 | 64.12 | 74.85 | 39.67 | 94.43 | 86.19 | 94.81 | 64.25 | 2.6 |
| | 1280 × 1280 | 85.91 | 77.64 | 86.26 | 50.62 | 92.91 | 87.55 | 94.91 | 64.70 | 2.6 |
| YOLOv8n | 640 × 640 | 83.51 | 67.82 | 78.27 | 44.28 | 92.65 | 89.71 | 95.67 | 67.95 | **1.7** |
| | 1280 × 1280 | 85.34 | 80.01 | 87.68 | 53.78 | 92.50 | 89.49 | 95.60 | 68.17 | 5.6 |

Table 7. Experimental results of wearable low-light pedestrian detection models trained on the TJU-PED-Campus Dataset and transfered to the HRBUST-LLPED Dataset.

| Method | Input Resolution | Trained and Tested on TJU-PED-Campus Dataset | | | | Transfered to and Tested on HRBUST-LLPED Dataset | | | | Infer Time (ms) |
|---|---|---|---|---|---|---|---|---|---|---|
| | | P (%) | R (%) | AP50 (%) | AP (%) | P (%) | R (%) | AP50 (%) | AP (%) | |
| YOLOv5s | 640 × 640 | 84.95 | 63.04 | 72.23 | 46.55 | **94.60** | 89.17 | 96.16 | 68.83 | 3.2 |
| | 1280 × 1280 | 89.57 | 73.77 | 84.36 | 57.33 | 93.91 | 90.43 | 96.27 | 69.19 | 3.2 |
| YOLOv8s | 640 × 640 | 88.13 | 64.71 | 74.39 | 51.02 | 93.35 | 90.64 | 96.17 | 69.73 | 3.4 |
| | 1280 × 1280 | **90.39** | **76.00** | **85.43** | **61.15** | 92.96 | **91.66** | **96.34** | **69.87** | 12.6 |
| YOLOv5n | 640 × 640 | 85.31 | 57.11 | 66.84 | 40.91 | 93.69 | 88.18 | 95.40 | 65.87 | 2.6 |
| | 1280 × 1280 | 87.40 | 69.97 | 80.25 | 51.93 | 93.91 | 87.35 | 95.24 | 66.01 | 2.6 |
| YOLOv8n | 640 × 640 | 85.97 | 59.92 | 69.49 | 46.07 | 93.06 | 89.82 | 95.73 | 68.77 | **1.7** |
| | 1280 × 1280 | 89.68 | 71.49 | 81.80 | 57.21 | 92.89 | 90.19 | 96.00 | 68.71 | 5.6 |

Table 8. Experimental results of wearable low-light pedestrian detection models trained on the TJU-PED Dataset and transfered to the HRBUST-LLPED Dataset.

| Method | Input Resolution | Trained and Tested on TJU-PED Dataset | | | | Transferred to and Tested on HRBUST-LLPED Dataset | | | | Infer Time (ms) |
|---|---|---|---|---|---|---|---|---|---|---|
| | | P (%) | R (%) | AP50 (%) | AP (%) | P (%) | R (%) | AP50 (%) | AP (%) | |
| YOLOv5s | 640 × 640 | 82.89 | 65.00 | 73.61 | 46.79 | **95.15** | 88.72 | 96.03 | 68.83 | 3.2 |
| | 1280 × 1280 | 89.20 | 74.13 | 84.87 | 57.05 | 94.56 | 89.19 | 96.06 | 68.97 | 3.2 |
| YOLOv8s | 640 × 640 | 87.27 | 66.56 | 75.88 | 51.04 | 93.46 | 90.66 | 96.24 | 69.59 | 3.4 |
| | 1280 × 1280 | **90.02** | **76.89** | **86.09** | **60.67** | 93.25 | **91.27** | **96.29** | **69.90** | 12.6 |
| YOLOv5n | 640 × 640 | 83.41 | 58.75 | 68.14 | 40.94 | 92.96 | 87.81 | 95.26 | 65.69 | 2.6 |
| | 1280 × 1280 | 87.24 | 70.49 | 80.96 | 51.84 | 93.19 | 88.68 | 95.36 | 65.81 | 2.6 |
| YOLOv8n | 640 × 640 | 85.35 | 60.88 | 70.80 | 45.82 | 92.67 | 90.27 | 95.75 | 68.53 | **1.6** |
| | 1280 × 1280 | 89.28 | 72.22 | 82.51 | 56.82 | 93.09 | 90.52 | 96.04 | 68.74 | 5.6 |

From Tables 4 and 5, it can be observed that pretraining on the KITTI-PED and KAIST-PED datasets leads to approximately a 1% improvement in accuracy, a 4–7% improvement in recall rate, a 3% improvement in AP@0.5, and a 5–7% improvement in AP for all four models. However, there is no improvement in inference speed. Training the models with an input resolution of 1280 × 1280 on the KITTI-PED dataset effectively improves model performance. On the other hand, when transferring the models to the HRBUST-LLPED dataset, it seems more beneficial to train them at a resolution of 640 × 640, as it improves the accuracy of the transferred models. This phenomenon corresponds to the resolution of the images in the dataset. Additionally, it is counterintuitive that changing the input resolution does not affect the inference speed of the YOLOv5 models, whereas it significantly slows down the YOLOv8 models by nearly four times.

As shown in Table 3, for the YOLOv5s model, features with a resolution of 640 × 640 are more effective for pedestrian recognition when trained on the LLVIP-PED dataset. This observation is supported by Figure 3, which indicates that the LLVIP-PED dataset contains significant annotations for large-scale pedestrians. Therefore, it seems that the YOLOv5s model is less sensitive to larger-sized objects.

Tables 6–8 further confirm the aforementioned conjecture, and we can draw the following preliminary conclusions:

- The model with the highest detection of precious is based on YOLOv5s pretrained on the TJU-PED dataset, with an accuracy of 95.15%. The model with the highest recall rate and AP@0.5 is based on YOLOv8s pretrained on the TJU-PED-Campus dataset, with 91.66% in terms of recall and 96.34% in terms of AP@0.5. The models with the highest AP@0.5:0.95 are based on YOLOv8s pretrained on the TJU-PED dataset, achieving a value of 69.90%. The fastest model is based on YOLOv8n, with an inference speed of 1.6 ms per image.
- Among the four selected models, YOLOv8s performs the best, achieving approximately a 3% higher for AP@0.5:0.95 than YOLOv5s. YOLOv8n and YOLOv5s have similar accuracies, but YOLOv8n is approximately 1.5 ms faster.
- For most pretrained datasets, training the model with pedestrian sizes closer to the target dataset leads to better performance when transferring the model.
- For the YOLOv5 models, the input image resolution does not affect the inference speed, whereas for the YOLOv8 models, the input image resolution impacts the model's speed.

*5.4. Further Analysis*

After obtaining the training results in Section 5.3, we will further analyze the impact of different pretrained datasets on transferring the models to the HRBUST-LLPED dataset, as well as the performance of our trained models during actual predictions in this section.

**The impact of different pretrained datasets on transferring the models to the HRBUST-LLPED dataset.**

In order to investigate this issue, we extracted the model weights with the highest AP@0.5:0.95 metric on each pretrained dataset, labeled as "best" for that dataset. We then set the input image resolution, starting from 256 × 256 square pixels, and increased it by 32 in stride, iterating up to 1280 × 1280 square pixels. We tested the models with different pixel inputs to assess their recognition speed and accuracy for different image sizes. All tests were conducted on an NVIDIA GeForce RTX 3080 Laptop GPU with 16GB of memory. The final results are shown in Figure 8, where each line is labeled in the format "study_Test Dataset_Train Dataset-best". First, it can be observed that training the models with other datasets before the target dataset improves the model's performance to some extent. Second, it can be seen that TJU-PED-Campus and TJU-PED are the most effective datasets for model training. This is likely because our dataset and these two datasets primarily focus on pedestrians within campus environments, making their distributions similar and facilitating model transfer. Additionally, LLVIP has the most negligible impact on model training, possibly because the LLVIP dataset contains many large objects that do not align well with the distribution of the HRBUST-LLPED dataset, requiring more time to train the model.

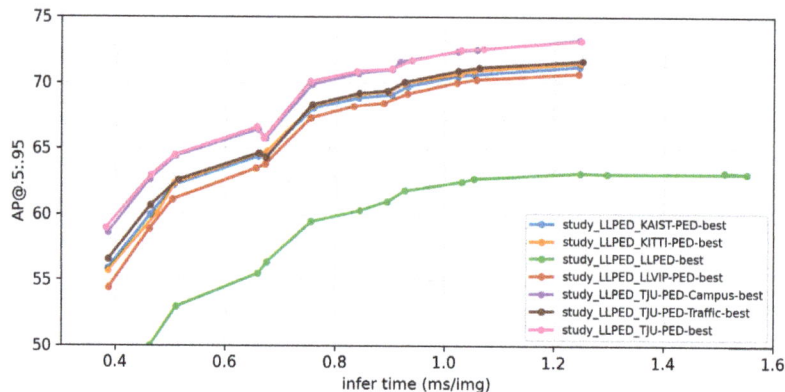

**Figure 8.** Model performance when trained on different datasets.

**The performance of our trained models during actual predictions.**

We have visualized several examples and their detection results to demonstrate each model's effectiveness in Figure 9. In the figure, from top to bottom, there are five test images, and from left to right, these show the original low-light image, ground truth labels, and predicted labels from YOLOv5s, YOLOv5n, YOLOv8s, and YOLOv8n, respectively. From the figure, it can be observed that all four wearable low-light pedestrian detection models can generally detect pedestrians present in the scenes. Additionally, the YOLOv5s-based and YOLOv8s-based wearable low-light pedestrian detection models can detect partially occluded pedestrians, such as the pedestrian in the far right of the first row, where only the upper body is visible. However, these four models still have room for improvement in detecting pedestrians with high overlap, as seen in the left-middle part of the last row, where all four models identify two overlapping individuals as single persons.

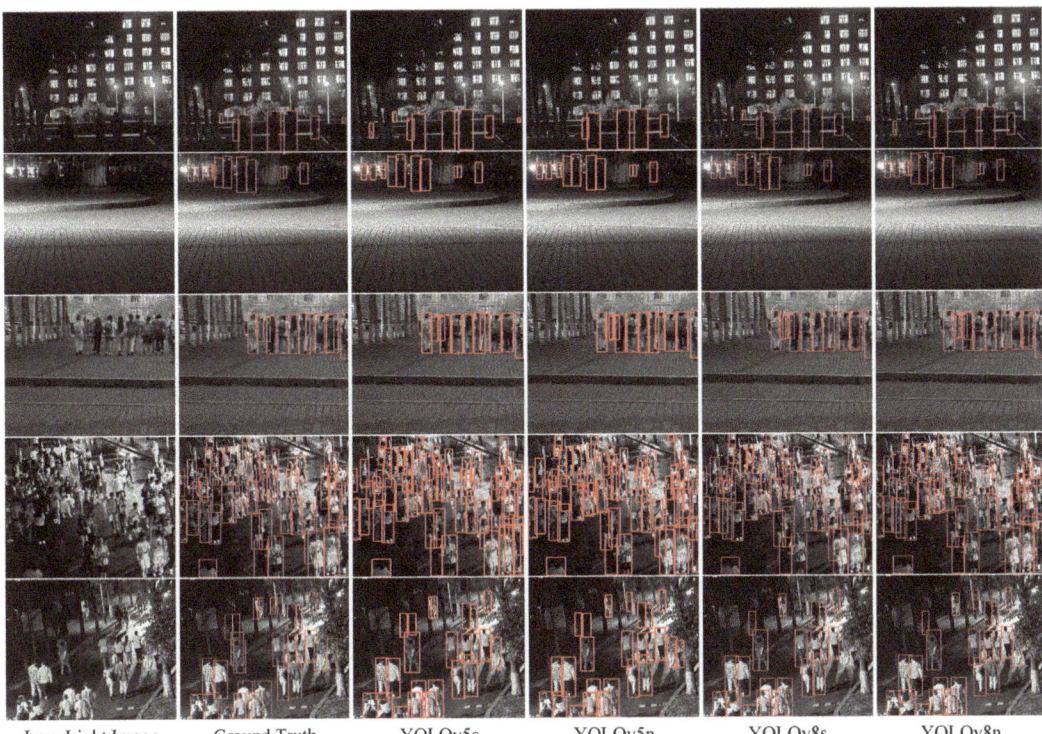

**Figure 9.** The qualitative result of different models on the HRBUST-LLPED dataset.

**The quantitative and qualitative analysis of the failure cases.**

In order to further analyze the characteristics of the HRBUST-LLPED dataset, we chose the models that pretrained on the TJU-PED dataset and were transferred to our dataset. In order to make the analysis results more accurate, we set the confidence threshold for detecting pedestrians to 0.3. We evaluated the model using three indicators: the missing rate, false detection rate, and incorrect localization rate. The smaller these indicators, the better the model. The experimental results are shown in Table 9. From the table, it can be found that the models with a pretrained resolution of $1280 \times 1280$ perform better. In addition, the YOLOv5s model has the lowest false negative rate, the YOLOv8n model has the lowest false positive rate, and the YOLOv8s model has more accurate localization.

**Table 9.** The missing rate (MR), false detection rate (FDR), and incorrect localization rate (ILR) of different models on the HRBUST-LLPED Dataset with the confidence threshold set to 0.3.

| Method   | Input Resolution    | MR (%)   | FDR (%)   | ILR (%)  |
|----------|---------------------|----------|-----------|----------|
| YOLOv5s  | $640 \times 640$    | 6.77     | 10.04     | 2.71     |
|          | $1280 \times 1280$  | **6.66** | 9.53      | 2.66     |
| YOLOv8s  | $640 \times 640$    | 7.66     | 9.27      | 2.92     |
|          | $1280 \times 1280$  | 6.98     | 8.72      | **2.51** |
| YOLOv5n  | $640 \times 640$    | 9.03     | 10.22     | 3.62     |
|          | $1280 \times 1280$  | 8.66     | 10.49     | 3.15     |
| YOLOv8n  | $640 \times 640$    | 8.13     | 9.03      | 2.92     |
|          | $1280 \times 1280$  | 7.88     | **8.67**  | 2.65     |

Subsequently, we selected the models with a pretrained resolution of 1280 × 1280 for qualitative analysis, as shown in Figure 10. The green bounding box indicates the ground truth, while the red indicates that predicted by the model. From the first row, it can be found that pedestrians in dark places with only a small part of the area exposed due to severe occlusion are prone to miss detection. From the second line, it can be found that areas with overexposure and severe occlusion are more prone to false detection. From the third line, it can be found that people who are obscured are more likely to be in an inaccurate location. Some pictures may cause multiple errors, as shown in the fourth line. Overall, our dataset contains occluded and poorly exposed images, posing challenges to pedestrian detection models.

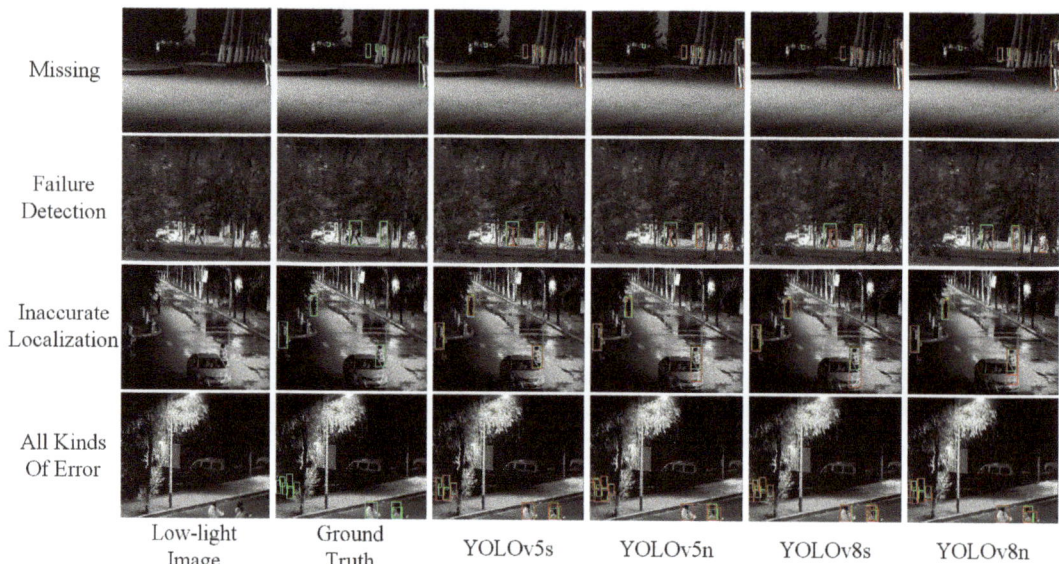

**Figure 10.** The cases of different kinds of failure.

## 6. Conclusions

In this paper, we have constructed a low-light pedestrian detection dataset specifically targeting campus scenes. By using low-light cameras to capture clear images during nighttime, our dataset is suitable for developing pedestrian detection algorithms in low-light environments. We have also provided wearable low-light pedestrian detection models based on the well-known YOLO architecture to test the validity of the HRBUST-LLEPD dataset. We pretrained the models on the public pedestrian datasets and fine-tuned them on our dataset. The experimental results provide a baseline performance for low-light pedestrian detection research. We hope this study will contribute to developing wearable nighttime pedestrian assistance systems.

**Author Contributions:** Conceptualization, T.L. and L.Y.; methodology, T.L.; software, T.L.; validation, T.L. and K.Z.; formal analysis, T.L.; investigation, T.L.; resources, T.L. and G.S.; data curation, T.L. and K.Z.; writing—original draft preparation, T.L.; writing—review and editing, T.L., G.S. and L.Y.; visualization, T.L.; supervision, G.S.; project administration, T.L.; funding acquisition, G.S. All authors have read and agreed to the published version of the manuscript.

**Funding:** This research was funded by the Key Research and Development Project of Heilongjiang Province, grant number 2022ZX01A34, and the 2020 Heilongjiang Province Higher Education Teaching Reform Project, grant number SJGY 20200320.

**Data Availability Statement:** The datasets generated and/or analyzed during the current study are available from the corresponding author upon reasonable request.

**Acknowledgments:** We would like to thank the Harbin Institute of Technology for providing technical support in terms of the wearable devices.

**Conflicts of Interest:** The authors declare no conflict of interest.

## References

1. Li, H.; Liu, H.; Li, Z.; Li, C.; Meng, Z.; Gao, N.; Zhang, Z. Adaptive Threshold Based ZUPT for Single IMU Enabled Wearable Pedestrian Localization. *IEEE Internet Things J.* **2023**, *10*, 11749–11760. [CrossRef]
2. Tang, Z.; Zhang, L.; Chen, X.; Ying, J.; Wang, X.; Wang, H. Wearable supernumerary robotic limb system using a hybrid control approach based on motor imagery and object detection. *IEEE Trans. Neural Syst. Rehabil. Eng.* **2022**, *30*, 1298–1309. [CrossRef] [PubMed]
3. Han, Y.; Yarlagadda, S.K.; Ghosh, T.; Zhu, F.; Sazonov, E.; Delp, E.J. Improving food detection for images from a wearable egocentric camera. *arXiv* **2023**, arXiv:2301.07861.
4. Li, X.; Holiday, S.; Cribbet, M.; Bharadwaj, A.; White, S.; Sazonov, E.; Gan, Y. Non-Invasive Screen Exposure Time Assessment Using Wearable Sensor and Object Detection. In Proceedings of the 2022 44th Annual International Conference of the IEEE Engineering in Medicine & Biology Society (EMBC), Glasgow, UK, 11–15 July 2022; pp. 4917–4920.
5. Kim, J.; Moon, N. Dog behavior recognition based on multimodal data from a camera and wearable device. *Appl. Sci.* **2022**, *12*, 3199. [CrossRef]
6. Park, K.B.; Choi, S.H.; Lee, J.Y.; Ghasemi, Y.; Mohammed, M.; Jeong, H. Hands-free human–robot interaction using multimodal gestures and deep learning in wearable mixed reality. *IEEE Access* **2021**, *9*, 55448–55464. [CrossRef]
7. Dimitropoulos, N.; Togias, T.; Michalos, G.; Makris, S. Operator support in human–robot collaborative environments using AI enhanced wearable devices. *Procedia CIRP* **2021**, *97*, 464–469. [CrossRef]
8. Zou, Z.; Chen, K.; Shi, Z.; Guo, Y.; Ye, J. Object detection in 20 years: A survey. *Proc. IEEE* **2023**, *111*, 257–276. [CrossRef]
9. Liu, W.; Anguelov, D.; Erhan, D.; Szegedy, C.; Reed, S.; Fu, C.Y.; Berg, A.C. Ssd: Single shot multibox detector. In Proceedings of the Computer Vision–ECCV 2016: 14th European Conference, Amsterdam, The Netherlands, 11–14 October 2016; Proceedings Part I 14; Springer: Berlin/Heidelberg, Germany, 2016; pp. 21–37.
10. Redmon, J.; Divvala, S.; Girshick, R.; Farhadi, A. You only look once: Unified, real-time object detection. In Proceedings of the IEEE Conference on Computer Vision and Pattern Recognition, Las Vegas, NV, USA, 27–30 June 2016; pp. 779–788.
11. Ren, S.; He, K.; Girshick, R.; Sun, J. Faster R-CNN: Towards Real-Time Object Detection with Region Proposal Networks. *IEEE Trans. Pattern Anal. Mach. Intell.* **2017**, *39*, 1137–1149. [CrossRef]
12. Lin, T.Y.; Dollár, P.; Girshick, R.; He, K.; Hariharan, B.; Belongie, S. Feature pyramid networks for object detection. In Proceedings of the IEEE Conference on Computer Vision and Pattern Recognition, Honolulu, HI, USA, 21–26 July 2017; pp. 2117–2125.
13. Dollar, P.; Wojek, C.; Schiele, B.; Perona, P. Pedestrian detection: An evaluation of the state of the art. *IEEE Trans. Pattern Anal. Mach. Intell.* **2011**, *34*, 743–761. [CrossRef]
14. Geiger, A.; Lenz, P.; Urtasun, R. Are we ready for autonomous driving? The kitti vision benchmark suite. In Proceedings of the 2012 IEEE Conference on Computer Vision and Pattern Recognition, Providence, RI, USA, 16–21 June 2012; pp. 3354–3361.
15. Zhang, S.; Benenson, R.; Schiele, B. Citypersons: A diverse dataset for pedestrian detection. In Proceedings of the IEEE Conference on Computer Vision and Pattern Recognition, Honolulu, HI, USA, 21–26 July 2017; pp. 3213–3221.
16. Shao, S.; Zhao, Z.; Li, B.; Xiao, T.; Yu, G.; Zhang, X.; Sun, J. Crowdhuman: A benchmark for detecting human in a crowd. *arXiv* **2018**, arXiv:1805.00123.
17. Zhang, S.; Xie, Y.; Wan, J.; Xia, H.; Li, S.Z.; Guo, G. Widerperson: A diverse dataset for dense pedestrian detection in the wild. *IEEE Trans. Multimed.* **2019**, *22*, 380–393. [CrossRef]
18. Braun, M.; Krebs, S.; Flohr, F.; Gavrila, D.M. Eurocity persons: A novel benchmark for person detection in traffic scenes. *IEEE Trans. Pattern Anal. Mach. Intell.* **2019**, *41*, 1844–1861. [CrossRef] [PubMed]
19. Pang, Y.; Cao, J.; Li, Y.; Xie, J.; Sun, H.; Gong, J. TJU-DHD: A diverse high-resolution dataset for object detection. *IEEE Trans. Image Process.* **2020**, *30*, 207–219. [CrossRef]
20. Davis, J.W.; Sharma, V. OTCBVS Benchmark Dataset Collection. 2007. Available online: https://vcipl-okstate.org/pbvs/bench/ (accessed on 2 September 2023).
21. Toet, A. The TNO multiband image data collection. *Data Brief* **2017**, *15*, 249–251. [CrossRef] [PubMed]
22. González, A.; Fang, Z.; Socarras, Y.; Serrat, J.; Vázquez, D.; Xu, J.; López, A.M. Pedestrian detection at day/night time with visible and FIR cameras: A comparison. *Sensors* **2016**, *16*, 820. [CrossRef] [PubMed]
23. Hwang, S.; Park, J.; Kim, N.; Choi, Y.; So Kweon, I. Multispectral pedestrian detection: Benchmark dataset and baseline. In Proceedings of the IEEE Conference on Computer Vision and Pattern Recognition, Boston, MA, USA, 7–12 June 2015; pp. 1037–1045.
24. Jia, X.; Zhu, C.; Li, M.; Tang, W.; Zhou, W. LLVIP: A visible-infrared paired dataset for low-light vision. In Proceedings of the IEEE/CVF International Conference on Computer Vision, Montreal, BC, Canada, 11–17 October 2021; pp. 3496–3504.
25. Papageorgiou, C.; Poggio, T. A trainable system for object detection. *Int. J. Comput. Vis.* **2000**, *38*, 15–33. [CrossRef]

26. Dalal, N.; Triggs, B. Histograms of oriented gradients for human detection. In Proceedings of the 2005 IEEE Computer Society Conference on Computer Cision and Pattern Recognition (CVPR'05), San Diego, CA, USA, 20–25 June 2005; Volume 1, pp. 886–893.
27. Enzweiler, M.; Gavrila, D.M. Monocular pedestrian detection: Survey and experiments. *IEEE Trans. Pattern Anal. Mach. Intell.* **2008**, *31*, 2179–2195. [CrossRef]
28. Wojek, C.; Walk, S.; Schiele, B. Multi-cue onboard pedestrian detection. In Proceedings of the 2009 IEEE Conference on Computer Vision and Pattern Recognition, Miami, FL, USA, 20–25 June 2009; pp. 794–801.
29. Cordts, M.; Omran, M.; Ramos, S.; Rehfeld, T.; Enzweiler, M.; Benenson, R.; Franke, U.; Roth, S.; Schiele, B. The cityscapes dataset for semantic urban scene understanding. In Proceedings of the IEEE Conference on Computer Vision and Pattern Recognition, Las Vegas, NV, USA, 27–30 June 2016; pp. 3213–3223.
30. Girshick, R.; Donahue, J.; Darrell, T.; Malik, J. Rich feature hierarchies for accurate object detection and semantic segmentation. In Proceedings of the IEEE Conference on Computer Vision and Pattern Recognition, Columbus, OH, USA, 23–28 June 2014; pp. 580–587.
31. Girshick, R. Fast R-CNN. In Proceedings of the IEEE International Conference on Computer Vision, Santiago, Chile, 7–13 December 2015; pp. 1440–1448.
32. Cai, Z.; Fan, Q.; Feris, R.S.; Vasconcelos, N. A unified multi-scale deep convolutional neural network for fast object detection. In Proceedings of the Computer Vision–ECCV 2016: 14th European Conference, Amsterdam, The Netherlands, 11–14 October 2016; Proceedings Part IV 14; Springer: Berlin/Heidelberg, Germany, 2016; pp. 354–370.
33. Redmon, J.; Farhadi, A. YOLO9000: Better, faster, stronger. In Proceedings of the IEEE Conference on Computer Vision and Pattern Recognition, Honolulu, HI, USA, 21–27 July 2017; pp. 7263–7271.
34. Redmon, J.; Farhadi, A. Yolov3: An incremental improvement. *arXiv* **2018**, arXiv:1804.02767.
35. Bochkovskiy, A.; Wang, C.Y.; Liao, H.Y.M. YOLOv4: Optimal Speed and Accuracy of Object Detection. *arXiv* **2020**, arXiv:2004.10934.
36. Jocher, G.; Chaurasia, A.; Stoken, A.; Borovec, J.; Kwon, Y.; Michael, K.; Fang, J.; Yifu, Z.; Wong, C.; Montes, D.; et al. ultralytics/yolov5: v3.0. Zenodo. Available online: https://ui.adsabs.harvard.edu/abs/2022zndo...3908559J/abstract (accessed on 6 November 2023)
37. Li, C.; Li, L.; Jiang, H.; Weng, K.; Geng, Y.; Li, L.; Ke, Z.; Li, Q.; Cheng, M.; Nie, W.; et al. YOLOv6: A single-stage object detection framework for industrial applications. *arXiv* **2022**, arXiv:2209.02976.
38. Wang, C.Y.; Bochkovskiy, A.; Liao, H.Y.M. YOLOv7: Trainable bag-of-freebies sets new state-of-the-art for real-time object detectors. In Proceedings of the IEEE/CVF Conference on Computer Vision and Pattern Recognition, Vancouver, BC, Canada, 18–22 June 2023; pp. 7464–7475.
39. Eckert, M.; Blex, M.; Friedrich, C.M. Object detection featuring 3D audio localization for Microsoft HoloLens. In Proceedings of the 11th International Joint Conference on Biomedical Engineering Systems and Technologies (BIOSTEC 2018), Funchal, Portugal, 19–21 January 2018; Volume 5, pp. 555–561.
40. Bahri, H.; Krčmařík, D.; Kočí, J. Accurate object detection system on hololens using yolo algorithm. In Proceedings of the 2019 International Conference on Control, Artificial Intelligence, Robotics and Optimization (ICCAIRO), Athens, Greece, 8–10 December 2019; pp. 219–224.
41. Park, K.B.; Kim, M.; Choi, S.H.; Lee, J.Y. Deep learning-based smart task assistance in wearable augmented reality. *Robot. Comput. Integr. Manuf.* **2020**, *63*, 101887. [CrossRef]
42. Arifando, R.; Eto, S.; Wada, C. Improved YOLOv5-Based Lightweight Object Detection Algorithm for People with Visual Impairment to Detect Buses. *Appl. Sci.* **2023**, *13*, 5802. [CrossRef]
43. Maya-Martínez, S.U.; Argüelles-Cruz, A.J.; Guzmán-Zavaleta, Z.J.; Ramírez-Cadena, M.d.J. Pedestrian detection model based on Tiny-Yolov3 architecture for wearable devices to visually impaired assistance. *Front. Robot. AI* **2023**, *10*, 1052509. [CrossRef] [PubMed]

**Disclaimer/Publisher's Note:** The statements, opinions and data contained in all publications are solely those of the individual author(s) and contributor(s) and not of MDPI and/or the editor(s). MDPI and/or the editor(s) disclaim responsibility for any injury to people or property resulting from any ideas, methods, instructions or products referred to in the content.

Article

# Smart-Data-Glove-Based Gesture Recognition for Amphibious Communication

Liufeng Fan [1], Zhan Zhang [1], Biao Zhu [2], Decheng Zuo [1,*], Xintong Yu [1] and Yiwei Wang [1]

[1] School of Computer Science and Technology, Harbin Institute of Technology, Harbin 150001, China; fanliufeng@ftcl.hit.edu.cn (L.F.); zhangzhan@hit.edu.cn (Z.Z.); yuxintong@ftcl.hit.edu.cn (X.Y.); 22s103226@stu.hit.edu.cn (Y.W.)

[2] Department of Electronic and Information Science, University of Science and Technology of China, Hefei 230052, China; zhubiao7678@163.com

* Correspondence: zuodc@hit.edu.cn

**Abstract:** This study has designed and developed a smart data glove based on five-channel flexible capacitive stretch sensors and a six-axis inertial measurement unit (IMU) to recognize 25 static hand gestures and ten dynamic hand gestures for amphibious communication. The five-channel flexible capacitive sensors are fabricated on a glove to capture finger motion data in order to recognize static hand gestures and integrated with six-axis IMU data to recognize dynamic gestures. This study also proposes a novel amphibious hierarchical gesture recognition (AHGR) model. This model can adaptively switch between large complex and lightweight gesture recognition models based on environmental changes to ensure gesture recognition accuracy and effectiveness. The large complex model is based on the proposed SqueezeNet-BiLSTM algorithm, specially designed for the land environment, which will use all the sensory data captured from the smart data glove to recognize dynamic gestures, achieving a recognition accuracy of 98.21%. The lightweight stochastic singular value decomposition (SVD)-optimized spectral clustering gesture recognition algorithm for underwater environments that will perform direct inference on the glove-end side can reach an accuracy of 98.35%. This study also proposes a domain separation network (DSN)-based gesture recognition transfer model that ensures a 94% recognition accuracy for new users and new glove devices.

**Keywords:** hand gesture recognition; smart data glove; underwater gesture recognition; amphibious communication; deep learning; transfer learning

## 1. Introduction

With the continuous development of wearable sensor technology, human–computer interaction (HCI) has become an important research area in computer science. As an essential branch of HCI, gesture recognition technology can be applied to various fields, such as smart homes [1], intelligent driving [2], sign language recognition [3], virtual reality [4], and drone control [5]. With the continuous improvements in gesture recognition technology, this technology can also be used in amphibious environments to complete some tasks, such as communication with divers and underwater operations [6].

Although traditional vision-based gesture recognition technology has matured, it has significant limitations in underwater environments [7,8]. The cost of underwater cameras is high, the underwater shooting environment is complex, and it is very easy to be disturbed by water flow, water bubbles, etc., which hinder the line of sight and make shooting difficult. Sensor-based gesture recognition technology has become popular for underwater gesture recognition because of its lower cost and higher stability (not easily affected by the underwater environment). It has become a research area that many researchers are interested in. However, sensor-based gesture recognition technologies still face many challenges in amphibious environments.

First, the environment could affect the sensors, leading to some discrepancies between the gesture data collected on land and underwater. In the underwater environment, factors such as water pressure, water flow, and water quality will affect the sensors, affecting the accuracy and integrity of data collection. Secondly, people will feel more resistance and pressure due to the increased water depth, resulting in slow or non-standard gestures. Thirdly, there needs to be more accuracy when using a pretrained recognition model to perform cross-user and cross-device gesture recognition. Fourthly, Bluetooth signals cannot transmit underwater, and how to collect data and recognize real-time gestures underwater is also a problem that needs to be solved. Finally, although several existing data gloves are used for gesture recognition, none are used for amphibious environments. The above challenges must be considered when designing and selecting a gesture recognition model to ensure accuracy, robustness, and reliability.

This paper addresses the above research gaps by developing a smart data glove integrating environmental sensors, five-channel capacitive flexible stretch sensors, and six-axis IMU (three-axis accelerometer and three-axis gyroscope) and proposing a novel hierarchical hand gesture recognition model. The proposed model introduces a novel SqueezeNet-BiLSTM algorithm, a large complex recognition algorithm designed for land gesture recognition, and another lightweight stochastic SVD-optimized algorithm designed for underwater gesture recognition, which will be directly applied to the glove-end side. Additionally, this study introduces the DSN-based transfer learning for gesture recognition to ensure the recognition accuracy of new users and new glove devices. This paper makes the following contributions:

- A new smart data glove integrating environmental sensors, five-channel capacitive flexible stretch sensors, and six-axis IMU (three-axis accelerometer and three-axis gyroscope).
- A novel amphibious hierarchical gesture recognition (AHGR) model that can adaptively switch the classification algorithm based on the environment (underwater and land) between
  - ○ a complex SqueezeNet-BiLSTM classification algorithm for land gesture recognition and
  - ○ a lightweight stochastic SVD-optimized spectral clustering classification algorithm for underwater gesture recognition.
- A domain separation network (DSN)-based gesture recognition transfer model to ensure the recognition accuracy of new users and new glove devices.

The rest of the paper is organized as follows: Section 2 provides a review of related work. Section 3 introduces this study's proposed smart data glove and predefined gesture set. Section 4 describes the proposed amphibious hierarchical gesture recognition model. Section 5 describes the proposed DSN-based gesture recognition transfer model. Section 6 presents the experimental results and analysis. Section 7 concludes this paper.

## 2. Related Work

### 2.1. Sensor-Based Gesture Recognition

Sensor-based gesture recognition can be roughly divided into the following four types: surface electromyography (sEMG) signal-based gesture recognition, IMU-based gesture recognition, stretch-sensor-based gesture recognition, and multi-sensor-based gesture recognition.

sEMG usually records the combined effect of the electromyographic signal of the surface muscle and the nerve trunk's electrical activity on the skin's surface. sEMG-based gesture recognition usually relies on surface electrodes deployed on the human arm or forearm to collect sensor signals [9–12]. However, sEMG-based gesture recognition also has some drawbacks. Firstly, the signals correlate strongly with the user's status, leading to unstable recognition results. Secondly, the collection of sEMG signals requires the electrodes to be tightly attached to the user's skin, and prolonged use is susceptible to the influence of oils and sweat produced by the user's skin and makes users uncomfortable.

IMU-based gesture recognition mainly uses one or more combinations of accelerometers, gyroscopes, and magnetometers to collect hand movement information in the space field [13]. Siddiqui and Chan [14] used the minimum redundancy and maximum correlation algorithm to study the optimal deployment area of the sensor, deployed the sensor on the user's wrist, and proposed a multimodal framework to solve the IMU sensing during the gesture movement bottleneck problem. Galka et al. [15] placed seven inertial sensors on the experimenter's upper arm, wrist, and finger joints, proposed and used a parallel HMM model, and reached a recognition accuracy of 99.75%. However, inertial sensors still have limitations, and they focus more on spatial dimension information, which is mainly used for coarse-grained gesture recognition of large gesture movements. It is challenging to perform finer-grained segmentation and recognition, such as recognition of the degree of bending of finger joints.

Flexible stretch-sensor-based gesture recognition is usually used to record changes in gesturing finger joints. Stretch sensors are often highly flexible, thinner, and more portable than other sensors [16,17]. Therefore, in recent years, research on gesture recognition technology based on stretch sensors has also received extensive attention from researchers. However, the limitations of flexible stretch sensors are also evident. First, they can only capture hand joint information but cannot capture the spatial motion characteristics of gestures. Second, stretch sensors are usually sensitive, so they are more prone to damage, and the data they generate are more prone to bias than those from other sensors.

Although the above three sensor-based gesture recognition methods can achieve remarkable gesture recognition accuracy, they all have some limitations, because they only use a single type of sensor. Multisensor gesture recognition can perfectly solve these problems by fusing multisensor data, thereby improving the recognition accuracy and recognizing more types of gestures. Plawiak et al. [16] used a DG5 VHand glove device, which consists of five finger flexion sensors and IMU, to identify 22 dynamic gestures, and the recognition accuracy rate reached 98.32%. Lu et al. [18] used the framework of acceleration signal and surface electromyography signal fusion, proposed an algorithm based on Bayesian and dynamic time warping (DTW), and realized a gesture recognition system that can recognize 19 predefined gestures with a recognition accuracy rate of 95.0%. Gesture recognition with multisensor fusion can avoid the limitations of a single sensor, learn from the strengths of multiple approaches, capture the characteristics of each dimension of gestures from multiple angles, and improve the accuracy of gesture recognition.

To date, all these studies are based on gesture recognition on land, and there is no related research on sensor-based gesture recognition underwater. This paper aims to fill this research gap by using a multi-sensor-based gesture recognition approach and developing a new smart data glove that incorporates environmental sensors, five-channel capacitive flexible stretch sensors, and a six-axis IMU (three-axis acceleration meter and three-axis gyroscope) mounted on the back of the hand.

*2.2. Sensor-Based Gesture Recognition Algorithm*

Sensor-based gesture recognition algorithms are generally divided into the following two types: traditional machine learning and deep learning.

Gesture recognition algorithms based on machine learning (ML) include DTW, support vector machine (SVM), random forest (RF), K-means, and K-nearest neighbors [16,19–21]. These methods are widely applicable and adaptable to various types of complex gesture data. At present, many researchers have conducted research on the improvement of related algorithms in sensor-based gesture recognition. Although the ML-based gesture recognition method is relatively simple to implement, the number of parameters generated is also lower than that of neural networks, and the requirements for the computing equipment are relatively low. However, with the increase in gesture types and gesture data sequences, the training data required for learning is also increasing. The accuracy and response time of the recognition algorithm will also be affected to a certain extent.

The basic model of deep learning (DL)-based gesture recognition mainly includes the convolutional neural network (CNN) [22], deep neural network (DNN) [23], and recurrent neural network (RNN) methods [24]. The DL model has become the mainstream classification method in gesture recognition due to its excellent performance, high efficiency in extracting data features, and ability to process sequential data. Fang et al. [25] designed a CNN-based SLRNet network to recognize sign language. This method used an inertial-sensors-based data glove with 36 IMUs to collect a user's arm and hand motion data, and the accuracy can reach 99.2%. Faisal et al. [26] developed a low-cost data glove deployed with flexible sensors and an IMU, and introduced a spatial projection method that improves upon classic CNN models for gesture recognition. However, the accuracy of this method for static gesture recognition is only 82.19%. Yu et al. [27] used a bidirectional gated recurrent unit (Bi-GRU) network to recognize dynamic gestures, realize real-time recognition on the end side (data glove), and reach a recognition accuracy of 98.4%. The limitation of this approach is that it is not possible to only use the smart glove, but external IMUs must be employed on the user's arm, which can cause discomfort to the user.

The selected model needs to be determined according to the type of task, requirements, and other factors. Due to the complex amphibious environment, the underwater and land environments are different, and the interference to the sensor is entirely different. It is difficult to transmit Bluetooth signals underwater, and it is difficult to send data to the host wirelessly. Therefore, choosing a gesture recognition model suitable for the amphibious environment is essential. This study addresses this gap by proposing a novel amphibious hierarchical gesture recognition (AHGR) model that adaptively switches classification algorithms according to environmental changes (underwater and land) to ensure recognition accuracy in amphibious scenarios. In addition, it is also challenging to ensure accuracy for cross-user and cross-device recognition using a pretrained DL model. Although some studies on gesture recognition across users and in different environments has made some progress [12], they were mainly focused on EMG-based gesture recognition, and there is a lack of research on cross-user gesture recognition using data gloves based on stretch sensors and IMUs. This study, then, introduces the transfer learning framework to the recognition model and proposes a DSN-based gesture recognition transfer model to solve this issue.

## 3. Smart Data Glove and Gesture Set

The following subsections describe in detail the proposed smart data gloves and the predefined gesture set.

### 3.1. Smart Data Glove

The smart glove developed in this study is shown in Figure 1. As shown in Figure 1a, the glove uses a five-channel flexible capacitive stretch sensor to collect the bending state of five fingers. The main control module located on the back of the hand is equipped with a Bluetooth communication module for wireless transmission of the collected gesture data, a six-axis IMU (three-axis accelerometer and three-axis gyroscope) for collecting hand spatial motion information, an environmental sensor for inferring the land and underwater environment, a microcontroller to process the collected gesture data and perform some simple computational tasks, and a battery to support electricity energy. The microcontroller used in the smart data glove is the Esp32-S3-DevKitC-1 development board [28]. This microcontroller is equipped with an ESP32-S3-WROOM-1 module, a general-purpose Wi-Fi+ low-power Bluetooth MCU, which has rich peripheral interfaces, powerful neural network computing and signal processing capabilities, and is specially designed for artificial intelligence (AI) and Internet of Things (IoT) market creation. It is equipped with 384 KB of ROM, 512 KB of SRAM, 16 KB of RTC SRAM, and a maximum of 8 MB of PSRAM to meet the experimental requirements. The detailed technical information of the proposed smart data glove is shown in Table 1.

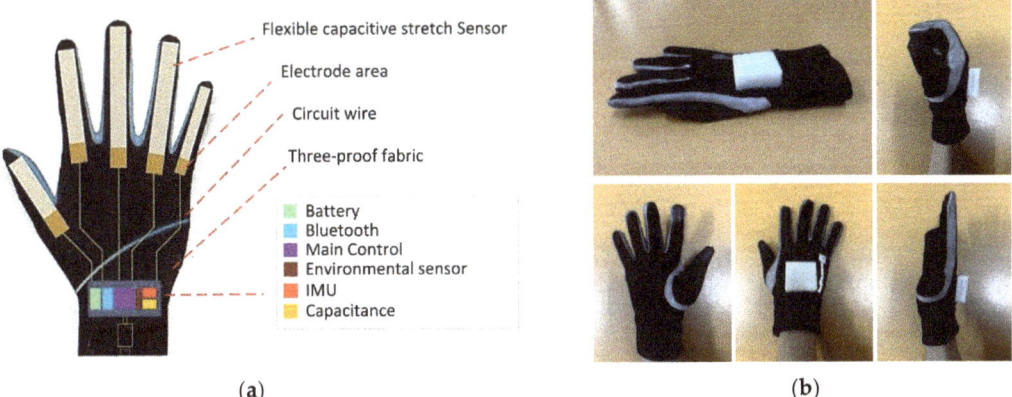

Figure 1. Proposed smart data glove: (a) structure diagram, (b) appearance.

Table 1. Detailed technical information of the proposed smart data glove.

| Indicator Name | Parameter |
| --- | --- |
| Stretch range | 0–50% |
| Minimum trigger strain | 0.05% |
| Response time | 15 ms |
| Stretch fatigue | >100,000 times |
| Data transmission | Bluetooth wireless data transmission. |
| Data collection frequency | 50 Hz |
| Microcontroller | Esp32-S3-DevKitC-1 |
| Battery capacity | 500 mah (when we deploy the proposed lightweight model on the glove side and conduct static gesture recognition and output signals in real time via Bluetooth. Through experimental testing, the battery power can last for 5–6 h.) |
| Battery type | Lithium polymer battery |
| Environmental sensor | Barometer Senor QMP6988 |
| IMUs | MPU6050 (3-axis accelerometer and gyroscope) |

*3.2. Gesture Set*

In the sensor-based gesture recognition technology, according to the characteristics of the stretch sensor and IMU loaded on the data glove, gestures can be divided into dynamic and static gestures according to the characteristics of the activity.

Static gestures are defined by the finger bending status. Since there are some difficult-to-operate gestures, some gestures were discarded, and 25 gestures were finally defined, as shown in Figure 2.

Dynamic gestures combine finger bending information (static gesture) with hand motion trajectories to characterize gesture types. We use the signal fluctuation of the motion sensor to distinguish the dynamic and static gestures. At the same time, the definition of the dynamic gesture set refers to the existing gesture sets, such as the sign language gesture set used by deaf–mute patients, and based on the distinguishability, operability, and understandability of the gesture design, 10 dynamic gestures are predefined, as shown in Figure 3.

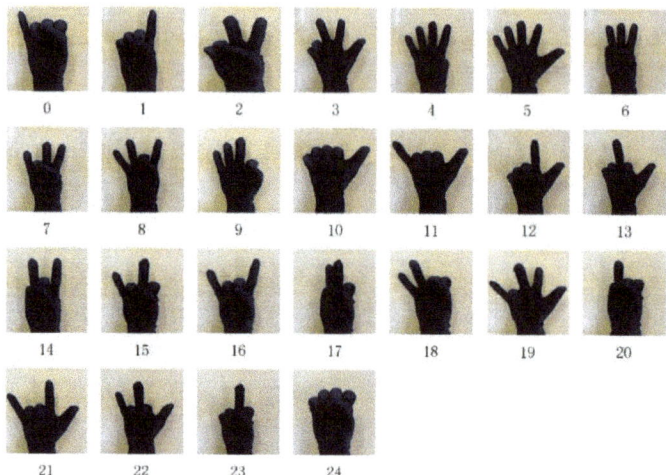

**Figure 2.** Static gesture set.

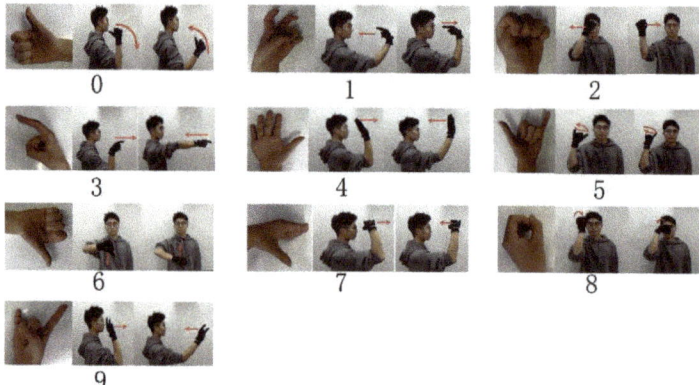

**Figure 3.** Dynamic gesture set.

In the face of different task environments, the gestures' meanings may differ. Therefore, this research does not assign specific meanings to static and dynamic gestures. It only describes them with serial numbers, where static gestures are assigned with serials from 0–24 and dynamic gestures with 0–9. Thus, users can assign meaning to gestures when dealing with different tasks. In the underwater environment, due to the influence of the water resistance and air pressure, the IMU data will be affected to a certain extent, resulting in data distortion. In contrast, stretch sensor data are very slightly affected by the environment. Secondly, users are easily affected by environmental factors such as the water flow, resulting in movement deviation and incomplete and non-standard gestures. This makes the data collected via IMU vary greatly for the same gesture, making training and testing difficult. Static gestures are less affected by the environment, and they can still be made accurately in an underwater environment. Finally, Bluetooth data are difficult to transmit underwater to the host, and underwater gestures must be recognized on the glove side. Static gesture recognition adopts a lightweight model that can be deployed on a microprocessor with limited computing power, so that static gestures can be recognized directly on the glove side. Based on the consideration of these factors, this study uses static gestures for underwater gesture recognition. The ground environment supports static and dynamic gesture recognition.

## 4. Amphibious Hierarchical Gesture Recognition Model

Due to the differences between underwater and land environments, this study proposes the AHGR model for gesture recognition in amphibious environments with a hierarchical structure. This section describes the details of the proposed AHGR model, including the hierarchical gesture recognition flow, the lightweight stochastic SVD-optimized spectral clustering algorithm for underwater gesture recognition, and the complex SqueezeNet-BiLSTM algorithm for land gesture recognition.

### 4.1. Hierarchical Gesture Recognition Flow

Affected by the underwater environment, it is difficult for users to make precise dynamic gestures underwater. The IMU signal will be greatly disturbed underwater, affected by water pressure, resistance, water flow, etc. Static gestures have no complex spatial motion, relying only on stretch sensor data to represent the gesture state information. Additionally, stretch sensors are less affected by the underwater environment. Thus, underwater gesture recognition only considers static gesture recognition using stretch sensor data. And since gesture recognition needs to be performed directly on the glove end in an underwater environment, choosing a recognition algorithm model with less recognition latency and less computing power requirements is necessary to ensure adequate gesture recognition performance in an underwater environment. Therefore, this study proposes a lightweight stochastic SVD-optimized spectral clustering algorithm to recognize underwater static gestures.

In the land environment, both static and dynamic gesture recognition are relatively easy to implement and acquire. There are still some challenges regarding dynamic gesture recognition on land. Although there is no interference from the water environment, the user will inevitably tremble to a certain extent when making gestures, which will cause fluctuations in sensor (IMUs) data and affect the recognition accuracy. The dynamic gesture recognition problem is a placement-independent problem with strong temporal characteristics, and a model capable of deep feature extraction in temporal and spatial dimensions is required. Thus, this study adopts the method of multisensor data fusion and proposes a complex SqueezeNet-BiLSTM algorithm for dynamic gesture recognition on land to ensure the effectiveness, robustness, and accuracy of the recognition results.

As shown in Figure 4, the detailed amphibious gesture recognition process of the AHGR model is as follows: The AHGR model first determines the recognition environment based on environmental sensors. The environmental sensor used in the AHGR model is a barometer sensor. According to the principles of hydrostatic pressure, when the air pressure sensor value is greater than the local standard atmospheric pressure plus 0.98 kpa (water depth is greater than 0.1 m), the current environment is underwater; otherwise, it is judged to be a land environment. If it is underwater, the AHGR model will switch to underwater gesture recognition and use the proposed lightweight stochastic SVD-optimized spectral clustering algorithm to recognize static gestures on the glove side. If it is on land, the AHGR model will first switch to land gesture recognition and determine the dynamic and static gestures through the fluctuations in the IMU data. If it is a static gesture, the land gesture recognition will directly output the result of the static gesture recognized using the lightweight stochastic SVD-optimized spectral clustering algorithm. If it is a dynamic gesture, land gesture recognition will use the SqueezeNet-BiLSTM algorithm to recognize dynamic gestures using multisensor data and encoded static gesture recognition results. The recognition results can be used to interact with or control devices in the IoT environment.

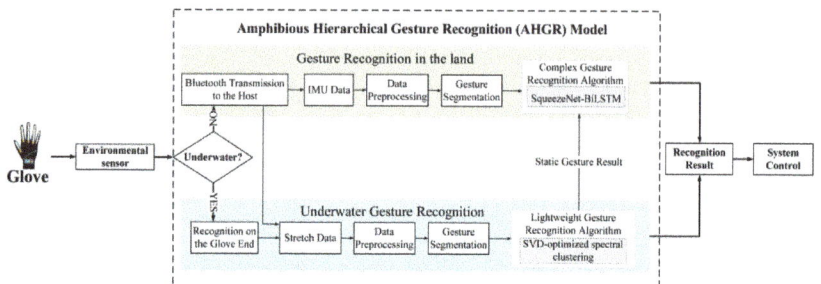

**Figure 4.** Amphibious hierarchical gesture recognition (AHGR) model.

*4.2. Stochastic SVD-Optimized Spectral Clustering Algorithm*

The spectral clustering algorithm is an algorithm evolved from graph theory [29]. Its main idea is to regard all data as points in the space, connect them with edges in the graph, calculate the weight by calculating the distance from the point to the edge, and finally realize clustering according to the weight. Although the spectral clustering algorithm can complete the clustering of high-dimensional data, the spectral clustering algorithm relies too heavily on the Laplacian matrix to complete the eigen decomposition. The calculation process requires extremely high space complexity and time complexity, and with the increase in data volume, the complexity also increases exponentially, seriously affecting the practical applications. Therefore, this study introduces the stochastic SVD [30] algorithm to accelerate the spectral clustering algorithm and reduce the computational cost.

SVD is a matrix decomposition method widely used in pattern recognition to reduce dimensions and solve ranks. The main process is to establish the connection between the large matrix and the small matrix and estimate the SVD result of the large matrix through the SVD decomposition result of the small matrix. This study considers using a stochastic SVD [31] algorithm. In this algorithm, an orthogonal matrix is established first and used as an orthogonal basis for the low-rank estimation of the original matrix. At the same time, the original matrix is projected to the subspace, the matrix formed in the subspace is subjected to SVD, and the decomposition result is mapped back to the original space. The detailed process is as follows:

Let the original matrix be $W \in R^{n \times n}$. First, select a standard Gaussian random matrix $\Omega$ of $n \times (k+p)$, where $k$ is the dimension of the low-rank estimate, and $p$ is the oversampling parameter, so that the rank of the random subspace is slightly larger than $k$. Let $Z = W\Omega$, and then find an orthogonal matrix $Q \in R^{n \times k}$ through QR decomposition to let $Z = QQ^T Z$. Map the original matrix $W$ to the subspace with $Q$ as the orthogonal basis, and obtain

$$B = Q^T W Q, \tag{1}$$

For the SVD decomposition of B, obtain

$$B = VMV^T, \tag{2}$$

Then, the k-rank estimation of the original matrix $W$ is obtained as

$$W \approx QBQ^T = QVMV^T Q^T, \tag{3}$$

Therefore, the estimated eigenvector of $W$ is $U = QV$. The stochastic SVD algorithm avoids direct SVD decomposition of large matrices by mapping high-dimensional matrices to low-dimensional subspaces. Hence, the information on the original matrix is almost completely preserved. The stochastic SVD-optimized spectral clustering algorithm is shown below as Algorithm 1.

**Algorithm 1:** SVD-optimized spectral clustering

**Input:** $X = \{x_1, x_2, \ldots, x_n\}, x_i \in R^N$
**Output:** Clustering result of $x_1, x_2, \ldots, x_n$

for i, j = 1, ..., n:
$s(x_i, x_j) \leftarrow \exp\left(\dfrac{*-d(x_i,x_j)^2}{2\sigma^2}\right)$
$A_{ij} = s(x_i, x_j)$
end
$[u, s, v]$ = Randomized_SVD(A)
# u, v is the left and right singular vector matrix of A
# s is the singular value matrix of A, s = diag$(\sigma_1, \sigma_2, \ldots, \sigma_n,)$
$U \leftarrow \{u_1, u_2, \ldots, u_l\} \in R^{n \times l}$, where ui is the i-th vector of u
$y_i \in R^l, i = 1, 2, \ldots, n$ is the i-th row vector of matrix U
$C_1, C_2, \ldots, C_k \leftarrow Kmeans(y_i)$
Create mapping $x_i \in R^N \vdash y_i \in R^l, i = 1, 2, \ldots, n$
Output the clustering results of $x_1, x_2, \ldots, x_n$

### 4.3. SqueezeNet-BiLSTM Algorithm

The proposed SqueezeNet-BiLSTM gesture classification algorithm first uses the Tucker decomposition algorithm to reduce the dimensionality and extract features of the preprocessed gesture data. After that, the SqueezeNet [32] network is used to extract in-depth data features and combined with the Bi-LSTM [33] network to extract the time series features of the gesture data to ensure the robustness of the gesture recognition model and improve the recognition accuracy. Tucker [20] decomposition is a high-dimensional data analysis method, especially suitable for dimensionality reduction and feature extraction of multidimensional data. It decomposes higher-order tensors into products of core tensors and some modality matrices. In this process, the dimensionality reduction of the data can be achieved by retaining the principal components of the core tensor, thereby removing irrelevant information and noise. The SqueezeNet [24] network adopts the idea of compression and expansion. Compared with the traditional convolutional neural network, it reduces the model parameters while ensuring the gesture recognition accuracy. A Bi-LSTM network, through the stacking of two layers of LSTM structure, solves the limitation that LSTM can only predict the output of the next moment based on the timing information of the previous moment. It can better combine the context for output and more effectively utilize the input gesture data's forward and backward feature information. The structure diagram of the proposed SqueezeNet-BiLSTM algorithm is shown in Figure 5.

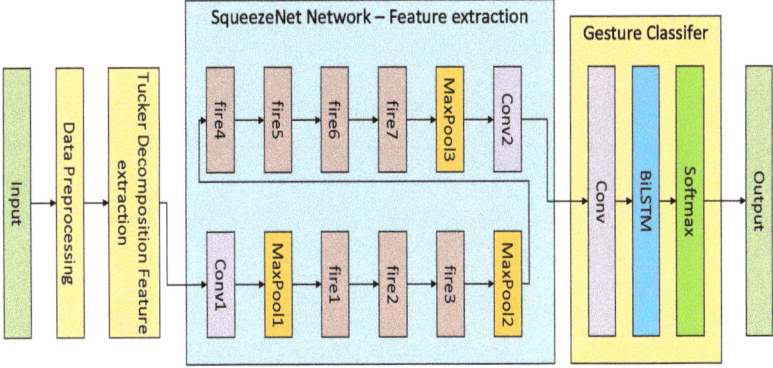

**Figure 5.** Structure diagram of SqueezeNet-BiLSTM algorithm.

The gesture recognition process of the SqueezeNet-BiLSTM model is as follows: For the gesture data collected by the smart data glove, the scale of the original sensor data is adjusted to a uniform length through operations such as sliding window, filter processing, standardization, normalization, data length normalization, and Turker decomposition [34]. The processed gesture feature data are input into the SqueezeNet network to obtain the corresponding feature vector through the multilayer convolution module, fire module, and maximum pooling layer, and then, the time series features are extracted from the gesture data through the BiLSTM network, and finally through the SoftMax to complete the gesture classification.

## 5. DSN-Based Gesture Recognition Transfer Model

During gesture recognition, the collected gesture data from the data gloves are subject to variations due to different users and different data gloves, leading to discrepancies that result in reduced recognition accuracy when incorporating new users or new data gloves into the recognition system. Employing user-specific model training during recognition requires substantial data from diverse users. While this approach may yield personalized gesture recognition models tailored to the unique characteristics of each user, it can potentially compromise the user experience for new users. Leveraging transfer learning facilitates the adaptation of existing gesture recognition models to acquire the distinctive gesture data features associated with new users and new data gloves. This approach enables the preservation of the intrinsic gesture recognition domain features while concurrently acquiring domain-specific features from the new context, thereby enhancing the recognition efficiency of the source model when confronted with novel data. Therefore, this study presents a novel DSN-based [35] gesture recognition transfer model, leveraging the principles of transfer learning. By collecting a small but representative dataset from the new domain, this model facilitates the transfer of the gesture recognition model, ensuring its effectiveness in accurately recognizing new data and enhancing the overall user experience.

### 5.1. Domain Separation Networks

Considering the inherent differences in gesture data among various users and different data gloves, it is acknowledged that the data space for gesture data is not entirely congruent. However, it is observed that certain common features exist alongside the distinct characteristics that are specific to each data domain. A transfer learning methodology utilizing DSN is considered to address this. This approach aims to uncover shared feature representations across users and data gloves while capturing domain-specific features simultaneously. During the transfer process, the source domain's private features are discarded, while the shared features are preserved, thereby ensuring the successful migration of the model.

The main work of DSNs [35] is divided into two parts: extracting common features of different domains and using common features for migration. The obtained DSN structure is shown in Figure 6.

A DSN is a "Decoder-Encoder" structure, which can be divided into five parts:

1. Target Domain Private Encoder $E_p^t(X^t)$: Used to extract private features of the target domain.
2. Source Domain Private Encoder $E_p^s(X^s)$: Used to extract private features of the source domain.
3. Shared Encoder $E_c(X)$: Used to extract the common features of the source and target domains.
4. Shared Decoder $D(E_c(X) + E_p(X))$: Used to decode samples composed of private features and shared features.
5. Classifier $G(E_c(X^s))$: The source domain samples are classified during training, and the classification is completed directly on the target domain when the training is completed.

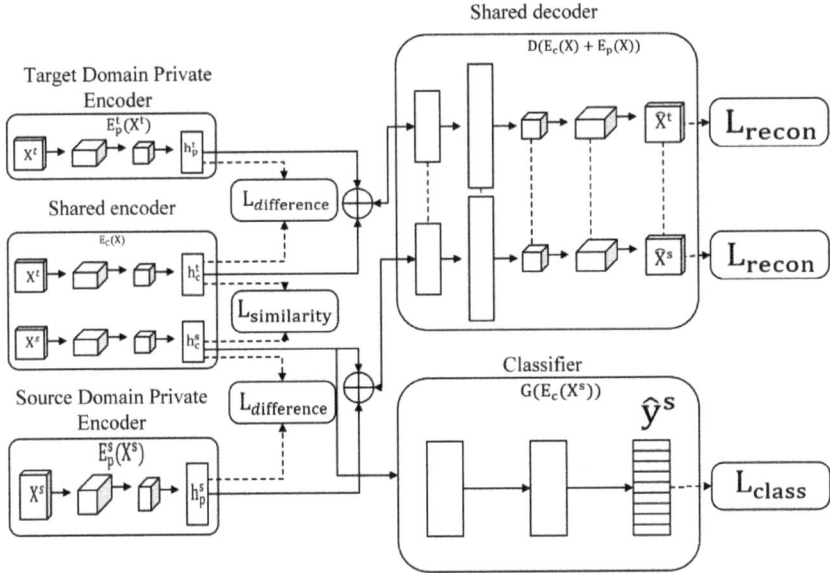

**Figure 6.** DSN structure diagram.

The overall work of the DSN is based on the original gesture recognition model structure, the model is used as an encoder, and the overall training goal is to minimize the difference loss $L_{difference}$:

$$L_{difference} = \left\| H_c^{sT} H_p^s \right\|_F^2 + \left\| H_c^{tT} H_p^t \right\|_F^2 \tag{4}$$

$L_{difference}$ calculates the similarity between $h_c^s$ and $h_p^s$ and $h_c^t$ and $h_p^t$. When $h_c^s = h_p^s$ and $h_c^t = h_p^t$, $L_{difference}$ is the largest, and when $h_c^s$ and $h_p^s$ are orthogonal (that is, completely different) and $h_c^t$ and $h_p^t$ are orthogonal, $L_{difference}$ is the smallest. Therefore, the purpose of completely separating $h_c^s$ from $h_p^s$ and $h_c^t$ from $h_p^t$ can be achieved by minimizing $L_{difference}$.

While ensuring that $h_c^s$ and $h_p^s$ and $h_c^t$ and $h_p^t$ are completely separated, it is necessary to ensure that $h_c^s$ s and $h_c^t$ can be transferred, meaningthat it is necessary to improve the similarity between the two, that is, to reduce the similarity loss $L_{similarity}$:

$$L_{similarity} = \frac{1}{(N^s)^2} \sum_{i,j=0}^{N^s} k\left(h_{ci}^s, h_{cj}^s\right) - \frac{1}{N^s N^t} \sum_{i,j=0}^{N^s,N^t} k\left(h_{ci}^s, h_{cj}^t\right) + \frac{1}{(N^t)^2} \sum_{i,j=0}^{N^t} k\left(h_{ci}^t, h_{cj}^t\right) \tag{5}$$

When the similarity loss $L_{similarity}$ is the smallest, $h_c^s$ and $h_c^t$ can be made the most similar or even become the same distribution. When the two distributions are similar, the classifier that is effective on $h_c^s$ can also work on $h_c^t$. While meeting the above conditions, it is also necessary to complete the measurement of the source domain data and perform target domain data assurance. Using the "encoder-decoder" structure, set the reconstruction loss $L_{recon}$:

$$L_{si\_mse}(x, \hat{x}) = \frac{1}{k} \|x - \hat{x}\|_2^2 - \frac{1}{k^2} ([x - \hat{x}] \cdot 1_k)^2 \tag{6}$$

$$L_{recon} = \sum_{i=1}^{N_s} L_{si\_mse}(x_i^s, \hat{x}_i^s) + \sum_{i=1}^{N_t} L_{si\_mse}(x_i^t, \hat{x}_i^t) \tag{7}$$

After extracting the shared features and their respective private features of the source domain and target domain samples, it is still necessary to classify the samples and set the classifier loss function $L_{task}$. After minimizing $L_{similarity}$, the distribution of the shared part of the source domain and the target domain is approximated. The classifier is effective in the common part of the source domain while ensuring that the common part of the target domain is also effective. Therefore, it only needs to use the labeled source domain data to train the classifier.

$$L_{task} = -\sum_{i=0}^{N_s} y_i^s \cdot \log \hat{y}_i^s \qquad (8)$$

*5.2. The Structure of the Gesture Recognition Model*

According to the DSN structure and basic principles, and based on the gesture recognition process, the small-sample gesture recognition transfer model proposed in this study is shown in Figure 7.

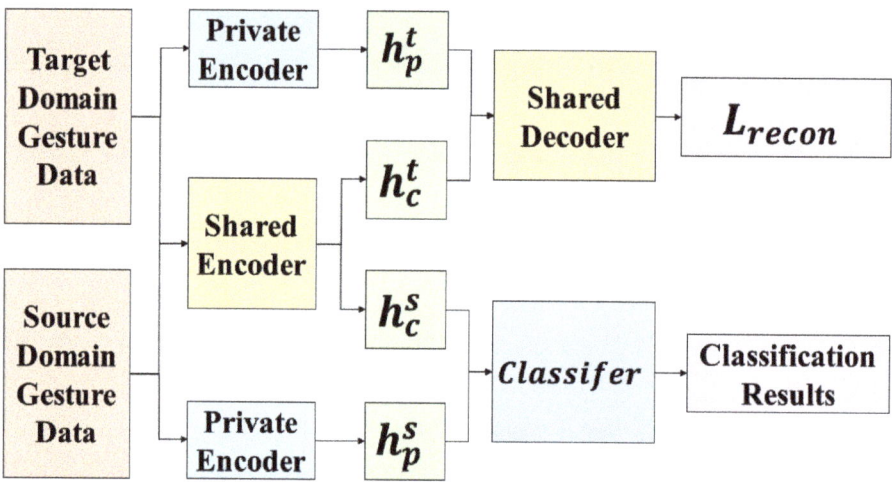

**Figure 7.** DSN-based small-sample gesture recognition transfer model.

The network recognition process is as follows: The labeled source domain gesture data are processed using private encoders and shared encoders to extract private features and shared features, respectively. Similarly, the unlabeled target domain gesture data are processed using private encoders and shared encoders to extract private features and shared features separately. By performing the computations outlined in Equations (4) and (5), the difference loss $L_{difference}$ and similarity loss $L_{similarity}$ are obtained. The shared features from the source and target domains, along with the target domain's private features, are fed into the shared decoder. This process involves the computations specified in Equations (5) and (6), resulting in the reconstruction loss $L_{recon}$. Furthermore, a classifier $L_{task}$ is constructed using the shared features from the source domain and the corresponding data labels. This entire procedure is repeated iteratively to minimize the overall loss function $L_{task} + \alpha L_{recon} + \beta L_{difference} + \gamma L_{similarity}$, where $\alpha$, $\beta$, and $\gamma$ are hyperparameters controlling the respective loss terms. Ultimately, the obtained classifier is utilized for recognizing gesture data collected from the target domain, i.e., new users with new data gloves. The network structure of the encoder and decoder is shown in Figure 8.

For the encoder part, we use a two-layer convolution structure to encode the gesture data. The first-layer convolution kernel size is set to three and passed through the ReLU layer to accelerate model convergence. At the same time, a maximum pooling layer with a kernel size of two is used to alleviate the convolution layer's sensitivity to positional relationships. The second-layer convolution kernel size is five in order to capture the data correlation characteristics of different areas. It then adopts a similar ReLU layer and maximum pooling layer, and then accesses the coding features obtained by the fully connected layer output operation.

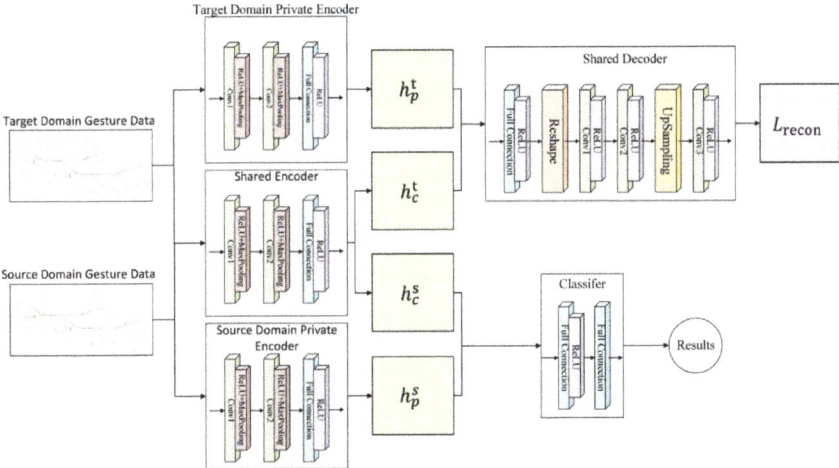

**Figure 8.** Proposed DSN network structure diagram.

For the shared decoder part, we first use the fully connected layer to decode the private features and public features and use the Reshape unit to modify the output of the fully connected layer to the size corresponding to the convolutional neural network. Then, we use two layers of convolution and ReLU layers with a convolution kernel of five and a UpSampling unit for deconvolution to restore the data. Finally, the restored data are operated through the convolution and ReLU layers to obtain the reconstruction loss $L_{recon}$.

## 6. Experimental Results and Analysis

This section will discuss the gesture data collection, experiments, and results to verify the effectiveness of the AHGR model proposed in this study.

### 6.1. Data Collection

Based on the amphibious environment, this study will collect and build hand gesture datasets in land and underwater environments. The gesture data collection setup is shown in Table 2.

**Table 2.** Gesture data collection setup.

| Parameters | On Land | Underwater |
|---|---|---|
| Gesture Type | 25 static gestures and 10 dynamic gestures | 25 static gestures |
| Data collected | 3-axis acceleration data, 3-axis gyroscope data, and 5-channel stretch data | 5-channel stretch sensor data |
| Data collection devices | Developed smart data glove and used Bluetooth to transfer data to Android devices | Esp32-S3-DevKitC-1 |
| Sensors | Accelerometer, gyroscope, and 5-channel flexible capacity stretch sensors | Smart data glove built-in 5-channel flexible capacity stretch sensors |
| Sampling rate | 50 HZ (0.02 s per reading) | 50 HZ (0.02 s per reading) |
| Time duration | 10 min for each gesture | 10 min for each gesture |
| Environment | 20 volunteers wear smart data gloves on land | 20 volunteers wear smart data gloves and put them under water |
| File format for storing data | .txt file | .txt file |

The land environment's gesture dataset includes dynamic and static gesture data. A total of 20 volunteers participated in the data collection experiments. During the data collection, the volunteers were asked to wear a data glove on their right hand and maintain a stable standing posture. After starting the gesture collection, volunteers had to make corresponding predefined dynamic and static gestures, and each gesture lasted for ten minutes. The land gesture dataset collected a total of 250,000 sets of static gesture data and 100,000 sets of dynamic gesture data, and each set of data comprises 60 data points, which is the window size.

The underwater gesture dataset is defined and constructed for the static gesture set, and the data collection flow diagram is shown in Figure 9. The underwater gesture data collection process is as follows: First, simulate the underwater environment and use a water-filled pool. Second, 20 volunteers put the smart data glove on their right hand, then put on a thin nitrile glove to make it waterproof. Third, volunteers put their hands into the water-filled pool, make the corresponding gesture, and then turn on the data glove's power. The fingers of the hand should be at least 0.15 m away from the bottom of the pool, and the elbow should be at least 0.5 m away from the water's surface. For each gesture, the volunteers had to remain underwater for at least 1 min. After a gesture data collection process is completed, the glove must be connected to the computer to export the gesture data saved on the glove side. According to the static gesture set, repeat the above steps until all 25 predefined static gesture data are collected. The underwater gesture dataset collected a total of 25,000 sets of static gestures, and each set of data comprises 60 data points.

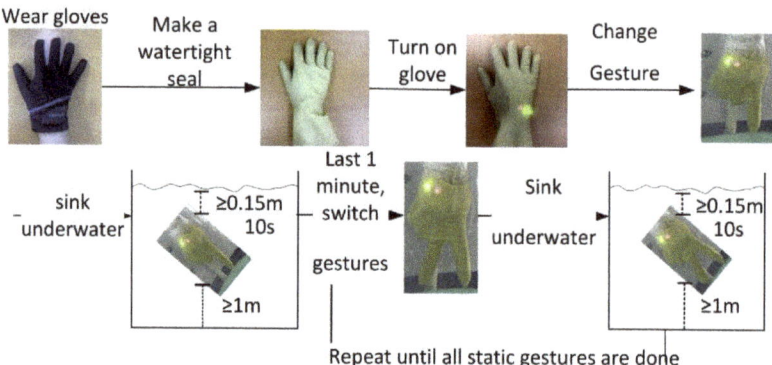

**Figure 9.** Underwater gesture data collection flow diagram.

*6.2. Evaluation of the Stochastic SVD-Optimized Spectral Clustering Algorithm*

Due to the usage of a static gesture set for underwater gestures, this research focuses solely on the gesture characteristics conveyed by the stretch sensors in the underwater data, while disregarding the data from the IMU. The comparison between the collected underwater gesture data and the corresponding land-based gesture data is illustrated in Figures 10 and 11. As shown in Figure 10, the upper part of the figure represents the underwater gesture data, while the lower part represents the gesture data captured on land. The figure displays three gestures, numbered 1, 2, and 6, from the predefined static gesture set depicted in Figure 2. As shown in Figure 11, the blue curve represents the gesture data collected underwater, and the orange curve represents the gesture data collected on land. A total of three dynamic gesture data points are compared in Figure 11, namely, dynamic gestures 0, 1, and 2 from the predefined dynamic gesture set depicted in Figure 3.

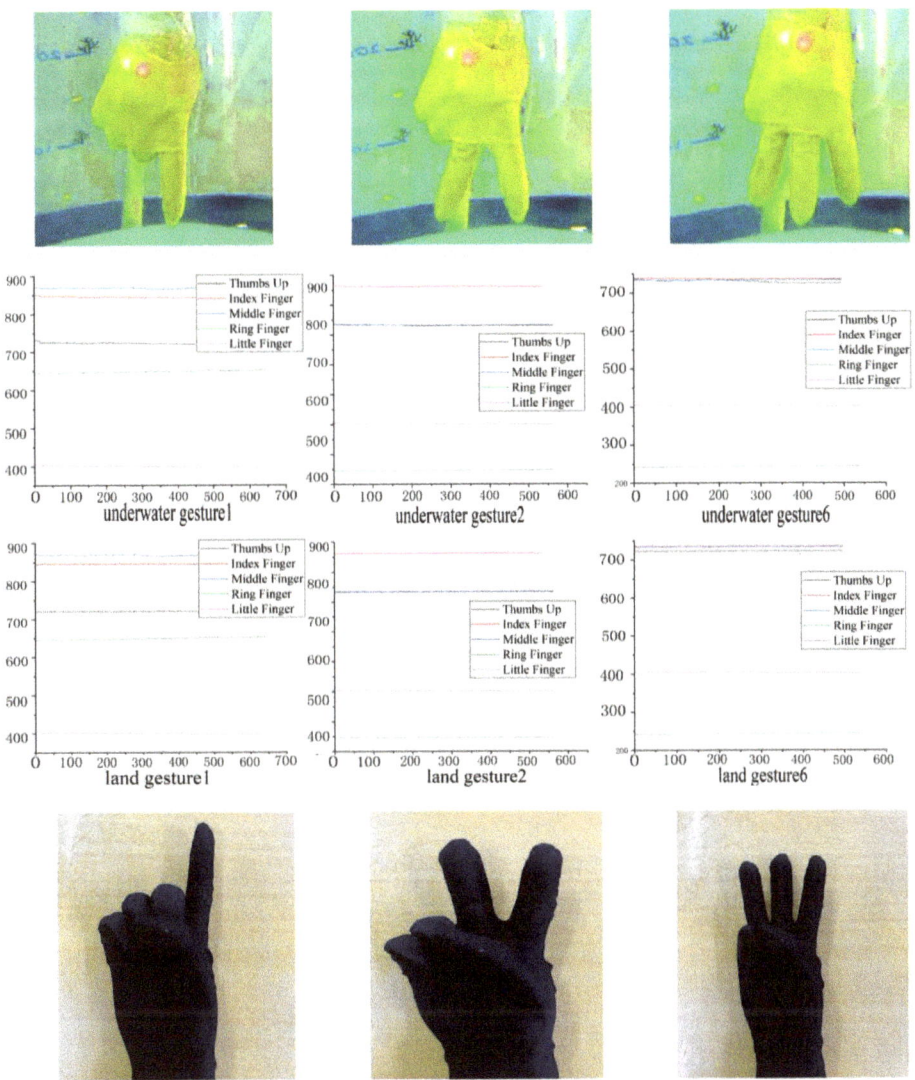

**Figure 10.** Comparison of static gesture data collected from underwater and land.

As illustrated in Figure 10, after undergoing gesture preprocessing and standardization normalization, the underwater gesture data captured by the stretch sensors exhibit similarity to the land-based gesture data collected by stretch sensors. The signal variations caused by the water pressure on the stretch sensors are found to be less than 1%. As illustrated in Figure 11, the dynamic gesture data show huge differences between underwater and on land, which can make pretrained dynamic gesture models difficult to use underwater. The above comparative results verify the feasibility of using static gestures underwater and the difficulty of using dynamic gestures. Moreover, since the underwater environment has little impact on the gesture data, the verification of underwater gesture recognition algorithms (stochastic SVD-optimized spectral clustering algorithm) can use on-land static gesture data as a reference.

A total of 25 static gesture data samples from 10 individuals were collected for experimentation. The collected data underwent preprocessing, normalization, and standardization procedures, with the application of a sliding window filtering technique to eliminate noise. Feature vectors were extracted from each gesture sample, and the extracted data were inputted into the stochastic SVD-optimized spectral clustering algorithm. The recognition accuracy and recognition time were recorded and compared with the performance of classic classifiers such as SVM, K-NN, and multilayer perceptron (MLP). The comparative results are summarized in Table 3.

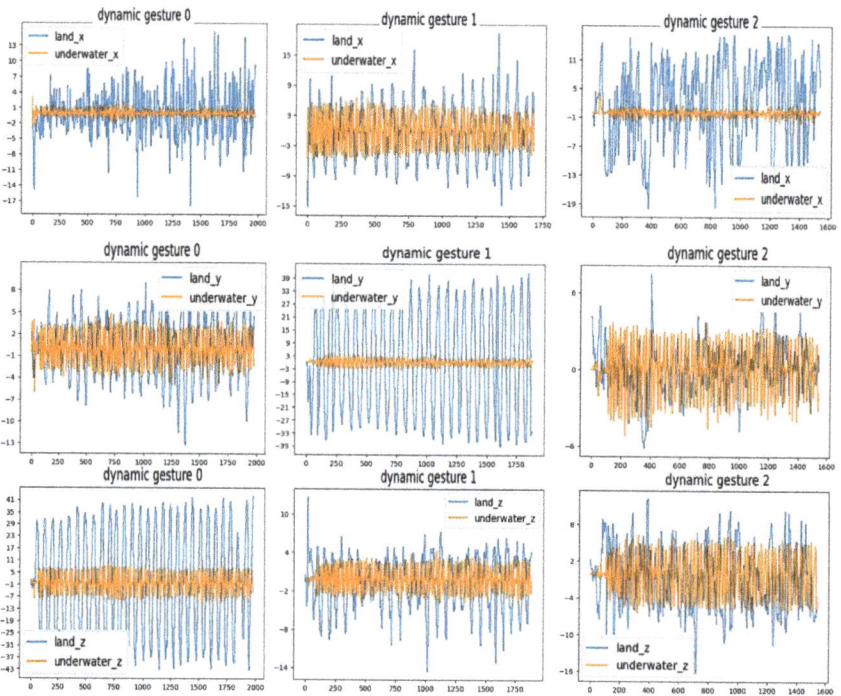

Figure 11. Comparison of dynamic gesture data collected from underwater and land.

Table 3. The performance results of different classification algorithms for underwater gestures.

|  | SVM | K-NN | Spectral Clustering | MLP | Stochastic SVD-Optimized Spectral Clustering |
|---|---|---|---|---|---|
| Person 1 | 0.9857 | 0.9667 | 0.9519 | 0.9651 | 0.9759 |
| Person 2 | 0.9333 | 0.9784 | 0.9484 | 0.9545 | 0.9828 |
| Person 3 | 0.9333 | 0.9667 | 0.9324 | 0.9456 | 0.9865 |
| Person 4 | 0.9996 | 0.9568 | 0.9349 | 0.9464 | 0.9794 |
| Person 5 | 0.9655 | 0.9935 | 0.9724 | 0.9652 | 0.9827 |
| Person 6 | 0.9666 | 0.9667 | 0.9282 | 0.9413 | 0.9863 |
| Person 7 | 0.8275 | 0.9382 | 0.9873 | 0.9613 | 0.9827 |
| Person 8 | 0.9667 | 0.9348 | 0.9932 | 0.9734 | 0.9932 |
| Person 9 | 0.7586 | 0.9655 | 0.9923 | 0.9426 | 0.9793 |
| Person 10 | 0.8621 | 0.8965 | 0.9838 | 0.9756 | 0.9862 |
| Average | 0.9199 | 0.9564 | 0.9625 | 0.9571 | 0.9835 |
| Inference time (ms) | 36.70 | 129.47 | 35.83 | 45.65 | 30.50 |
| Training time (s) | 320 | 240 | 153 | 356 | 135 |

The experimental validation revealed that the gesture recognition algorithm employed in this study achieved an average recognition accuracy that was approximately 7% higher than for SVM, 3% higher than for K-NN, and around 2% higher than for spectral clustering. Furthermore, the inference time and training time of the proposed algorithm were comparatively shorter than those of the other algorithms. These results provide empirical evidence of the effectiveness of the adopted stochastic SVD-optimized spectral clustering algorithm for underwater gesture recognition.

### 6.3. Evaluation of the SqueezeNet-BiLSTM Algorithm

To evaluate the proposed SqueezeNet-BiLSTM algorithm, this study conducted comparative experiments on several sensor-based deep learning gesture recognition algorithms, including convolution neural network (CNN)-LSTM, BiLSTM, CNN-LSTM, and SqueezeNet-LSTM.

The CNN-LSTM [36] network is a classic DL network model. It uses a CNN to extract the features of gestures in the spatial dimension and LSTM to extract the features of gesture data in the time dimension. The BiLSTM [37] network uses bidirectional LSTM network units to realize the two-way feature in the time dimension of gesture data extraction. The CNN-BiLSTM network [38,39] combines the classic CNN and BiLSTM to compare and verify the impact of the SqueezeNet network architecture on the accuracy of gesture recognition. The SqueezeNet-LSTM network connects the SqueezeNet network and the LSTM network to show the characteristics of the bidirectional time feature extraction of the gesture data via the bidirectional LSTM network.

The dataset collected in Section 6.1 is used to train the model. The ratio of training and testing sets is 7:3. The window size is set up to be 60. The loss and accuracy curves of the five selected algorithms are shown in Figure 12. As the epoch increases, the loss rate of the training model gradually approaches 0, and the accuracy rate approaches 1. Although the trends of all training models tend to be consistent in the end, the loss and accuracy curve of the SqueezeNet-BiLSTM algorithm is smoother, and it converges faster than those of the other four selected algorithms, which shows that the performance of the SqueezeNet-BiLSTM algorithm is more suitable for the current situation. The performance results of the proposed SqueezeNet-BiLSTM algorithm and the other four selected algorithms are shown in Table 4.

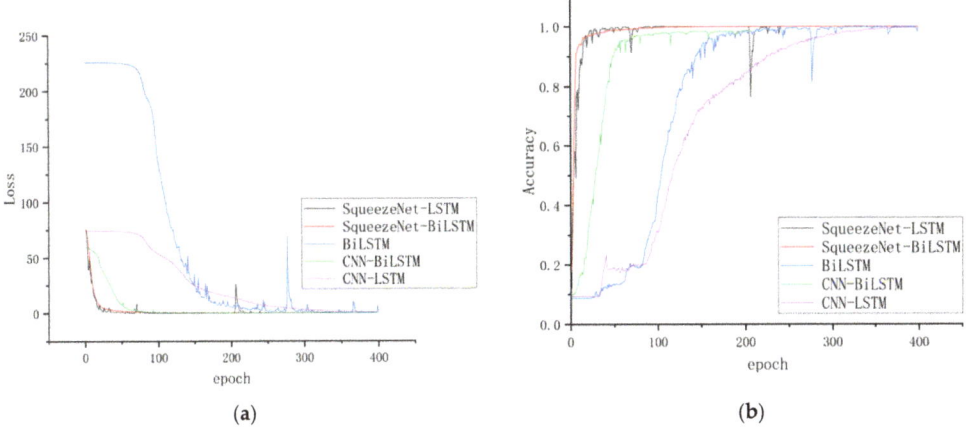

**Figure 12.** Loss and accuracy curve for selected algorithm: (**a**) loss curve; (**b**) accuracy curve.

**Table 4.** The performance results of different gesture recognition algorithms.

|  | Accuracy | Precision | Recall | F1 Score | Inference Time (ms) | Training Time (s) |
|---|---|---|---|---|---|---|
| SqueezeNet-BiLSTM | 98.94% | 97.34% | 98.21% | 97.21% | 85.4 | 638 |
| CNN-LSTM | 94.59% | 94.72% | 94.59% | 94.69% | 235.6 | 678 |
| BiLSTM | 92.34% | 94.70% | 92.34% | 92.34% | 73.6 | 436 |
| CNN-BiLSTM | 94.68% | 95.06% | 93.32% | 93.32% | 386.8 | 896 |
| SqueezeNet-LSTM | 95.97% | 96.01% | 97.08% | 95.97% | 65.3 | 563 |

According to the above experimental results, the recognition accuracy of the gesture recognition based on the BiLSTM network is the worst compared with other algorithms and can only reach 92.3%. Its network structure only pays attention to the information characteristics of the gesture sequence in the time dimension, ignoring the character of the gesture data in the spatial dimenstion, and the recognition accuracy is relatively low. The recognition accuracy of the gesture recognition algorithm based on the CNN-LSTM network structure and the CNN-BiLSTM network structure is higher than that of the gesture recognition algorithm based on the BiLSTM network. This is because its network structure fully integrates the characteristics of CNN and LSTM networks and fully extracts the attributes of gesture data in various dimensions. The recognition accuracies obtained by the CNN-LSTM and the CNN-BiLSTM network are close. The reason is that the two network structures are similar, and the difference mainly lies in the Bi-LSTM network structure used by the latter.

Compared with the other four selected classification algorithms, the gesture recognition algorithm based on the SqueezeNet-BiLSTM network proposed in this study has the best recognition accuracy, and its recognition accuracy, precision, recall, and F1 score reach 98.94%, 97.34%, 98.21%, and 97.21%, respectively. Its training time and inference time are at a medium level compared with the state-of-the-arts algorithms. This is an acceptable result, because although SqueezeNet is a lightweight convolutional neural network, whose training time and inference time are usually short, when the BiLSTM layer is connected behind SqueezeNet, as the complexity of the model increases, the recognition accuracy increases, and the training time and inference time inevitably increase.

*6.4. Evaluation of DSN-Based Gesture Recognition Transfer Model*

The experiment employed the gesture data of two volunteers to validate the efficacy of the proposed DSN-based gesture recognition transfer model. The experiment randomly selected four volunteers as UserA, UserB, UserC, and UserD. Their gesture data were excluded from the collected dataset, and the remaining data were utilized to train the SqueezeNet-BiLSTM source model. Following the completion of training, the model was tested by inputting the gesture data of these four users and the remaining data. The obtained average recognition accuracy is presented in Table 5, while the confusion matrix of users A and B is shown in Figure 13. The outcomes reflected in Table 5 underscore the substantial dissimilarities among the gesture data of different users, with the source gesture recognition model failing to extract the distinctive features of the novel users' gesture data, leading to a diminished accuracy in recognizing new users' gestures.

**Table 5.** Comparison of the recognition accuracy of SqueezeNet-BiLSTM on new users' data.

|  | UserA | UserB | UserC | UserD | Remaining Data |
|---|---|---|---|---|---|
| Accuracy | 66.1% | 63.6% | 63.0% | 61.4% | 96.57% |

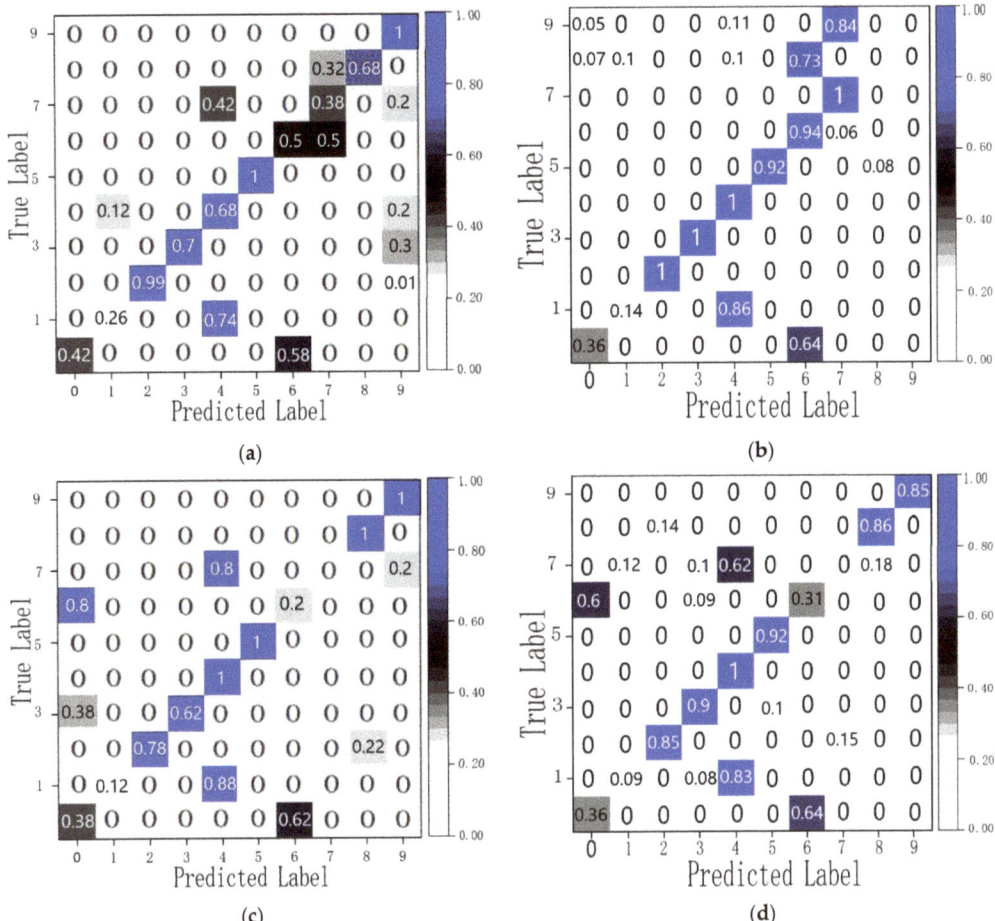

**Figure 13.** Confusion matrix of new users' recognition: (**a**) UserA; (**b**) UserB; (**c**) UserC; (**d**) UserD.

Figure 13 demonstrates that certain gesture recognition accuracies, such as gestures 0, 1, 6, and 7, are notably low. Gesture 0 and gesture 6 are often misrecognized for each other. This may be because the finger bending state is the same in the two gestures, and the hand movements are also similar. This leads to mutual misrecognition when user actions are not very standardized. Gestures 1 and 7 are always recognized as gesture 4. This may be due to the similar hand movements of these gestures and the non-standard bending of the user's fingers. In Figure 13, some special cases arise; in the test results of User B, gestures 8 and 9 show recognition problems, which may be caused by non-standard bending movements of the user's fingers or ill-fitting gloves. Since our gloves only come in one size, people with small hands cannot fit the gloves perfectly when wearing them, making it difficult to obtain accurate stretch sensor data, ultimately leading to inaccurate recognition. For other relatively small identification problems, these can be attributed to variances in personal hand size, movement patterns, and sensor data from the glove, resulting in significant disparities between certain gesture data and the data employed during training. To avoid these problems, we will first perform bending and stretching calibration in the early stage of gesture recognition to minimize recognition errors caused by palm size. Secondly, in the data preprocessing stage, filtering algorithms are used to reduce data noise and then put through data normalization, as well as data up-sampling and down-sampling, to reduce dynamic gesture recognition errors caused by personal hand movement habits. Although a series of measures have been taken to ensure the accuracy of identification, everyone's behavioral habits still vary greatly. In practical environments, it is still difficult to obtain good recognition accuracy using untrained data.

The experiment performed a model transfer test regarding small-sample data, using gesture data of varying scales. Specifically, the experiments collected samples of 5, 10, 20, 30, 40, and 50 instances for each gesture category. To verify the superiority of our proposed -DSNbased gesture recognition transfer model, we also selected several state-of-the-art transfer learning models for comparison, including generative adversarial network (GAN)- [40] and conditional generative adversarial networks (CGAN)-based [41] transfer learning models. The transfer process involved utilizing our proposed DSN-based gesture recognition transfer model and selected state-of-the-art transfer learning models, with incremental updates applied to enhance the model's performance. Subsequently, the experiment conducted tests using the gesture data of UserA, UserB, UserC, and UserD to evaluate the recognition accuracy of the transferred gesture recognition model. The results depicting the recognition accuracy for each user are illustrated in Figure 14.

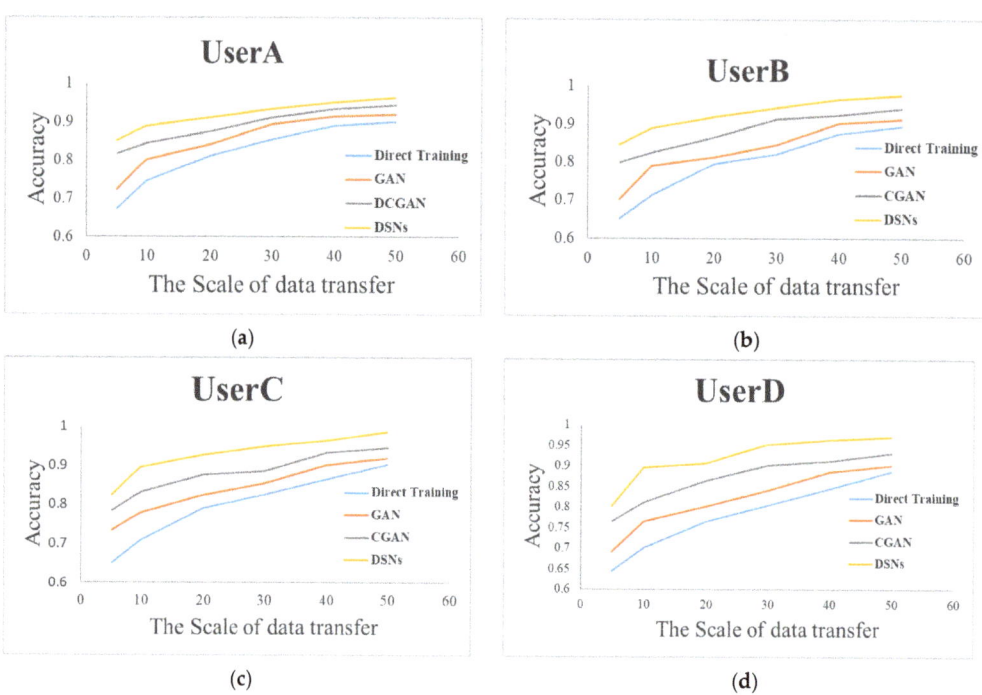

**Figure 14.** Transfer experiment of new user gesture recognition based on DSNs. (**a**) UserA; (**b**) UserB; (**c**) UserC; (**d**) UserD.

As shown in Figure 14, it can be observed that the recognition accuracy for new users increases with the growth of the data scale. During data transfer training with the same sample size, the accuracy of the proposed DSN-based gesture recognition transfer model is significantly better than the state-of-the-art algorithms. When using the novel DSN gesture recognition transfer model in the target domain, the model effectively extracts the domain-generalizable features from the source domain data and applies them to the target domain. As a result, the recognition accuracy is significantly improved compared with direct training when conducting small-scale data transfer training. Therefore, new users only need to provide a small amount of training data to ensure the accuracy of the recognition model, thereby effectively enhancing the user experience.

## 7. Conclusions and Future Work

This study developed a smart data glove with five-channel flexible capacitive stretch sensors, accelerometers, and gyroscopes for gesture recognition in an amphibious environment. To ensure recognition accuracy, this study also proposed a novel AHGR model, which can adaptively change the gesture recognition model to adopt an amphibious environment. This model contains two classification algorithms, the SqueezeNet-BiLSTM algorithm for land gesture recognition and the stochastic SVD-optimized spectral clustering algorithm for underwater gesture recognition. The accuracy of the SqueezeNet-BiLSTM algorithm and the stochastic SVD-optimized spectral clustering algorithm can reach 98.94% and 98.35%, respectively. This study also introduces a DSN-based gesture recognition transfer model, so that new users and new devices only need small-scale data transferring and training to ensure that the recognition accuracy reaches 94%.

In future work, we plan to conduct more professional underwater hand gesture testing, such as hiring divers to test in deeper water. We also plan to develop a waterproof smart data glove that can be used directly underwater and add an acoustic modem to transmit gesture data wirelessly. In addition, we plan to analyze the energy consumption of different models running on the gloves and optimize the design model to reduce energy consumption while ensuring high accuracy.

**Author Contributions:** Conceptualization, L.F.; Methodology, L.F. and X.Y.; Software, L.F. and Y.W.; Validation, L.F.; Formal analysis, L.F.; Investigation, L.F. and X.Y.; Resources, L.F.; Data curation, L.F., X.Y. and Y.W.; Writing—original draft, L.F. and X.Y.; Writing—review & editing, L.F., Z.Z. and D.Z.; Visualization, L.F.; Supervision, L.F., Z.Z. and D.Z.; Project administration, Z.Z. and D.Z.; Funding acquisition, Z.Z., B.Z. and D.Z. All authors have read and agreed to the published version of the manuscript.

**Funding:** This research is supported by National Natural Science Foundation of China under Grant No. 62171155.

**Acknowledgments:** We thank the members and staff in the Research Center of Fault-Tolerant and Mobile Computing for supporting this work. We also thank our study volunteers for taking the time to help us collect the gesture data.

**Conflicts of Interest:** The authors declare no conflict of interest.

## References

1. Muneeb, M.; Rustam, H.; Jalal, A. *Automate Appliances via Gestures Recognition for Elderly Living Assistance*; IEEE: New York, NY, USA, 2023; pp. 1–6.
2. Miao, Y.; Shi, E.; Lei, M.; Sun, C.; Shen, X.; Liu, Y. *Vehicle Control System Based on Dynamic Traffic Gesture Recognition*; IEEE: New York, NY, USA, 2022; pp. 196–201.
3. Lee, M.; Bae, J. Real-Time Gesture Recognition in the View of Repeating Characteristics of Sign Languages. *IEEE Trans. Ind. Inform.* **2022**, *18*, 8818–8828. [CrossRef]
4. Tan, P.; Han, X.; Zou, Y.; Qu, X.; Xue, J.; Li, T.; Wang, Y.; Luo, R.; Cui, X.; Xi, Y. Self-Powered Gesture Recognition Wristband Enabled by Machine Learning for Full Keyboard and Multicommand Input. *Adv. Mater.* **2022**, *34*, 2200793. [CrossRef] [PubMed]
5. Bello, H.; Suh, S.; Geißler, D.; Ray, L.; Zhou, B.; Lukowicz, P. CaptAinGlove: Capacitive and Inertial Fusion-Based Glove for Real-Time on Edge Hand Gesture Recognition for Drone Control. In Proceedings of the 2023 ACM International Joint Conference on Pervasive and Ubiquitous Computing & the 2023 ACM International Symposium on Wearable Computing, Cancún, Mexico, 8–12 October 2023.
6. Antillon, D.W.O.; Walker, C.R.; Rosset, S.; Anderson, I.A. Glove-Based Hand Gesture Recognition for Diver Communication. *IEEE Trans. Neural Netw. Learning Syst.* **2022**, *1*, 1–13. [CrossRef] [PubMed]
7. Kvasic, I.; Miskovic, N.; Vukic, Z. Convolutional Neural Network Architectures for Sonar-Based Diver Detection and Tracking. In Proceedings of the OCEANS 2019—Marseille, Marseille, France, 17–20 June 2019; IEEE: Marseille, France, 2019; pp. 1–6.
8. Rautaray, S.S.; Agrawal, A. Vision Based Hand Gesture Recognition for Human Computer Interaction: A Survey. *Artif. Intell. Rev.* **2015**, *43*, 1–54. [CrossRef]

9. Geng, W.; Du, Y.; Jin, W.; Wei, W.; Hu, Y.; Li, J. Gesture Recognition by Instantaneous Surface EMG Images. *Sci. Rep.* **2016**, *6*, 36571. [CrossRef]
10. Hu, Y.; Wong, Y.; Wei, W.; Du, Y.; Kankanhalli, M.; Geng, W. A Novel Attention-Based Hybrid CNN-RNN Architecture for sEMG-Based Gesture Recognition. *PLoS ONE* **2018**, *13*, e0206049. [CrossRef]
11. Milosevic, B.; Farella, E.; Benatti, S. Exploring Arm Posture and Temporal Variability in Myoelectric Hand Gesture Recognition. In Proceedings of the 2018 7th IEEE International Conference on Biomedical Robotics and Biomechatronics (Biorob), Enschede, The Netherlands, 26–29 August 2018; IEEE: Enschede, The Netherlands, 2018; pp. 1032–1037.
12. Duan, D.; Yang, H.; Lan, G.; Li, T.; Jia, X.; Xu, W. EMGSense: A Low-Effort Self-Supervised Domain Adaptation Framework for EMG Sensing. In Proceedings of the 2023 IEEE International Conference on Pervasive Computing and Communications (PerCom), Atlanta, GA, USA, 13–17 March 2023; IEEE: Atlanta, GA, USA, 2023; pp. 160–170.
13. Kim, M.; Cho, J.; Lee, S.; Jung, Y. IMU Sensor-Based Hand Gesture Recognition for Human-Machine Interfaces. *Sensors* **2019**, *19*, 3827. [CrossRef]
14. Siddiqui, N.; Chan, R.H.M. Multimodal Hand Gesture Recognition Using Single IMU and Acoustic Measurements at Wrist. *PLoS ONE* **2020**, *15*, e0227039. [CrossRef]
15. Galka, J.; Masior, M.; Zaborski, M.; Barczewska, K. Inertial Motion Sensing Glove for Sign Language Gesture Acquisition and Recognition. *IEEE Sens. J.* **2016**, *16*, 6310–6316. [CrossRef]
16. Plawiak, P.; Sosnicki, T.; Niedzwiecki, M.; Tabor, Z.; Rzecki, K. Hand Body Language Gesture Recognition Based on Signals From Specialized Glove and Machine Learning Algorithms. *IEEE Trans. Ind. Inf.* **2016**, *12*, 1104–1113. [CrossRef]
17. Preetham, C.; Ramakrishnan, G.; Kumar, S.; Tamse, A.; Krishnapura, N. Hand Talk-Implementation of a Gesture Recognizing Glove. In Proceedings of the 2013 Texas Instruments India Educators' Conference, Bangalore, India, 4–6 April 2013; IEEE: Bangalore, India, 2013; pp. 328–331.
18. Lu, Z.; Chen, X.; Li, Q.; Zhang, X.; Zhou, P. A Hand Gesture Recognition Framework and Wearable Gesture-Based Interaction Prototype for Mobile Devices. *IEEE Trans. Human-Mach. Syst.* **2014**, *44*, 293–299. [CrossRef]
19. Agab, S.E.; Chelali, F.Z. New Combined DT-CWT and HOG Descriptor for Static and Dynamic Hand Gesture Recognition. *Multimed Tools Appl.* **2023**, *82*, 26379–26409. [CrossRef] [PubMed]
20. Khan, A.M.; Tufail, A.; Khattak, A.M.; Laine, T.H. Activity Recognition on Smartphones via Sensor-Fusion and KDA-Based SVMs. *Int. J. Distrib. Sens. Netw.* **2014**, *10*, 503291. [CrossRef]
21. Trigueiros, P.; Ribeiro, F.; de Azurém, C.; Reis, L.P. A Comparison of Machine Learning Algorithms Applied to Hand Gesture Recognition. In Proceedings of the 7th Iberian Conference on Information Systems and Technologies (CISTI 2012), Madrid, Spain, 20–23 June 2012.
22. LeCun, Y.; Bengio, Y.; Hinton, G. Deep Learning. *Nature* **2015**, *521*, 436–444. [CrossRef]
23. Murad, A.; Pyun, J.-Y. Deep Recurrent Neural Networks for Human Activity Recognition. *Sensors* **2017**, *17*, 2556. [CrossRef]
24. Hammerla, N.Y.; Halloran, S.; Ploetz, T. Deep, Convolutional, and Recurrent Models for Human Activity Recognition Using Wearables. *arXiv* **2016**, arXiv:1604.08880.
25. Fang, B.; Lv, Q.; Shan, J.; Sun, F.; Liu, H.; Guo, D.; Zhao, Y. Dynamic Gesture Recognition Using Inertial Sensors-Based Data Gloves. In Proceedings of the 2019 IEEE 4th International Conference on Advanced Robotics and Mechatronics (ICARM), Toyonaka, Japan, 3–5 July 2019; IEEE: Toyonaka, Japan, 2019; pp. 390–395.
26. Faisal, M.A.A.; Abir, F.F.; Ahmed, M.U.; Ahad, A.R. Exploiting Domain Transformation and Deep Learning for Hand Gesture Recognition Using a Low-Cost Dataglove. *Sci. Rep.* **2022**, *12*, 21446. [CrossRef]
27. Yu, C.; Fan, S.; Liu, Y.; Shu, Y. End-Side Gesture Recognition Method for UAV Control. *IEEE Sens. J.* **2022**, *22*, 24526–24540. [CrossRef]
28. Espressif ESP32 WROOM 32E Datasheet 2023. Available online: https://www.espressif.com.cn/sites/default/files/documentation/esp32-wroom-32e_esp32-wroom-32ue_datasheet_en.pdf (accessed on 29 October 2023).
29. von Luxburg, U. A Tutorial on Spectral Clustering. *Stat. Comput.* **2007**, *17*, 395–416. [CrossRef]
30. De Lathauwer, L.; De Moor, B.; Vandewalle, J. A Multilinear Singular Value Decomposition. *SIAM J. Matrix Anal. Appl.* **2000**, *21*, 1253–1278. [CrossRef]
31. Halko, N.; Martinsson, P.G.; Tropp, J.A. Finding Structure with Randomness: Probabilistic Algorithms for Constructing Approximate Matrix Decompositions. *SIAM Rev.* **2011**, *53*, 217–288. [CrossRef]
32. Iandola, F.N.; Han, S.; Moskewicz, M.W.; Ashraf, K.; Dally, W.J.; Keutzer, K. SqueezeNet: AlexNet-Level Accuracy with 50x Fewer Parameters and <0.5 MB Model Size. *arXiv* **2016**, arXiv:1602.07360.
33. Lefebvre, G.; Berlemont, S.; Mamalet, F.; Garcia, C. BLSTM-RNN Based 3D Gesture Classification. In *Artificial Neural Networks and Machine Learning—ICANN 2013*; Mladenov, V., Koprinkova-Hristova, P., Palm, G., Villa, A.E.P., Appollini, B., Kasabov, N., Eds.; Lecture Notes in Computer Science; Springer: Berlin/Heidelberg, Germany, 2013; Volume 8131, pp. 381–388, ISBN 978-3-642-40727-7.
34. Kolda, T.G.; Bader, B.W. Tensor Decompositions and Applications. *SIAM Rev.* **2009**, *51*, 455–500. [CrossRef]
35. Bousmalis, K.; Trigeorgis, G.; Silberman, N.; Krishnan, D.; Erhan, D. Domain Separation Networks. In Proceedings of the 30th Conference on Neural Information Processing Systems, Barcelona, Spain, 5–10 December 2016.

36. Khatun, M.A.; Abu Yousuf, M.; Ahmed, S.; Uddin, M.Z.; Alyami, S.A.; Al-Ashhab, S.; Akhdar, H.F.; Khan, A.; Azad, A.; Moni, M.A. Deep CNN-LSTM With Self-Attention Model for Human Activity Recognition Using Wearable Sensor. *IEEE J. Transl. Eng. Health Med.* **2022**, *10*, 1–16. [CrossRef]
37. Li, Y.; Wang, L. Human Activity Recognition Based on Residual Network and BiLSTM. *Sensors* **2022**, *22*, 635. [CrossRef]
38. Rhanoui, M.; Mikram, M.; Yousfi, S.; Barzali, S. A CNN-BiLSTM Model for Document-Level Sentiment Analysis. *Mach. Learn. Knowl. Extr.* **2019**, *1*, 832–847. [CrossRef]
39. Challa, S.K.; Kumar, A.; Semwal, V.B. A Multibranch CNN-BiLSTM Model for Human Activity Recognition Using Wearable Sensor Data. *Vis. Comput.* **2022**, *38*, 4095–4109. [CrossRef]
40. Bousmina, A.; Selmi, M.; Ben Rhaiem, M.A.; Farah, I.R. A Hybrid Approach Based on GAN and CNN-LSTM for Aerial Activity Recognition. *Remote Sens.* **2023**, *15*, 3626. [CrossRef]
41. Jimale, A.O.; Mohd Noor, M.H. Fully Connected Generative Adversarial Network for Human Activity Recognition. *IEEE Access* **2022**, *10*, 100257–100266. [CrossRef]

**Disclaimer/Publisher's Note:** The statements, opinions and data contained in all publications are solely those of the individual author(s) and contributor(s) and not of MDPI and/or the editor(s). MDPI and/or the editor(s) disclaim responsibility for any injury to people or property resulting from any ideas, methods, instructions or products referred to in the content.

Article

# Health Monitoring System from Pyralux Copper-Clad Laminate Film and Random Forest Algorithm

Chi Cuong Vu [1], Jooyong Kim [2] and Thanh-Hai Nguyen [1,*]

[1] Faculty of Electrical and Electronics Engineering, Ho Chi Minh City University of Technology and Education, 01 Vo Van Ngan Street, Linh Chieu Ward, Ho Chi Minh City 700000, Vietnam; cuongvc@hcmute.edu.vn

[2] Department of Materials Science and Engineering, Soongsil University, Seoul 156-743, Republic of Korea; jykim@ssu.ac.kr

* Correspondence: nthai@hcmute.edu.vn

**Abstract:** Sensor technologies have been core features for various wearable electronic products for decades. Their functions are expected to continue to play an essential role in future generations of wearable products. For example, trends in industrial, military, and security applications include smartwatches used for monitoring medical indicators, hearing devices with integrated sensor options, and electronic skins. However, many studies have focused on a specific area of the system, such as manufacturing processes, data analysis, or actual testing. This has led to challenges regarding the reliability, accuracy, or connectivity of components in the same wearable system. There is an urgent need for studies that consider the whole system to maximize the efficiency of soft sensors. This study proposes a method to fabricate a resistive pressure sensor with high sensitivity, resilience, and good strain tolerance for recognizing human motion or body signals. Herein, the sensor electrodes are shaped on a thin Pyralux film. A layer of microfiber polyesters, coated with carbon nanotubes, is used as the bearing and pressure sensing layer. Our sensor shows superior capabilities in respiratory monitoring. More specifically, the sensor can work in high-humidity environments, even when immersed in water—this is always a big challenge for conventional sensors. In addition, the embedded random forest model, built for the application to recognize restoration signals with high accuracy (up to 92%), helps to provide a better overview when placing flexible sensors in a practical system.

**Keywords:** soft pressure sensor; water resistance; Pyralux copper-clad laminate; random forest; respiratory monitoring

## 1. Introduction

The sensor technology industry is expected to make great strides in the coming decades by adopting a wide range of wearable devices in everyday life [1,2]. Augmented and virtual reality devices [3–5], such as VR, AR, or MR, combine cameras, depth sensors, and tension or pressure sensors, allowing users to interact with specific content. Product categories such as smart clothing and other related products are all based on a core sensor, where communication between the human body and surrounding environments is a key function of the product. Various types of sensors are specific to different technologies and application contexts, in which the most significant proportion are chemical, optical, and electromechanical sensors [6–11].

Going deeper into the aspects and applications, we highlight some examples. Chemical sensors provide an alternative to needles [12]; they allow people with diabetes to monitor blood sugar without pricking a finger. Today, commercial devices are still mainly needles. Therefore, the quest to find new wearable sensors that are less invasive to the skin is ongoing. Motion sensors are used for pedometer applications [13,14]; most small wearables are set up with built-in accelerometers. Many motion-sensing components are included in clinical trials or track the movements of athletes to analyze and provide the most appropriate

and quick solutions. Optical sensors are used in heart rate detection [15]. We are familiar with smartwatches with a light cluster on the back of the device, which is used to obtain heart rate data or blood oxygen, thereby enabling deeper analysis into calories consumed or sleep quality. Many companies are competing for the lead in commercializing blood pressure monitors [16]. Other companies are targeting devices in the form of 'small clinics on the wrist' that will replace conventional tests in hospitals [17]. The electrodes allow the monitoring of heart rate, muscles, and brain signals [18]. Incorporating conductive materials into wearable technology is a simple and established concept. However, this is the principle behind countless types of wearable sensors today, including soft, wet electrodes glued to the skin to measure heart rate, dry electrodes in headphones to analyze brain electrical signals, and microneedle tips in skin patches to quantify muscle movements. This also creates a wide range of applications for electrode sensors [19–21], from vital sign monitoring and sleep analysis for healthcare to emotional response or stress monitoring.

Among the flexible wearable sensors, the pressure sensor [22,23] has become an integral part of the tasks of detecting movements or muscle changes. The new generation of soft pressure sensors [24,25] exhibits distinct advantages such as being lightweight, small, hypoallergenic, comfortable to wear on the body, and suitable for many different locations and ranges. The above properties are achieved by the materials that make up the sensor, the most important of which are nanomaterials such as metal nanowires (NWs), carbon nanotubes (CNTs), conductive polymers, and nanoparticles (NPs). Flexible pressure sensors fabricated from CNTs and graphene oxide (GO) are attracting much attention from research groups worldwide [26,27]. Their applications appear in various research areas, including patient body signal recognition, athlete tracking, human–robot interface, and electronic textiles [28–30]. There has been a lot of research on prominent soft pressure sensors over the years. These studies have focused on one or a few aspects of flexible wearable sensors, such as materials or construction. The above studies lack evaluations when placing sensors in an overall wearable system. In addition, studies also lack efficient and intelligent signal processing models that can optimize sensor performance.

From the above approach, this study constructed a system to monitor human health signals through a specific application: respiratory detection masks. In this study, a flexible sensor was developed and considered when connected with other components of the wearable system, consisting of a processing board and electrode connections. On the other hand, an embedded machine-learning model was built to observe the data obtained from the system. The model helps to provide a better overview of when the flexible sensor is used in an actual healthcare application. In more detail, the study presents a pressure sensor that uses a Pyralux film for electrodes, and the sensing layer is wipe-coated with carbon nanotubes (CNTs). Here, the sensors are thin (0.26 mm) and highly sensitive at $0.2\ \text{kPa}^{-1}$ (under 6 kPa)/$0.05\ \text{kPa}^{-1}$ (over 8 kPa). There are some highlights of this work. Firstly, the simple process helps the sensor retain the benefits of conventional resistive sensors and expands the limits of sensors in healthcare or electronic control devices (thin and high performance). Secondly, the sensor is water-resistant. This capability would suit many working environments and applications, such as body tracking systems, rain gear, and robots. Thirdly, the sensor structure can be hand-drawn without any design software intervention. More importantly, body signals are classified based on a combination of a multifunctional design, consisting of a new-generation flexible pressure sensor and an embedded machine-learning model built on the collected data. The sensors are expected to be applied in practice in clinics or hospitals. This is also a potential investment direction for companies or businesses in the near future.

## 2. Materials and Methods

Pyralux copper-clad laminate film was purchased from Dupont Ltd., Wilmington, DE, USA. The film consists of Kapton polyimide and is available in sheet form as a single-sided clad. The structure archives high flexibility and compatibility. The carbon nanotube (CNT) inks were 0.1 wt% solutions from KH Chemicals Company, Seoul City, Republic of

Korea. The wet-etching material (sodium persulfate ($Na_2S_2O_8$)) was obtained from SME Company, and the wet wipes were from SsangYong Company, Seoul, Republic of Korea. A spacer layer was constructed from a hot-melt adhesive film (PU film). The final sensor was protected and secured using a thin single-sided sticky film from 3M Technology Ltd., Seoul, Republic of Korea.

The manufacturing process of the flexible pressure sensors is shown in Figure 1a–c. There were three main steps: process the electrode layer, process the sensing layer, and assemble the final sensor. From a nonwoven master roll, wet wipes were fabricated via a converting process and contained multiple microfibers with a dimension of 10 μm arranged randomly together. They were dried at 100 °C for 10 min to remove the water, then dipped in conductive inks, squeezed, and dried again to obtain the dry wipe-coated CNTs. The CNT particles were attached to microfibers using the binder inside the inks and the interfiber force (friction). However, this adhesion is weak, and the CNTs could fall out when working in particle applications (large deformations). These CNT wipes had a sheet resistance of 27–30 Ω/sq and a thickness of 0.16 mm.

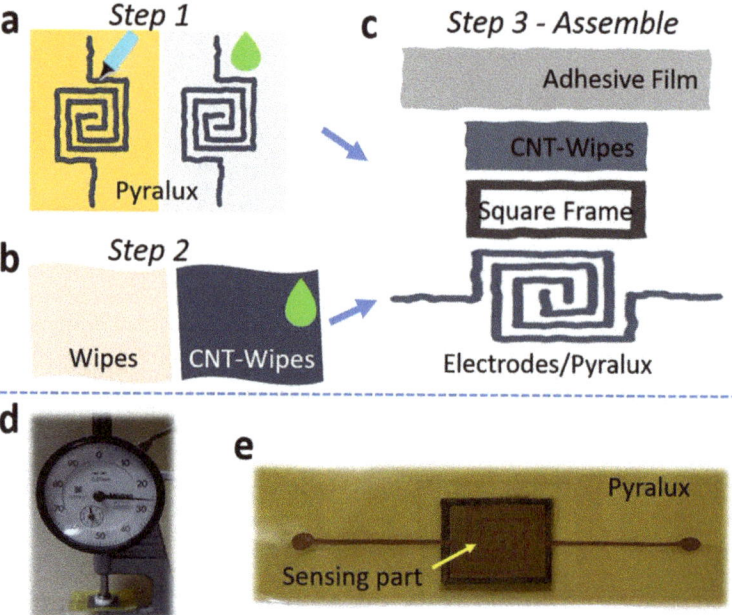

**Figure 1.** Manufacturing process of the flexible pressure sensors, consisting of (**a**) preparing the electrode layer, (**b**) preparing the sensing layer, (**c**) assembling sensor, (**d**) thickness of the sensor, and (**e**) real image of the final flexible sensor.

Pyralux film is a type of flexible printed circuit board (PCB) with a thickness of 0.032 mm. This film has two sides; one side is polyimide, and another is a copper layer. The electrodes were shaped with a permanent marker pen, and the distance between each electrode was about 1 mm. Then, the Pyralux film was dipped in etching liquid to remove nonelectrode parts for 25 min. After the wet-etching process, we used acetone liquid to clean the marker lines. Finally, the sensor was assembled from many parts. Figure 1c describes the structure, consisting of electrodes (Pyralux), a sensing layer (CNT wipes), a spacer film (double-sided hot melt adhesive film), and a cover layer (one-sided adhesive thin film). The total thickness was about 0.26 mm, and the size was 15 × 15 mm, as seen in Figure 1d,e. Obviously, the above structure has the potential to be applied in real-life cases.

## 3. Results and Discussion

Field-emission scanning electron microscope (FE-SEM) images of each layer are shown in Figure 2. The CNT wipes are a random structure of many fibers. There are many empty spaces between the fibers or bundle fibers, and each single fiber has a diameter of 10 μm. The conductive particles (CNTs) cover the single fiber or create some random film on the pristine surface of the wipes. Following the principle of the pressure sensor, our sensor converts the mechanical pressure value into an electrical resistance signal. As shown in Figure 3a, the thickness of the sensing layer (CNT wipes) decreases when applying force. The CNT arrays on the microfiber polyesters come closer together, increasing the conductivity of the sensing layer so that the overall resistance decreases. At that time, the conductivity between the electrode lines (copper) on the Pyralux layer increase. After removing the force, the CNT-wipe layer is released. The microfiber polyesters move further, causing the overall resistance to recover to its original value.

**Figure 2.** Scanning electron microscope images of the sensor: (**a**) sensing layer at 100 μm and 50 μm before dipping in CNTs, (**b**) sensing layer at 100 μm and 50 μm after dipping CNTs, and (**c**) assembled sensor at the top view and bottom view.

The working performances of the sensors were tested with a universal testing machine (UTM) (Figures 3b and S1), which included a controlling computer, a force load cell (DN-FGA-20), and an LCR meter (Keysight E4980AL). The compression speed of the UTM system was 0.02 mm/s, and the resolution of the force gauge (FGA-K2) was 0.0098 N/cm$^2$ with an error range of ±0.2%. All testing was performed at a room temperature of 25 °C. The sensor samples were located on the sole plate, and two electrodes were connected to two crocodile clips (soldering method) for reading the resistance signal. Figure 3c describes the current–voltage graphical curves (I–V) at different pressure levels from 0 to 50 kPa.

These curves indicate an Ohmic behavior at voltages from −3 to 3 V. This parameter demonstrates good linearity for practical applications. Figure 3d,e show the resistance change in the sensor with increasing/decreasing force. In addition, the sensitivity is calculated as $S = \delta(\triangle R/R_0)/\delta P$, where $R_0$ is the initial resistance, $R$ is the resistance under pressure, and the $\delta P$ is the pressure change.

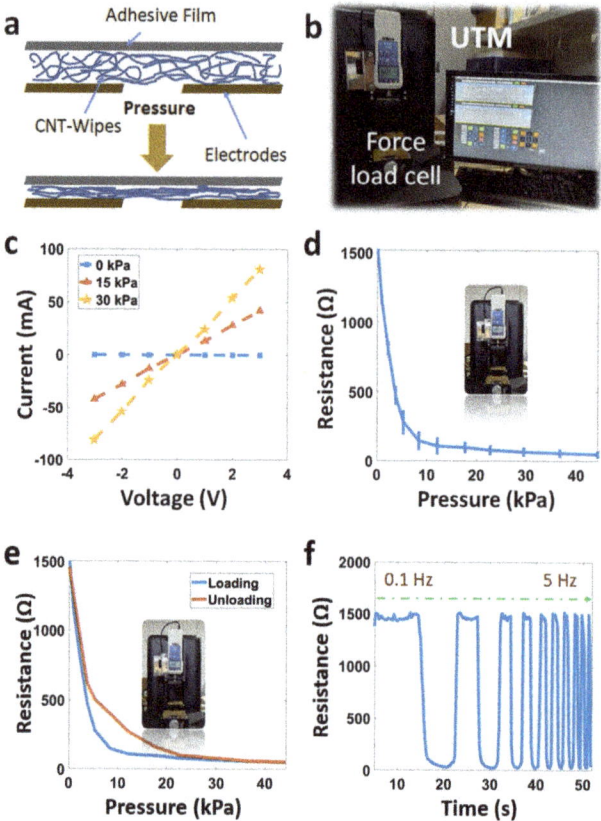

**Figure 3.** Characteristics of the sensor: (**a**) working principle, (**b**) universal testing machine, (**c**) current–voltage graphical curves, (**d**) resistance change under pressure, (**e**) hysteresis, and (**f**) resistance change at different frequencies.

It is clear that the sensitivity was about 0.2 kPa$^{-1}$ in the low-force range under 6 kPa and about 0.05 kPa$^{-1}$ in the compression force range was over 8 kPa. This is due to the saturated contacting area at high pressure. Figure 3d also shows the sensor has a small standard deviation error (<7%). Hysteresis is defined as the deviation between the response/release resistance lines and dependence of the history states. Two hysteresis lines can be seen in Figure 3e, with a slight difference for pressure from 0 to 45 kPa. In Figure 3f, the sensor show excellent dynamic performance over a wide mechanical frequency (0.1–5 Hz).

Figure 4 and Figures S2–S5 show essential characteristics, such as response/recovery time, resistance change at different frequencies, bending angles, temperatures. Due to the benefits of the chosen flexible materials, the sensor performed well in the working test at three different levels of radii (0–15–22.5 mm). Figure 4a describes the resistance change under bending deformations. Here, the workability diminishes with curvature change but is generally still suitable for the considered radii. The main reason for the delay is the connection time between CNT coating layers when applying pressure and the viscosity/elasticity of the fibers. We observed a short value at a response time of 70 ms and at a recovery time of 50 ms (Figure 4b). These above values ensure the electrical properties and the working potential of the sensor in monitoring and controlling cases.

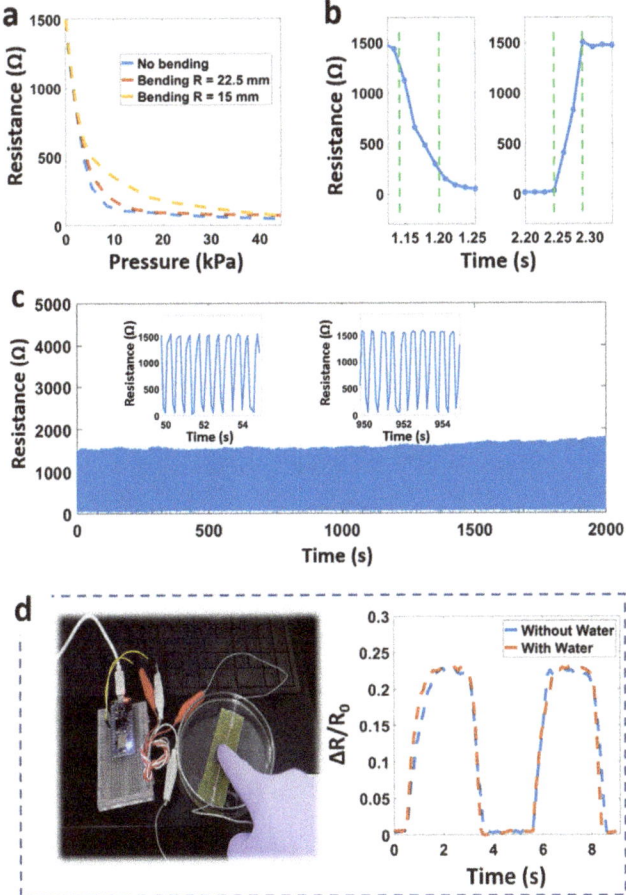

**Figure 4.** (a) Resistance change at different bending radii, (b) response and recovery times, (c) durability under loading/unloading cycles, and (d) working when dipped under water.

Durability represents the ability of the sensor to remain functional when faced with many loading/unloading cycles. In this work, the sensor had stable responses and mechanical integrity under 3000 cycles. We obtained a small difference (maximum resistance) with 5% after 1000 cycles and about 8% after 2500 cycles. These changes were caused by the permanent deformation in the structure of the sensor when working, leading to a change in the maximum resistance. The CNT particles can fall out of the fibers, and cracks appear in the CNT-coated layers during operation. We recommend a solution for this issue: replacing the CNT inks with Ag or Au pastes. However, this approach (Ag/Au) will raise costs and change the sensor's characteristics. Another sensor feature is the working capacity in high-humidity environments or directly underwater. As shown in Figure 4d, the sensor was immersed in water to a depth of about 1 cm (sensing part only). The result shows a slight change in the output signal. Figure S2 describes stable operation when immersed in water for different times (0, 6, or 12 h). The reason for this is that the sensing layer (CNT wipes) is entirely enclosed by two water-resistant layers, polyimide (from Pyralux), and

3M adhesive. Additionally, Table 1 summarizes some studies on flexible sensors, including principle, sensitivity, response/recovery time, and water resistance [22,31–36]. Our sensor clearly demonstrates good capabilities such as water resistance, short response time, and good sensitivity for a wide range of wearable applications.

**Table 1.** Comparison of some developed flexible pressure sensors.

| Ref. | Principle | Thickness (μm) | Response Time (ms) | Sensitivity (kPa$^{-1}$) | Water Resistance |
|---|---|---|---|---|---|
| [31] | Resistive | - | 200 | ~7.12 | Yes |
| [22] | Capacitive | ~290 | ~41 | 0.23 | Yes |
| [32] | Resistive | 150 | - | ~0.001 | No |
| [33] | Capacitive | ~110 | 180/120 | ~0.14 | No |
| [34] | Resistive | 1000 | - | ~0.13 | No |
| [35] | Capacitive | >1000 | ~100 | 0.18 | No |
| [36] | Capacitive | >1000 | ~100 | 0.0124 | No |
| Ours | Resistive | 260 | 70/50 | 0.2 | Yes |

## 4. Embedded Health-Monitoring System with Flexible Sensor and Random Forest Algorithm

To demonstrate the ability of the fabricated sensor, we propose a method to monitor the breathing rate of humans (Figure 5). This indicator is significant for applications to analyze daily clinical health after exercise or for respiratory disease. The system consists of three main parts: a smart mask with a flexible pressure sensor, a processing circuit board, and a lipo-battery (3.7 V). An embedded circuit collects the signals with an nRF52 module, including an analog/digital converter (ADC), a microcontroller (MCU), and a Bluetooth low-energy (BLE) part. This module is a completely embedded board that can run a small AI model. Here, a random forest model was built to classify the types of respiration: normal breath, deep breath, rapid breath, cough, and breath holding. As described, we collected the signals via sampling, digitizing, and converting them into digital signals at a rate of 1000 signals per min from the system. One filter removes unusual vibrations and noises. Then, the features are extracted from the time- and frequency-domain breathing samples, including signal amplitude, standard deviation, breathing cycle, mean value, variance, root mean square, and unbiased estimation. These features are used for the random forest model [37,38]. Finally, the classified respiration data are sent via Bluetooth to phones, tablets, or smartwatches.

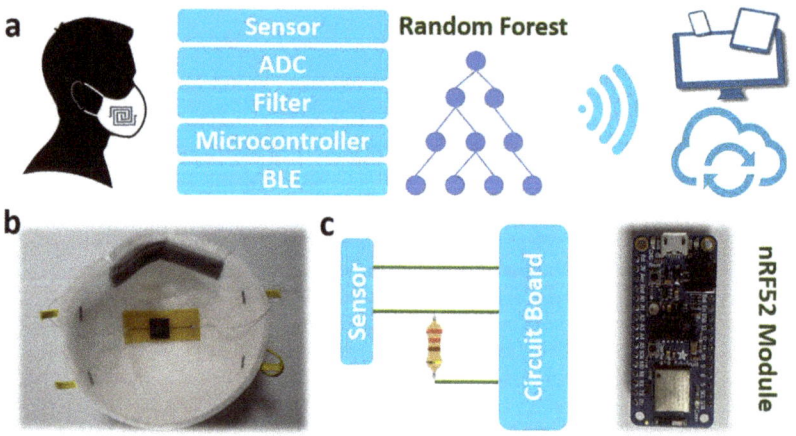

**Figure 5.** (**a**) Respiration checking with the embedded system, (**b**) smart mask with the flexible sensor, and (**c**) signal processing board with nRF52 module.

Random forest (RF) is a type of machine-learning algorithm based on decision trees and is commonly used for regression or classification problems. RF builds many decision trees, and each tree is unique. The final classifications are from the aggregation of these decision trees. This algorithm has proven particularly powerful for tasks with data from soft sensors. The application shows an overview of how the flexible sensor can be placed in a complete system containing many components. Here, we evaluated two main problems with flexible sensors when used in a system without considering the manufacturing process. Firstly, there is a connection between the sensors and the electrical wires. Soft sensors are not created from heat-resistant materials, so soldering the metal wires directly onto the electrode layers is not reasonable. The best practice is to secure part of the wire with strong adhesive and fix the rest of these wire ends with a laminating layer. Secondly, a suitable AI model is necessary for manipulating the obtained data. Wearables have an issue in that the average deep- or machine-learning models are too large to be applicable on a small embedded board. Therefore, choosing a suitable model for each application is extremely necessary. This leads to a concept—tiny machine learning for wearable devices. Basically, it takes a lot of experiments and tweaking of the system and signal acquisition method on the embedded board (programming) to obtain the best combination.

Figure 6a,b show different breathing signals and a complete system, including a smart mask, flexible pressure sensor, and circuit board. Figure 6c describes the initial signal and smoothing signal. This paper considers some filters, such as Kalman, Butterworth, and average filters, to reduce the noise in the raw data. Kalman filtering is an efficient optimal estimator given the measurements observed over time. The Butterworth filter finds the best compromise between phase response and attenuation. The average filter calculates the average output sample from a finite number of input samples. We concluded that the average filter was sufficient for obtaining a smoothing signal in this work.

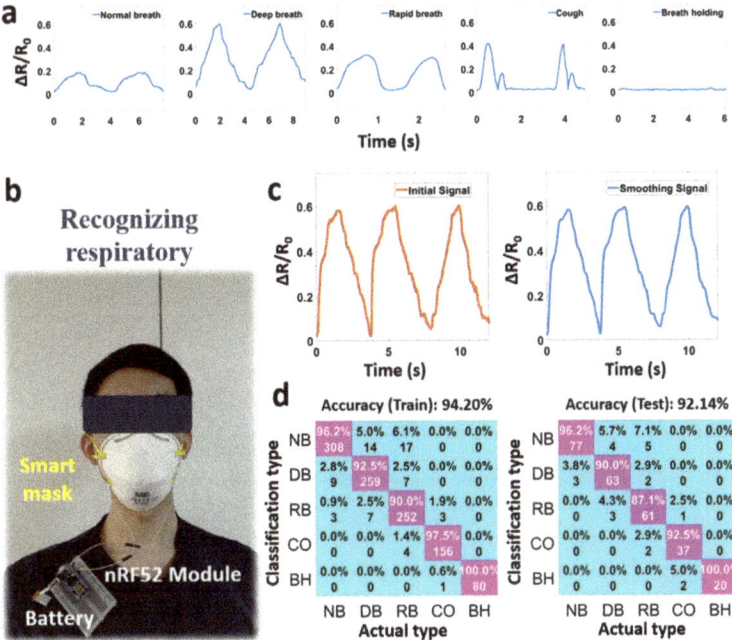

**Figure 6.** Respiratory monitoring system, consisting of (**a**) different breathing signals, (**b**) smart mask for recognizing respiratory, (**c**) initial and smoothing signal, and (**d**) confusion matrices.

The system recorded 1400 data samples, of which 80% were used for the training dataset and 20% for the testing dataset. The performance of the models was evaluated through confusion matrices. We randomly selected 280 breathing samples (20%), consisting of 80 normal breath samples, 70 deep breath samples, 70 rapid breath samples, 40 cough samples, and 20 breath-holding samples as a test dataset to evaluate the model. Figure 6d shows that the recognition accuracy was up to 94% for the training and 92% for the testing datasets. In these matrices, NB represents normal breathing, DB represents deep breathing, RB represents rapid breathing, CO represents cough, and BH represents breath-holding. The cough and breath-holding types were the easiest to distinguish, while the rapid and deep breath types were easily confused with the normal type. Among them, the accuracy of the breath-holding type was the highest at 100%. The accuracy of the rapid breath type was the lowest at 87%. Through the above application, the sensor demonstrated potential for daily health-monitoring tasks.

## 5. Conclusions

In summary, we have proposed a respiratory monitoring system from a flexible resistive pressure sensor, an embedded module, and a random forest algorithm. The sensor was assembled from Pyralux film, CNT wipes, and thin adhesive layers. This thin structure demonstrated superior performance with a sensitivity of 0.2 $kPa^{-1}$ (<6 kPa) and high water resistance. The potential of the sensor was evaluated in a practical system with various components. Among the challenges a system needs to address, the connections (between the flexible sensors and the electrical wires) and the signal processing model (which analyzes the resulting data) are two problems that need to be considered in detail. Here, the random forest algorithm was applied to recognize the different types of respiration, such as normal breath, deep breath, rapid breath, cough, and breath-holding. A model achieved a good accuracy of up to 92%, opening some new directions for studies in the future. The soft sensing system and machine-learning combination will become a universal platform for various wearable applications.

**Supplementary Materials:** The following supporting information can be downloaded at: https://www.mdpi.com/article/10.3390/mi14091726/s1, Figure S1: Universal testing machine (UTM), consisting of a computer, a force load cell, an LCR meter, and a sole plate; Figure S2: Signal of the sensor when immersed in water for different durations (0, 6, or 12 h); Figure S3: SEM picture of Pyralux film after etching solution; Figure S4: Durability of the sensor (in the bending state with R = 22.5 mm) after 2000 cycles; Figure S5: The resistance change in the sensor at different temperatures.

**Author Contributions:** Conceptualization, T.-H.N., J.K. and C.C.V.; methodology, C.C.V.; software, T.-H.N.; validation, C.C.V., T.-H.N. and J.K.; formal analysis, T.-H.N.; investigation, J.K. and C.C.V.; resources, C.C.V.; data curation, C.C.V.; writing—original draft preparation, J.K. and C.C.V.; writing—review and editing, C.C.V.; visualization, T.-H.N. and C.C.V.; supervision, T.-H.N. and C.C.V.; project administration, C.C.V. and J.K.; funding acquisition, C.C.V. All authors have read and agreed to the published version of the manuscript.

**Funding:** The research was supported by The Youth Incubator for Science and Technology Programme, managed by the Youth Promotion Science and Technology Center—Ho Chi Minh Communist Youth Union and Department of Science and Technology of Ho Chi Minh City; the contract number is "41/2022/HĐ-KHCNT-VU" signed on 30 December 2022. We would like to thank Ho Chi Minh City University of Technology and Education (HCMUTE) for supporting our group.

**Institutional Review Board Statement:** Not applicable.

**Informed Consent Statement:** Not applicable.

**Data Availability Statement:** The datasets generated and/or analyzed during the current study are available from the corresponding author upon reasonable request.

**Conflicts of Interest:** The authors declare no conflict of interest.

## References

1. Mishra, S.; Khouqeer, G.A.; Aamna, B.; Alodhayb, A.; Ali Ibrahim, S.J.; Hooda, M.; Jayaswal, G. A Review: Recent Advancements in Sensor Technology for Non-Invasive Neonatal Health Monitoring. *Biosens. Bioelectron. X* **2023**, *14*, 100332. [CrossRef]
2. Zhang, Z.; Wen, F.; Sun, Z.; Guo, X.; He, T.; Lee, C. Artificial Intelligence-Enabled Sensing Technologies in the 5G/Internet of Things Era: From Virtual Reality/Augmented Reality to the Digital Twin. *Adv. Intell. Syst.* **2022**, *4*, 2100228. [CrossRef]
3. Holt, S. Virtual Reality, Augmented Reality and Mixed Reality: For Astronaut Mental Health; and Space Tourism, Education and Outreach. *Acta Astronaut.* **2023**, *203*, 436–446. [CrossRef]
4. Zhou, Y.; Chen, J.; Wang, M. A Meta-Analytic Review on Incorporating Virtual and Augmented Reality in Museum Learning. *Educ. Res. Rev.* **2022**, *36*, 100454. [CrossRef]
5. Zhu, J.; Ji, S.; Yu, J.; Shao, H.; Wen, H.; Zhang, H.; Xia, Z.; Zhang, Z.; Lee, C. Machine Learning-Augmented Wearable Triboelectric Human-Machine Interface in Motion Identification and Virtual Reality. *Nano Energy* **2022**, *103*, 107766. [CrossRef]
6. Singh, M.; Chauhan, M.; Mishra, Y.K.; Wallen, S.L.; Kaur, G.; Kaushik, A.; Chaudhary, G.R. Novel Synthesis of Amorphous CP@HfO2 Nanomaterials for High-Performance Electrochemical Sensing of 2-Naphthol. *J. Nanostruct. Chem.* **2023**, *13*, 423–438. [CrossRef]
7. Olorunyomi, J.F.; Teng Geh, S.; Caruso, R.A.; Doherty, C.M. Metal–Organic Frameworks for Chemical Sensing Devices. *Mater. Horiz.* **2021**, *8*, 2387–2419. [CrossRef]
8. Arabi, M.; Chen, L. Technical Challenges of Molecular-Imprinting-Based Optical Sensors for Environmental Pollutants. *Langmuir* **2022**, *38*, 5963–5967. [CrossRef]
9. Venketeswaran, A.; Lalam, N.; Wuenschell, J.; Ohodnicki, P.R., Jr.; Badar, M.; Chen, K.P.; Lu, P.; Duan, Y.; Chorpening, B.; Buric, M. Recent Advances in Machine Learning for Fiber Optic Sensor Applications. *Adv. Intell. Syst.* **2022**, *4*, 2100067. [CrossRef]
10. Huang, Y.; Liu, B.; Zhang, W.; Qu, G.; Jin, S.; Li, X.; Nie, Z.; Zhou, H. Highly sensitive active-powering pressure sensor enabled by integration of double-rough surface hydrogel and flexible batteries. *NPJ Flex. Electron.* **2022**, *6*, 92. [CrossRef]
11. Yuan, J.; Li, Q.; Ding, L.; Shi, C.; Wang, Q.; Niu, Y.; Xu, C. Carbon Black/Multi-Walled Carbon Nanotube-Based, Highly Sensitive, Flexible Pressure Sensor. *ACS Omega* **2022**, *7*, 44428–44437. [CrossRef]
12. Manasa, G.; Mascarenhas, R.J.; Shetti, N.P.; Malode, S.J.; Mishra, A.; Basu, S.; Aminabhavi, T.M. Skin Patchable Sensor Surveillance for Continuous Glucose Monitoring. *ACS Appl. Bio Mater.* **2022**, *5*, 945–970. [CrossRef]
13. Lv, P.; Qian, J.; Yang, C.; Liu, T.; Wang, Y.; Wang, D.; Huang, S.; Cheng, X.; Cheng, Z. Flexible All-Inorganic Sm-Doped PMN-PT Film with Ultrahigh Piezoelectric Coefficient for Mechanical Energy Harvesting, Motion Sensing, and Human-Machine Interaction. *Nano Energy* **2022**, *97*, 107182. [CrossRef]
14. Gupta, N.; Adepu, V.; Tathacharya, M.; Siraj, S.; Pal, S.; Sahatiya, P.; Kuila, B.K. Piezoresistive Pressure Sensor Based on Conjugated Polymer Framework for Pedometer and Smart Tactile Glove Applications. *Sens. Actuators A* **2023**, *350*, 114139. [CrossRef]
15. Han, P.; Li, L.; Zhang, H.; Guan, L.; Marques, C.; Savović, S.; Ortega, B.; Min, R.; Li, X. Low-Cost Plastic Optical Fiber Sensor Embedded in Mattress for Sleep Performance Monitoring. *Opt. Fiber Technol.* **2021**, *64*, 102541. [CrossRef]
16. Schutte, A.E.; Kollias, A.; Stergiou, G.S. Blood pressure and its variability: Classic and novel measurement techniques. *Nat. Rev. Cardiol.* **2022**, *19*, 643–654. [CrossRef]
17. Niu, X.; Gao, X.; Liu, Y.; Liu, H. Surface bioelectric dry Electrodes: A review. *Measurement* **2021**, *183*, 109774. [CrossRef]
18. Nelson, E.C.; Verhagen, T.; Noordzij, M.L. Health empowerment through activity trackers: An empirical smart wristband study. *Comput. Hum. Behav.* **2016**, *62*, 364–374. [CrossRef]
19. Ambaye, A.D.; Kefeni, K.K.; Mishra, S.B.; Nxumalo, E.N.; Ntsendwana, B. Recent Developments in Nanotechnology-Based Printing Electrode Systems for Electrochemical Sensors. *Talanta* **2021**, *225*, 121951. [CrossRef]
20. Mostafiz, B.; Bigdeli, S.A.; Banan, K.; Afsharara, H.; Hatamabadi, D.; Mousavi, P.; Hussain, C.M.; Keçili, R.; Ghorbani-Bidkorbeh, F. Molecularly Imprinted Polymer-Carbon Paste Electrode (MIP-CPE)-Based Sensors for the Sensitive Detection of Organic and Inorganic Environmental Pollutants: A Review. *Trends Environ. Anal. Chem.* **2021**, *32*, e00144. [CrossRef]
21. Liu, C.; Wang, X.; Zhang, H.J.; You, X.; Yue, O. Self-Healable, High-Strength Hydrogel Electrode for Flexible Sensors and Supercapacitors. *ACS Appl. Mater. Interfaces* **2021**, *13*, 36240–36252. [CrossRef]
22. Vu, C.C.; Kim, J. Simultaneous Sensing of Touch and Pressure by Using Highly Elastic e-Fabrics. *Appl. Sci.* **2020**, *10*, 989. [CrossRef]
23. Zhang, J.; Zhang, Y.; Li, Y.; Wang, P. Textile-Based Flexible Pressure Sensors: A Review. *Polym. Rev.* **2022**, *62*, 65–94. [CrossRef]
24. Huang, C.-Y.; Yang, G.; Huang, P.; Hu, J.-M.; Tang, Z.-H.; Li, Y.-Q.; Fu, S.-Y. Flexible Pressure Sensor with an Excellent Linear Response in a Broad Detection Range for Human Motion Monitoring. *ACS Appl. Mater. Interfaces* **2023**, *15*, 3476–3485. [CrossRef] [PubMed]
25. Pierre Claver, U.; Zhao, G. Recent Progress in Flexible Pressure Sensors Based Electronic Skin. *Adv. Eng. Mater.* **2021**, *23*, 2001187. [CrossRef]
26. Vu, C.C.; Kim, J. Waterproof, Thin, High-Performance Pressure Sensors-Hand Drawing for Underwater Wearable Applications. *Sci. Technol. Adv. Mater.* **2021**, *22*, 718–728. [CrossRef] [PubMed]
27. Ai, J.; Cheng, S.-R.; Miao, Y.-J.; Li, P.; Zhang, H.-X. Graphene/Electrospun Carbon Nanofiber Sponge Composites Induced by Magnetic Particles for Mutil-Functional Pressure Sensor. *Carbon* **2023**, *205*, 454–462. [CrossRef]

28. Hou, Y.; Wang, L.; Sun, R.; Zhang, Y.; Gu, M.; Zhu, Y.; Tong, Y.; Liu, X.; Wang, Z.; Xia, J.; et al. Crack-Across-Pore Enabled High-Performance Flexible Pressure Sensors for Deep Neural Network Enhanced Sensing and Human Action Recognition. *ACS Nano* **2022**, *16*, 8358–8369. [CrossRef]
29. Shen, J.; Guo, Y.; Fu, T.; Yao, S.; Zhou, J.; Wang, D.; Bi, H.; Zuo, S.; Wu, X.; Shi, F.; et al. Skin-Inspired Hierarchical Structure Sensor for Ultrafast Active Human–Robot Interaction. *Adv. Mater. Technol.* **2023**, *8*, 2202008. [CrossRef]
30. Choudhry, N.A.; Shekhar, R.; Rasheed, A.; Arnold, L.; Wang, L. Effect of Conductive Thread and Stitching Parameters on the Sensing Performance of Stitch-Based Pressure Sensors for Smart Textile Applications. *IEEE Sens. J.* **2022**, *22*, 6353–6363. [CrossRef]
31. Han, Z.; Li, H.; Xiao, J.; Song, H.; Li, B.; Cai, S.; Chen, Y.; Ma, Y.; Feng, X. Ultralow-Cost, Highly Sensitive, and Flexible Pressure Sensors Based on Carbon Black and Airlaid Paper for Wearable Electronics. *ACS Appl. Mater. Interfaces* **2019**, *11*, 33370–33379. [CrossRef] [PubMed]
32. Zhu, Z.; Zhang, H.; Xia, K.; Xu, Z. Hand-Drawn Variable Resistor and Strain Sensor on Paper. *Microelectron. Eng.* **2018**, *191*, 72–76. [CrossRef]
33. Liu, Y.-Q.; Zhang, Y.-L.; Jiao, Z.-Z.; Han, D.-D.; Sun, H.-B. Directly Drawing High-Performance Capacitive Sensors on Copying Tissues. *Nanoscale* **2018**, *10*, 17002–17006. [CrossRef]
34. Gilanizadehdizaj, G.; Aw, K.C.; Stringer, J.; Bhattacharyya, D. Facile Fabrication of Flexible Piezo-Resistive Pressure Sensor Array Using Reduced Graphene Oxide Foam and Silicone Elastomer. *Sens. Actuators A* **2022**, *340*, 113549. [CrossRef]
35. Hwang, J.; Kim, Y.; Yang, H.; Oh, J.H. Fabrication of Hierarchically Porous Structured PDMS Composites and Their Application as a Flexible Capacitive Pressure Sensor. *Compos. Part B* **2021**, *211*, 108607. [CrossRef]
36. Ji, B.; Zhou, Q.; Chen, G.; Dai, Z.; Li, S.; Xu, Y.; Gao, Y.; Wen, W.; Zhou, B. In Situ Assembly of a Wearable Capacitive Sensor with a Spine-Shaped Dielectric for Shear-Pressure Monitoring. *J. Mater. Chem. C* **2020**, *8*, 15634–15645. [CrossRef]
37. Zhang, C.; Zhang, L.; Tian, Y.; Bao, B.; Li, D. A Machine-Learning-Algorithm-Assisted Intelligent System for Real-Time Wireless Respiratory Monitoring. *Appl. Sci.* **2023**, *13*, 3885. [CrossRef]
38. Vu, C.C.; Kim, J. Human motion recognition using SWCNT textile sensor and fuzzy inference system based smart wearable. *Sens. Actuators A.* **2018**, *283*, 263–272. [CrossRef]

**Disclaimer/Publisher's Note:** The statements, opinions and data contained in all publications are solely those of the individual author(s) and contributor(s) and not of MDPI and/or the editor(s). MDPI and/or the editor(s) disclaim responsibility for any injury to people or property resulting from any ideas, methods, instructions or products referred to in the content.

Article

# Flexible Pressure Sensors and Machine Learning Algorithms for Human Walking Phase Monitoring

Thanh-Hai Nguyen, Ba-Viet Ngo, Thanh-Nghia Nguyen and Chi Cuong Vu *

Faculty of Electrical and Electronics Engineering, Ho Chi Minh City University of Technology and Education, 01 Vo Van Ngan Street, Linh Chieu Ward, Ho Chi Minh City 700000, Vietnam; nthai@hcmute.edu.vn (T.-H.N.); vietnb@hcmute.edu.vn (B.-V.N.); nghiant@hcmute.edu.vn (T.-N.N.)
* Correspondence: cuongvc@hcmute.edu.vn

**Abstract:** Soft sensors are attracting much attention from researchers worldwide due to their versatility in practical projects. There are already many applications of soft sensors in aspects of life, consisting of human-robot interfaces, flexible electronics, medical monitoring, and healthcare. However, most of these studies have focused on a specific area, such as fabrication, data analysis, or experimentation. This approach can lead to challenges regarding the reliability, accuracy, or connectivity of the components. Therefore, there is a pressing need to consider the sensor's placement in an overall system and find ways to maximize the efficiency of such flexible sensors. This paper proposes a fabrication method for soft capacitive pressure sensors with spacer fabric, conductive inks, and encapsulation glue. The sensor exhibits a good sensitivity of $0.04\ kPa^{-1}$, a fast recovery time of 7 milliseconds, and stability of 10,000 cycles. We also evaluate how to connect the sensor to other traditional sensors or hardware components. Some machine learning models are applied to these built-in soft sensors. As expected, the embedded wearables achieve a high accuracy of 96% when recognizing human walking phases.

**Keywords:** soft capacitive pressure sensors; healthcare; connectivity; random forest; flexible embedded system

**Citation:** Nguyen, T.-H.; Ngo, B.-V.; Nguyen, T.-N.; Vu, C.C. Flexible Pressure Sensors and Machine Learning Algorithms for Human Walking Phase Monitoring. *Micromachines* **2023**, *14*, 1411. https://doi.org/10.3390/mi14071411

Academic Editors: Lei Jing, Yoshinori Matsumoto and Zhan Zhang

Received: 21 June 2023
Revised: 8 July 2023
Accepted: 10 July 2023
Published: 13 July 2023

**Copyright:** © 2023 by the authors. Licensee MDPI, Basel, Switzerland. This article is an open access article distributed under the terms and conditions of the Creative Commons Attribution (CC BY) license (https:// creativecommons.org/licenses/by/ 4.0/).

## 1. Introduction

Nowadays, intelligent devices use a large number of different sensors. Their applications are found in manufacturing, environmental protection, biotechnology, medical diagnostics, marine exploration, and space development. Large amounts of signals are collected in each of the above unique environments. New sensor technology has evolved along the following trends: developing novel materials and processes, realizing sensor integration and intelligence, and realizing the hardware system.

Wearable technology refers to electronic devices worn on the user's body. These devices exist in various forms, including accessories, jewelry, clothing, and medical devices. The complexity of wearable devices varies. The most complex examples include Google Glass, artificial intelligence (AI) hearing aids, Microsoft's HoloLens, or holographic computers—virtual reality (VR) headsets. Less sophisticated products are disposable skin patches with sensors that transmit patient data to monitoring equipment in a healthcare facility. Wearable devices contain built-in sensors that allow monitoring of body signals, location tracking, or biometric recognition. Most of these devices are attached to clothing/footwear and operate with or without direct contact with the human body. Smart cards and smartphones are portable and track users' movements. Some wearable devices use remote sensors and accelerometers to recognize human motion and speed. Others use optical or chemical sensors to measure glucose, blood oxygen, and heart rates. One thing these devices have in common is that they all operate in real-time.

Among the new types of sensors, flexible soft sensors emerge as the ideal technology for the future of wearables. Flexible soft sensors are a general concept for sensors that can be applied to a variety of soft surfaces with different irregular shapes or spaces, such as human skin and clothing [1–4]. These sensors are often capable of bending or stretching while exhibiting their electromechanical properties. For example, He et al. [5] propose a facile capacitive pressure sensor optimized by a low-cost nylon netting, showing a good response sensitivity (0.33 kPa$^{-1}$) for monitoring the pulses and clicks. This structure achieves operational stability after 1000 cycles and a response time of 20 ms. The group of Kim [6] presents hand-drawing pressure sensors using carbon nanotubes (CNTs), wet tissues, and pyralux films. The sensors are flexible, thin, water-resistant, and high-performance. With a sensitivity of 0.2 kPa$^{-1}$, these structures are suitable for touch detection when controlling a mobile phone or tablet. In another work, Zhang et al. [7] show a capacitive pressure sensor that is capable of making measurements at low pressure (0.2 g soybean). This research uses an elastic metalized sponge (nickel-plated polyurethane sponge) as the elastic porous electrode. The sensor is applied to the grasping robot. It is clear that the most significant benefit of soft sensors is their usability in different environments, flexibility, safety, and comfort for the wearer [8–10]. On the other hand, the main challenges of these sensors are signal collection and stable operation in an application [11–13].

In recent years, flexible sensors have been extensively studied in various fields, such as medicine [14,15], healthcare [16,17], the environment [18,19], and biology [20,21]. There are some commercial products, consisting of motion-tracking clothing [22,23], medical care devices [24,25], or health alarms [26,27]. However, these studies only approach a small part of the entire system. Considering a system, we need connections, signal acquisition, processing circuits, data analysis techniques, and other components/sensors with tasks that the flexible sensors cannot yet perform. From the above approach, the paper models a comprehensive system for human health monitoring through pressure gloves and another example of a walking signal classification/tracking system. In this model, the soft capacitive pressure sensors are based on thin spacer fabric [28,29], single-walled carbon nanotubes (SWCNTs) [30], stretchable silver paste [31], and encapsulating glue [32]. This structure can generate a 3D response from the sensor under deformation. The fabricated sensor has shown good properties in response time of 7 ms, resiliency, and durability after 10,000 loading/unloading cycles. More specifically, the sensors are considered when connecting with other system components, such as processing circuits and traditional sensors. This system can classify the human walking phases using two machine learning models, including a simple deep neural network [33] and a random forest [34]. The deep neural network is a model based on a neural network containing many hidden layers, and the random forest is a classification model based on multiple decision trees. The work will provide an overview of flexible sensors in a complete embedded system.

## 2. Materials and Methods

### 2.1. Materials

Spacer fabrics are unique fabrics comprising two separate layers connected by fiber (polyester). This 3D structure creates 3D microclimates between the layers, leading to outstanding properties such as durability, flexibility, and breathability for the spacer fabric. Two outer stretch fabrics are woven from a primary blend of synthetic polyester (polyethylene terephthalate) and spandex with a ratio of 76/24 from SNT Co., Ltd., Seoul, Republic of Korea. This method of co-weaving provides excellent elasticity for many particular applications. After surface coating the polyester/spandex fibers, these fabrics will form conductive layers that are able to change the conductivity through deformations. Surface coating ink was purchased from KH Chemicals Co., Seoul, Republic of Korea; silver paste and encapsulation glue were prepared from Dycotec Materials Ltd. (Calne, UK).

### 2.2. Fabrication Method

The fabrication processes of the capacitive pressure sensors are a combination of three phases: compositing the conductive layer with SWCNT and silver paste, shaping the sensors by laser cutting machines, and injecting the encapsulation glue into the spacer layer.

Firstly, an ultrasonic stirring machine was used to remove the air bubble inside SWCNT ink at 0.1 wt%. The stirring machine worked for 2 h at a stirring frequency of 20 Hz, a temperature of 75 °C, and a speed of 1000 rpm. In addition, the silver paste was kept in a fridge at 4 °C in a tightly sealed bottle. Using screen-printing technology at a speed of 20 mm/s, SWCNT layers were coated on each surface of the spacer fabric (Figure 1a). The excess water inside the ink was removed with a two-way dryer (Figure 1b). This step was set up at a temperature of 200 °C, a time of 2 min, and a fan speed of 1500 rpm. Then, we applied the above process with silver paste to the two sides of the spacer fabric, as shown in Figure 1c,d. At that time, the parameters underwent a slight change. The dryer worked at a temperature of 130 °C for 15 min to remove the solvents. Secondly, the pressure sensors could achieve their own customizable shapes with a laser cutter (Figure 1e). In this step, we shaped the sensors into a square of 10 mm × 10 mm. Thirdly, the encapsulation glue was inserted inside the spacer layer by a small injection needle, as described in Figure 1f. The dryer worked at a temperature of 130 °C for 10 min to fix these glue positions. Finally, the obtained sensor is shown in Figure 1g, consisting of two electrode layers of the polyester/spandex fabrics coated with SWCNT/silver. These fabric layers achieve high conductivity, stretchability, flexibility, and quick response/recovery under pressure deformations for practical applications. Between two electrode layers, a dielectric layer was formed by encapsulation glue and polyester fibers. This structure guarantees that the sensors have strong and constant elasticity even under large forces. In addition, the thin spacer fabric would fit many small spaces for different wearable devices.

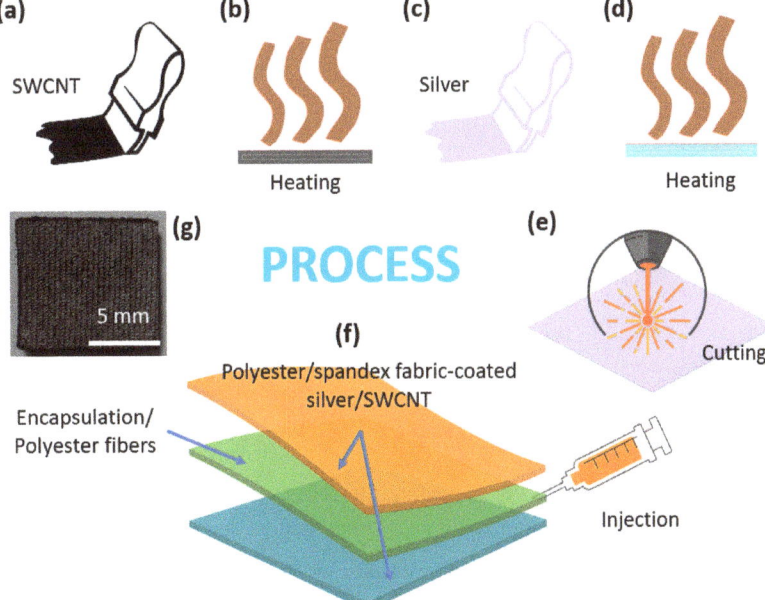

**Figure 1.** The fabrication process of a capacitive pressure sensor, including (**a**) Screen printing SWCNT, (**b**) Heating in the first, (**c**) Screen printing silver paste, (**d**) Heating in the second, (**e**) Laser cutting, (**f**) Injeting encapsulation glue, and (**g**) Real capacitive pressure sensor.

## 3. Results and Discussion

The sensor's characteristics are checked by a universal testing machine (UTM), as shown in Figure 2a. Figure 2b describes the working principle of the fabricated capacitive pressure sensor. Accordingly, the dielectric layers from encapsulation and polyester fibers are compressed when loading a force. The distance between the two electrode layers (silver/SWCNT) becomes closer, which changes the capacitance of the sensor. One challenge with soft capacitive pressure is resiliency and noise under deformations. Here, the polyester filaments help ensure resilience quickly. The dielectric layer ensures minimal interference in flexible applications.

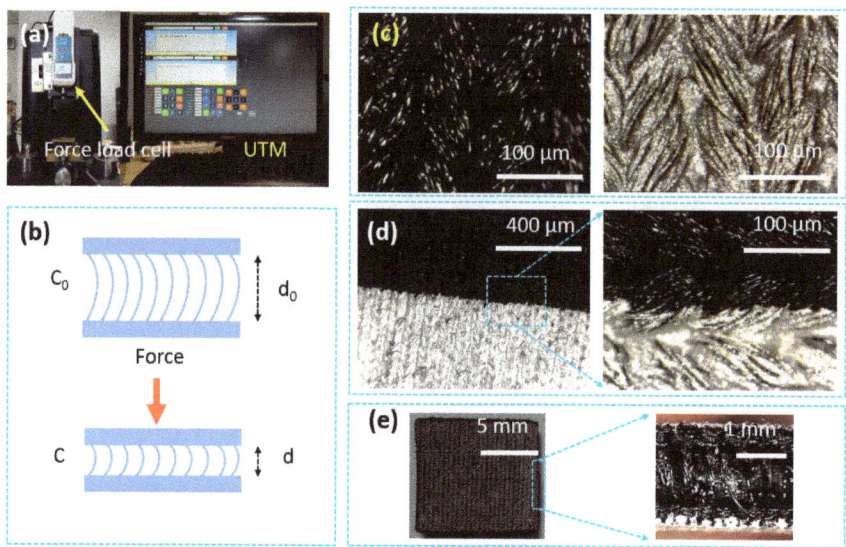

**Figure 2.** (**a**) Universal testing machine (UTM); (**b**) Working principle; (**c**) Scanning electron microscope (SEM) figures of the sensor at 100 μm; (**d**) Scanning electron microscope (SEM) figures of the sensor at the edge printing; (**e**) The final sensor and the cross view.

Capacitance change is calculated according to Equation (1), where C represents the capacitance at the compressed force, $\varepsilon_0$ represents the constant for the dielectric permittivity of vacuum, $\varepsilon_r$ represents the permittivity of the dielectric layer, A represents the area of the electrode layer, and $d_0$ represents the distance between two electrode layers, respectively. The sensitivity (S) can be defined from Equation (2), where $C_0$ is the initial capacitance, $\Delta C$ is the capacitance change, and P is the loading pressure.

$$C_{sensor} = \varepsilon_0 \varepsilon_r \frac{A}{d_0} \qquad (1)$$

$$S = \frac{\Delta C/C}{C_0} \qquad (2)$$

From Equation (1), the capacitance change can be raised by decreasing the distance d and increasing the dielectric constant ε. This ε of the dielectric layer will be defined in Equation (3), where $V_{air}$ represents the volume of the air, $V_{PET}$ is the volume of the PET fibers, $V_{encapsulation}$ represents the volume of the encapsulation glue, $\varepsilon_{air}$ represents the permittivity of the air, $\varepsilon_{PET}$ is the permittivity of the PET fibers, and $\varepsilon_{encapsulation}$ is the permittivity of the encapsulation glue. It is demonstrated that the decrease in the volume of air gaps contributes significantly to the increase in capacitance obtained when compressing and releasing the force.

$$\varepsilon_r = \left( \%V_{air} \cdot \varepsilon_{air} + \%V_{PET} \cdot \varepsilon_{PET} + \%V_{encapsulation} \cdot \varepsilon_{encapsulation} \right) \qquad (3)$$

Scanning electron microscopy (SEM) was used to analyze the surface morphology of the sensor. Figure 2c–e shows the knitted structure of two electrode layers of polyester/spandex fabrics (before and after the screen printing process). This structure consists of interconnected loops that create extremely high stretchability. The diameter of the single fibers is about 10 μm, and there are many gaps between the fibers or bundles. After the printing processes, the filaments are coated with silver/SWCNT, covering about 90% of the printed area. Figure 2c,d shows that the surface is illustrated at different steps in the overall process. Encapsulation glue is inserted with an injection needle between the gaps inside the spacer fabric. As shown in Figure 2e, the air volume is minimized to increase the dielectric constant.

Figure 3 describes the capacitance change at the different levels of compression. The pressure sensitivity is about $4 \times 10^{-2}$ kPa$^{-1}$. Thanks to the construction of the spacer fabric, the sensor can withstand pressures of up to 1000 kPa, as seen in Figure 3b–d. It demonstrates that the sensor has an extensive working range and is suitable for many applications. The resilience of the sensor package depends mainly on the viscosity of the polyester pile fibers and the encapsulation glue. Resilience is vitally essential for practical applications that require a rapid response. Here, the response and recovery times can be seen in Figure 3e, with 7 ms at 100 kPa pressure. The main reason for this good value is the ability of the polyester fibers to release two parallel electrodes quickly. Additionally, we consider the dynamic performance of the sensor in Figure 3f. The result shows that the sensor has a stable response at different frequencies, from 0.1 to 2 Hz.

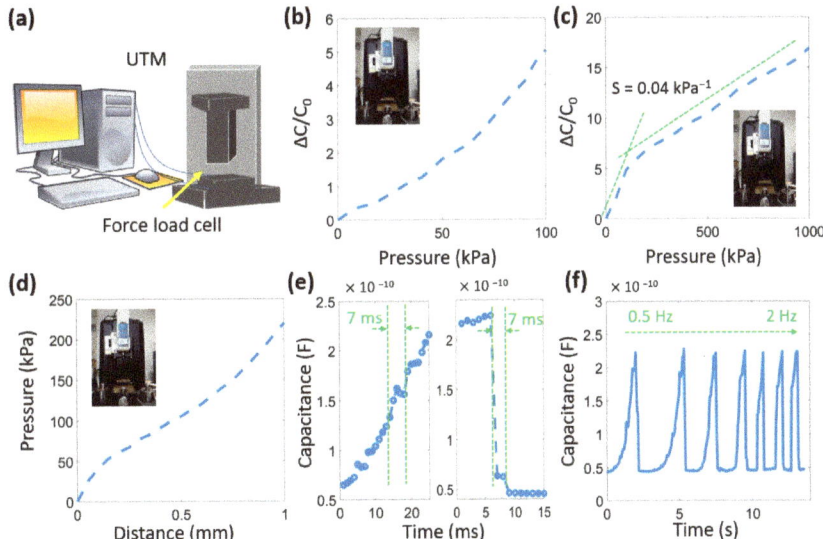

**Figure 3.** Properties of the sensor, consisting of (**a**) Diagram of UTM machine; (**b**) Capacitance change under pressure from 0–100 kPa; (**c**) Capacitance change under pressure from 0–1000 kPa; (**d**) Distance change under pressure; (**e**) Response and recovery time; (**f**) Capacitance change at different frequencies.

In the event of undesirable deformations, the electrode layers can still maintain high conductivity, as demonstrated in Figure 4a,b. The resistance change is under 0.5 Ω after stretching, pressing, bending, or twisting. This is due to the high stretchability of the stretchable silver paste layer. Another highlight is the ability to maintain capacitance after washing. As shown in Figure 4c, the maximum capacitance of the sensor has a slight change of 8%. Figure 4d describes the testing results to evaluate the durability of the sensor under many working cycles. Accordingly, our sensor exhibits a stable electrical function and excellent mechanical integrity (<7% change) at 100 kPa after 10,000 loading/releasing cycles. This variation in dynamic durability is mainly due to fatigue and plastic deformation of the dielectric layer (encapsulation and polyester fibers), causing permanent structural deformation. These above characteristics are significant for practical wearable systems. In addition, Table 1 summarizes some studies on flexible sensors, including their principles, thickness, sensitivity, and response time [32–38]. Our sensor clearly demonstrates good capabilities, such as a fast response time and small thickness, for a wide range of wearable applications.

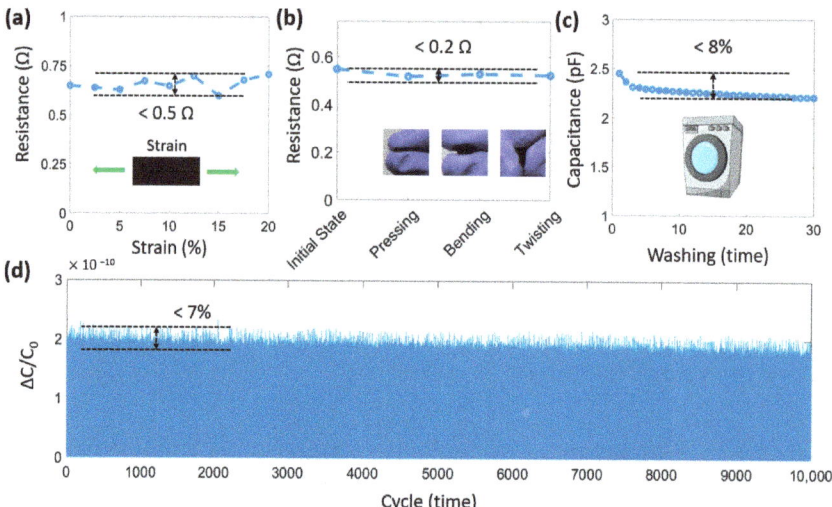

**Figure 4.** Properties of the sensor, consisting of (**a**) Resistance change under strain from 0–20%; (**b**) Resistance change when pressing, bending, and twisting; (**c**) Capacitance change after washing; and (**d**) Capacitance change under 10,000 loading/unloading cycles.

**Table 1.** A comparison of some flexible pressure sensors.

| Ref. | Principle | Thickness | Response Time | Sensitivity |
| --- | --- | --- | --- | --- |
| [5] | Capacitive | 2.698 mm | 20 ms | 0.33 kPa$^{-1}$ |
| [7] | Capacitive | >4 mm | >50 ms | - |
| [35] | Capacitive | 1.185 mm | - | 0.283 kPa$^{-1}$ |
| [36] | Capacitive | 1.96 mm | 7 ms | 0.0121 kPa$^{-1}$ |
| [6] | Resistive | 0.26 mm | 70 ms | 0.2 kPa$^{-1}$ |
| [37] | Resistive | 3 mm | - | 0.05 kPa$^{-1}$ |
| [38] | Resistive | 0.15 mm | - | 0.001 kPa$^{-1}$ |
| Ours | Capacitive | 1.85 mm | 7 ms | 0.04 kPa$^{-1}$ |

## 4. Recognizing Walking Phases and Wearable Embedded System

To demonstrate the ability of the fabricated sensors and wearable systems based on flexible sensors, we propose a method to monitor the overall indicators of human exercise (Figure 5a). This system will track signals such as phase, heart rate, and SPO$_2$ ratio during different walking phases. These indicators are significant for applications to analyze daily clinical health after exercise or for patients recovering from injury. To achieve this aim, the system will consist of three main parts: the smart sock-added four flexible pressure sensors, the MAX30105 module, and the Arduino Nano 33 BLE Sense module. The MAX30105 board contains two LEDs that are used to detect pulse oximetry and heart rate signals. The Arduino Nano 33 BLE Sense is a completely tiny, embedded board with many sensors that can run a small AI model. Some machine learning models are built to classify walking phases from the received signals, such as simple deep neural networks and random forests.

Human walking activity can be divided into four main phases, including heel-strike (HS), hell-off (HO), toe-off (TO), and mid-swing (MS). The amplitude change of the signals is the lowest in the MS phase and the highest in the HS phase. As described in Figure 5b,c, we collect signals at a rate of 1000 samples/min from the wearable system. The analog signals are then converted to digital signals and filtered to remove noise or unusual vibrations. Some features extracted from the walking samples are applied to machine learning models. Finally, the classified phases are sent via Bluetooth to phones, tablets, or smartwatches.

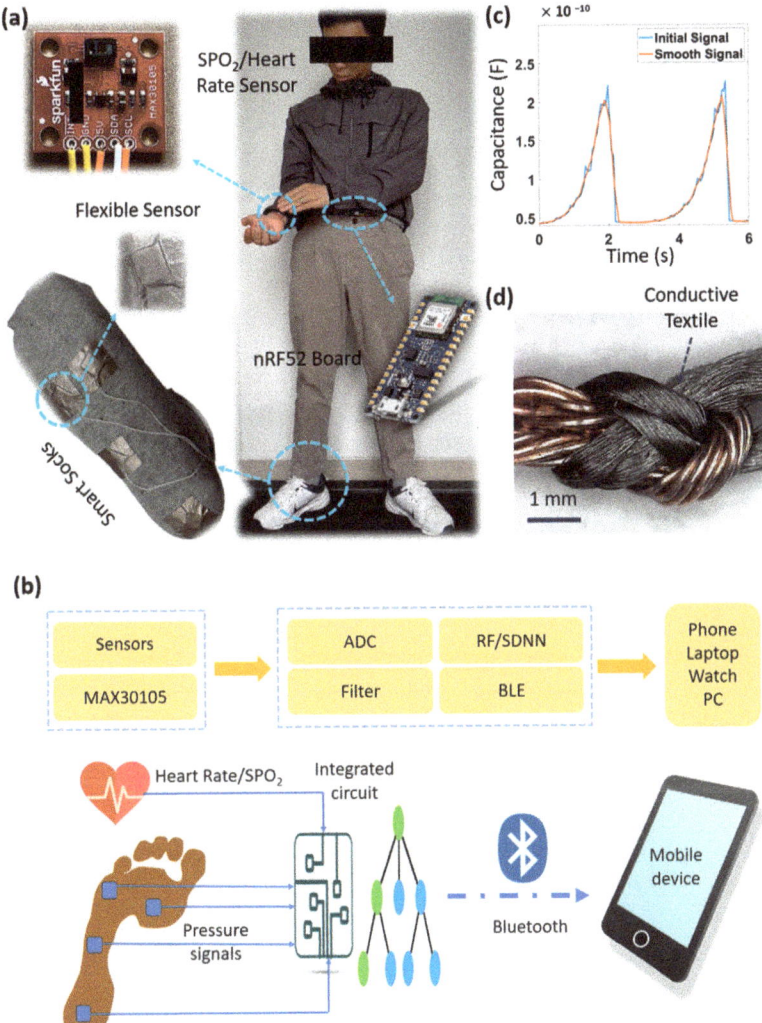

**Figure 5.** (a) Structure of the wearable monitoring system, including four capacitive pressure sensors (smart socks), MAX30105, and Arduino Nano 33 BLE Sense; (b) Working diagram of the smart wearable system; (c) Smooth signal; (d) Connection (sheet bend knot) between copper wire and conductive textile wire.

Deep learning is a neural network with many hidden layers between the input and output layers. These neural networks attempt to simulate the behavior of the human brain, allowing it to learn and observe exciting patterns between data patterns. Deep learning appears in many AI applications and services, which help collect data and improve analysis without human intervention. However, the data required for a deep learning model must be large enough and take a long time to train. Due to the characteristics of the existing dataset, we chose a simple deep neural network (SDNN) model with three hidden and fully connected layers. Each layer consists of multiple nodes (neurons) and builds upon the previous layer to refine and optimize the classifier. The output of one node will become the input of the next node. Data will be passed from one layer to the next (a feedforward neural network). The total number is about 35 nodes (neurons) in hidden layers. We also evaluated the system with a random forest (RF) model. Random forest is a type of supervised machine learning algorithm for classification and regression problems. This algorithm combines the output of multiple decision trees to reach a single result. RF will build many decision trees, and each tree is unique. The prediction results are based on the aggregation of the decision trees. The RF algorithm has proven particularly powerful for tasks involving the data of soft sensors. The paper will show an overview of the capabilities of two SDNN and RF algorithms with a system that integrates flexible pressure sensors.

We evaluated the three main problems of flexible sensors when placed in an overall system without considering the complex manufacturing process. Firstly, there is a connection between the sensors and the electrical wires. The soft sensors are not created from heat-resistance materials, so soldering the metal wires directly onto the electrode layers is not reasonable. Here, the best practice is to secure part of the wire with strong adhesive and fix the rest of the wire ends with a laminating layer, as shown in Figure 5. Secondly, there is a difference in working phase between the soft sensors and other sensors in the system. These sensors can be other soft sensors or traditional sensors. The reason could be due to the flexibility of the new sensors or loose connections. Some studies used conductive fabric-type electrical wires, which further increase the possibility of noise causing instability in the received signal. The connection between the metal wire and the wire based on fabric is also an issue. We suggest twisting the two ends of the wire together, such as in a sheet bend knot, and fixing them with instant glue (Figure 5d). Then, the connected position is wrapped by a heat shrink tube. This method ensures the signal's transmission capacity and protects the connection. A new AI model will require a lot of experimentation and tweaking in the system and signal processing method to get the best combination. However, wearables have the issue that the average deep learning or machine learning model is too large to be applicable on a small embedded board. Therefore, choosing a suitable model for each application is extremely necessary.

Figure 6a describes the walking motion with the wearable system, including four flexible sensors and the MAX30105 module. Signals from the system are recorded with 1500 data samples, and each phase of walking motion will have 375 samples, 80% for training and 20% for testing. The performance of the models will be evaluated through the confusion matrixes, as shown in Figure 6b,c. The confusion matrix calculates the model's metrics, such as accuracy and recall. Each column of the matrix represents the actual walking phases. In contrast, each row of the matrix represents the predicted walking phases. Each cell measures how many times a phase was correctly and incorrectly classified. High values on the diagonal line show the performance of the model. After the training process, the average accuracy reaches 91.92% for SDNN and 96.5% for RF. However, the average accuracy of SDNN is only 88.33% when testing new data. This index of RF is much better, at 96.33%. The accuracy of each phase (HS, HO, TO, and MS) is 97.4%, 94.7%, 94.7%, and 98.6%, respectively. We observe that the HS and MS phases are the easiest to distinguish, while the HO and TO phases are easily confused with each other. It is clear that the RF model is suitable for the existing dataset. Among them, the highest accuracy is obtained when classifying the MS phase (98%), and the lowest accuracy is obtained when recognizing the HO phase (94%). The practical application has demonstrated that the built-in flexible capacitive pressure sensor has a high potential for wearable devices in tracking and recognizing body movements. However, the sensor still has a limitation. There is a difference in the working ability of each position on the socks. This problem leads to asynchronous signals. At that time, the system could not comprehensively evaluate the flexible sensors. We recommend several different structures of the spacer fabric for different positions. This direction will be carried out in the subsequent studies.

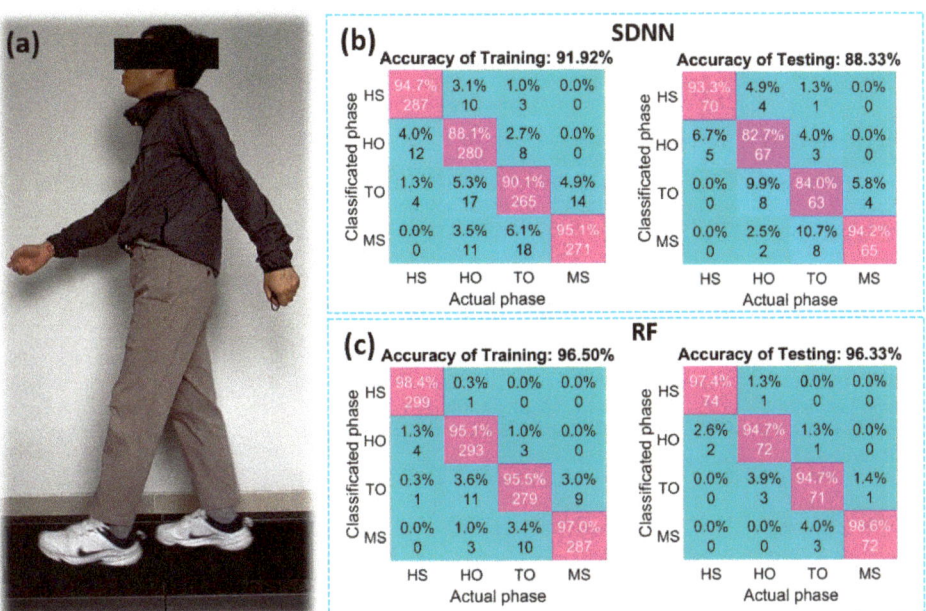

**Figure 6.** (**a**) Human walking exercise; (**b**) Confusion matrixes of simple deep neural networks; (**c**) Confusion matrixes of random forests.

## 5. Conclusions

This paper proposes a method to fabricate a flexible capacitive pressure sensor based on spacer fabric, polyester/spandex fabric, encapsulation glue, SWCNT, and silver paste. Thanks to the unique 3D structure of the spacer fabric containing polyester piles, the sensor achieves good sensitivity and extreme resilience for a variety of applications. In addition, the stretchability of polyester/spandex fabric and conductive silver paste has increased the durability of the sensor when subjected to unwanted deformations. The combination of two printing layers (SWCNT/silver) ensures the high conductivity of the electrodes. More importantly, we evaluated the capability of a practical system with various components, such as embedded boards, other soft sensors, or traditional sensors. Among the challenges a system needs to address, the connections (between the flexible sensors and the electrical wires) and the signal processing model (which analyzes the resulting data) are two problems that need to be considered in detail. On testing with two machine learning models, RF gave better results and will be used for further studies. The combination of the soft sensing system and machine learning will become a universal platform for various wearable applications in the future.

**Author Contributions:** Conceptualization, T.-H.N. and C.C.V.; methodology, B.-V.N.; software, T.-N.N.; validation, B.-V.N., T.-H.N. and C.C.V.; formal analysis, T.-N.N.; investigation, B.-V.N.; resources, C.C.V.; data curation, C.C.V.; writing—original draft preparation, T.-N.N.; writing—review and editing, C.C.V.; visualization, T.-H.N.; supervision, T.-H.N.; project administration, C.C.V.; funding acquisition, T.-H.N. All authors have read and agreed to the published version of the manuscript.

**Funding:** This research was supported by Ho Chi Minh City University of Technology and Education (HCMUTE) under Grant No. T2023-42.

**Data Availability Statement:** The datasets generated and/or analyzed during the current study are available from the corresponding author on reasonable request.

**Acknowledgments:** We would like to thank Ho Chi Minh City University of Technology and Education (HCMUTE), Vietnam under Grant No. T2023-42.

**Conflicts of Interest:** The authors declare no conflict of interest.

## References

1. Li, W.-D.; Ke, K.; Jia, J.; Pu, J.-H.; Zhao, X.; Bao, R.-Y.; Liu, Z.-Y.; Bai, L.; Zhang, K.; Yang, M.-B.; et al. Recent Advances in Multiresponsive Flexible Sensors towards E-Skin: A Delicate Design for Versatile Sensing. *Small* **2022**, *18*, 2103734. [CrossRef]
2. Lu, Y.; Fujita, Y.; Honda, S.; Yang, S.-H.; Xuan, Y.; Xu, K.; Arie, T.; Akita, S.; Takei, K. Wireless and Flexible Skin Moisture and Temperature Sensor Sheets toward the Study of Thermoregulator Center. *Adv. Healthc. Mater.* **2021**, *10*, 2100103. [CrossRef] [PubMed]
3. Li, J.-W.; Huang, C.-Y.; Zhou, B.-H.; Hsu, M.-F.; Chung, S.-F.; Lee, W.-C.; Tsai, W.-Y.; Chiu, C.-W. High Stretchability and Conductive Stability of Flexible Hybrid Electronic Materials for Smart Clothing. *Chem. Eng. J. Adv.* **2022**, *12*, 100380. [CrossRef]
4. Chiu, C.-W.; Huang, C.-Y.; Li, J.-W.; Li, C.-L. Flexible Hybrid Electronics Nanofiber Electrodes with Excellent Stretchability and Highly Stable Electrical Conductivity for Smart Clothing. *ACS Appl. Mater. Interfaces* **2022**, *14*, 42441–42453. [CrossRef] [PubMed]
5. He, Z.; Chen, W.; Liang, B.; Liu, C.; Yang, L.; Lu, D.; Mo, Z.; Zhu, H.; Tang, Z.; Gui, X. Capacitive Pressure Sensor with High Sensitivity and Fast Response to Dynamic Interaction Based on Graphene and Porous Nylon Networks. *ACS Appl. Mater. Interfaces* **2018**, *10*, 12816–12823. [CrossRef]
6. Vu, C.C.; Kim, J. Waterproof, Thin, High-Performance Pressure Sensors-Hand Drawing for Underwater Wearable Applications. *Sci. Technol. Adv. Mater.* **2021**, *22*, 718–728. [CrossRef]
7. Zhang, Y.; Lin, Z.; Huang, X.; You, X.; Ye, J.; Wu, H. Highly Sensitive Capacitive Pressure Sensor with Elastic Metallized Sponge. *Smart Mater. Struct.* **2019**, *28*, 105023. [CrossRef]
8. Zhao, C.; Wang, Y.; Tang, G.; Ru, J.; Zhu, Z.; Li, B.; Guo, C.F.; Li, L.; Zhu, D. Ionic Flexible Sensors: Mechanisms, Materials, Structures, and Applications. *Adv. Funct. Mater.* **2022**, *32*, 2110417. [CrossRef]
9. Jumet, B.; Bell, M.D.; Sanchez, V.; Preston, D.J. A Data-Driven Review of Soft Robotics. *Adv. Intell. Syst.* **2022**, *4*, 2100163. [CrossRef]
10. Singh, K.R.; Nayak, V.; Singh, J.; Singh, R.P. Nano-Enabled Wearable Sensors for the Internet of Things (IoT). *Mater. Lett.* **2021**, *304*, 130614. [CrossRef]
11. Zhang, J.; Zhang, Y.; Li, Y.; Wang, P. Textile-Based Flexible Pressure Sensors: A Review. *Polym. Rev.* **2022**, *62*, 65–94. [CrossRef]
12. Htwe, Y.Z.N.; Mariatti, M. Printed Graphene and Hybrid Conductive Inks for Flexible, Stretchable, and Wearable Electronics: Progress, Opportunities, and Challenges. *J. Sci. Adv. Mater. Devices* **2022**, *7*, 100435. [CrossRef]
13. Vu, C.C.; Kim, S.J.; Kim, J. Flexible Wearable Sensors—An Update in View of Touch-Sensing. *Sci. Technol. Adv. Mater.* **2021**, *22*, 26–36. [CrossRef]
14. Chen, J.; Zhang, J.; Hu, J.; Luo, N.; Sun, F.; Venkatesan, H.; Zhao, N.; Zhang, Y. Ultrafast-Response/Recovery Flexible Piezoresistive Sensors with DNA-Like Double Helix Yarns for Epidermal Pulse Monitoring. *Adv. Mater.* **2022**, *34*, 2104313. [CrossRef]
15. Wang, J.; Zhu, Y.; Wu, Z.; Zhang, Y.; Lin, J.; Chen, T.; Liu, H.; Wang, F.; Sun, L. Wearable Multichannel Pulse Condition Monitoring System Based on Flexible Pressure Sensor Arrays. *Microsyst. Nanoeng.* **2022**, *8*, 16. [CrossRef] [PubMed]
16. Geng, D.; Chen, S.; Chen, R.; You, Y.; Xiao, C.; Bai, C.; Luo, T.; Zhou, W. Tunable Wide Range and High Sensitivity Flexible Pressure Sensors with Ordered Multilevel Microstructures. *Adv. Mater. Technol.* **2022**, *7*, 2101031. [CrossRef]
17. Sun, G.; Wang, P.; Jiang, Y.; Sun, H.; Meng, C. Intrinsically Flexible and Breathable Supercapacitive Pressure Sensor Based on MXene and Ionic Gel Decorating Textiles for Comfortable and Ultrasensitive Wearable Healthcare Monitoring. *ACS Appl. Electron. Mater.* **2022**, *4*, 1958–1967. [CrossRef]
18. Duan, Z.; Jiang, Y.; Huang, Q.; Yuan, Z.; Zhao, Q.; Wang, S.; Zhang, Y.; Tai, H. A Do-It-Yourself Approach to Achieving a Flexible Pressure Sensor Using Daily Use Materials. *J. Mater. Chem. C* **2021**, *9*, 13659–13667. [CrossRef]
19. Lei, D.; Zhang, Q.; Liu, N.; Su, T.; Wang, L.; Ren, Z.; Gao, Y. An Ion Channel-Induced Self-Powered Flexible Pressure Sensor Based on Potentiometric Transduction Mechanism. *Adv. Funct. Mater.* **2022**, *32*, 2108856. [CrossRef]
20. Kim, J.-S.; So, Y.; Lee, S.; Pang, C.; Park, W.; Chun, S. Uniform Pressure Responses for Nanomaterials-Based Biological on-Skin Flexible Pressure Sensor Array. *Carbon* **2021**, *181*, 169–176. [CrossRef]
21. Pan, H.; Lee, T.-W. Recent Progress in Development of Wearable Pressure Sensors Derived from Biological Materials. *Adv. Healthc. Mater.* **2021**, *10*, 2100460. [CrossRef] [PubMed]
22. Hou, Y.; Wang, L.; Sun, R.; Zhang, Y.; Gu, M.; Zhu, Y.; Tong, Y.; Liu, X.; Wang, Z.; Xia, J.; et al. Crack-Across-Pore Enabled High-Performance Flexible Pressure Sensors for Deep Neural Network Enhanced Sensing and Human Action Recognition. *ACS Nano* **2022**, *16*, 8358–8369. [CrossRef] [PubMed]

23. Zeng, Y.; Xiang, H.; Zheng, N.; Cao, X.; Wang, N.; Wang, Z.L. Flexible Triboelectric Nanogenerator for Human Motion Tracking and Gesture Recognition. *Nano Energy* **2022**, *91*, 106601. [CrossRef]
24. Wang, H.; Li, Z.; Liu, Z.; Fu, J.; Shan, T.; Yang, X.; Lei, Q.; Yang, Y.; Li, D. Flexible Capacitive Pressure Sensors for Wearable Electronics. *J. Mater. Chem. C* **2022**, *10*, 1594–1605. [CrossRef]
25. Yu, Z.; Cai, G.; Liu, X.; Tang, D. Pressure-Based Biosensor Integrated with a Flexible Pressure Sensor and an Electrochromic Device for Visual Detection. *Anal. Chem.* **2021**, *93*, 2916–2925. [CrossRef]
26. Sun, J.; Xiu, K.; Wang, Z.; Hu, N.; Zhao, L.; Zhu, H.; Kong, F.; Xiao, J.; Cheng, L.; Bi, X. Multifunctional Wearable Humidity and Pressure Sensors Based on Biocompatible Graphene/Bacterial Cellulose Bioaerogel for Wireless Monitoring and Early Warning of Sleep Apnea Syndrome. *Nano Energy* **2023**, *108*, 108215. [CrossRef]
27. Kang, K.; Park, J.; Kim, K.; Yu, K.J. Recent Developments of Emerging Inorganic, Metal and Carbon-Based Nanomaterials for Pressure Sensors and Their Healthcare Monitoring Applications. *Nano Res.* **2021**, *14*, 3096–3111. [CrossRef]
28. Vu, C.C.; Kim, J. Highly Elastic Capacitive Pressure Sensor Based on Smart Textiles for Full-Range Human Motion Monitoring. *Sens. Actuators A Phys.* **2020**, *314*, 112029. [CrossRef]
29. Yu, S.; Dong, M.; Jiang, G.; Ma, P. Compressive Characteristics of Warp-Knitted Spacer Fabrics with Multi-Layers. *Compos. Struct.* **2021**, *256*, 113016. [CrossRef]
30. Yang, F.; Zhao, H.; Li, R.; Liu, Q.; Zhang, X.; Bai, X.; Wang, R.; Li, Y. Growth Modes of Single-Walled Carbon Nanotubes on Catalysts. *Sci. Adv.* **2022**, *8*, eabq0794. [CrossRef]
31. Ibrahim, N.; Akindoyo, J.O.; Mariatti, M. Recent Development in Silver-Based Ink for Flexible Electronics. *J. Sci. Adv. Mater. Devices* **2022**, *7*, 100395. [CrossRef]
32. Li, H.; Ma, Y.; Huang, Y. Material Innovation and Mechanics Design for Substrates and Encapsulation of Flexible Electronics: A Review. *Mater. Horiz.* **2021**, *8*, 383–400. [CrossRef]
33. Samek, W.; Montavon, G.; Lapuschkin, S.; Anders, C.J.; Müller, K.-R. Explaining Deep Neural Networks and Beyond: A Review of Methods and Applications. *Proc. IEEE* **2021**, *109*, 247–278. [CrossRef]
34. Bai, J.; Li, Y.; Li, J.; Yang, X.; Jiang, Y.; Xia, S.-T. Multinomial Random Forest. *Pattern Recognit.* **2022**, *122*, 108331. [CrossRef]
35. Wu, R.; Ma, L.; Patil, A.; Hou, C.; Zhu, S.; Fan, X.; Lin, H.; Yu, W.; Guo, W.; Liu, X.Y. All-Textile Electronic Skin Enabled by Highly Elastic Spacer Fabric and Conductive Fibers. *ACS Appl. Mater. Interfaces* **2019**, *11*, 33336–33346. [CrossRef] [PubMed]
36. Atalay, O.; Atalay, A.; Gafford, J.; Walsh, C. A Highly Sensitive Capacitive-Based Soft Pressure Sensor Based on a Conductive Fabric and a Microporous Dielectric Layer. *Adv. Mater. Technol.* **2018**, *3*, 1700237. [CrossRef]
37. Gilanizadehdizaj, G.; Aw, K.C.; Stringer, J.; Bhattacharyya, D. Facile Fabrication of Flexible Piezo-Resistive Pressure Sensor Array Using Reduced Graphene Oxide Foam and Silicone Elastomer. *Sens. Actuators A Phys.* **2022**, *340*, 113549. [CrossRef]
38. Zhu, Z.; Zhang, H.; Xia, K.; Xu, Z. Hand-Drawn Variable Resistor and Strain Sensor on Paper. *Microelectron. Eng.* **2018**, *191*, 72–76. [CrossRef]

**Disclaimer/Publisher's Note:** The statements, opinions and data contained in all publications are solely those of the individual author(s) and contributor(s) and not of MDPI and/or the editor(s). MDPI and/or the editor(s) disclaim responsibility for any injury to people or property resulting from any ideas, methods, instructions or products referred to in the content.

MDPI AG
Grosspeteranlage 5
4052 Basel
Switzerland
Tel.: +41 61 683 77 34

*Micromachines* Editorial Office
E-mail: micromachines@mdpi.com
www.mdpi.com/journal/micromachines

Disclaimer/Publisher's Note: The statements, opinions and data contained in all publications are solely those of the individual author(s) and contributor(s) and not of MDPI and/or the editor(s). MDPI and/or the editor(s) disclaim responsibility for any injury to people or property resulting from any ideas, methods, instructions or products referred to in the content.

www.ingramcontent.com/pod-product-compliance
Lightning Source LLC
LaVergne TN
LVHW070725100526
838202LV00013B/1172